LOUIS

Q
143
P2
D78
1986

Dubos, Rene J. (Rene Jules), 1901-

Louis Pasteur, free lance of science.

15401

$13.95

NWACC Library
One College Drive
Bentonville, AR 72712

BAKER & TAYLOR BOOKS

THE DA CAPO SERIES IN SCIENCE

LOUIS PASTEUR

FREE LANCE OF SCIENCE

RENÉ DUBOS

A DA CAPO PAPERBACK

Library of Congress Cataloging in Publication Data

Dubos, René Jules, 1901–
Louis Pasteur, free lance of science.

(The Da Capo series in science)
A Da Capo paperback)
Bibliography: p.
1. Pasteur, Louis, 1822–1895. 2. Scientists—France—Biography. I. Title. II. Series.
Q143.P2D78 1986 509.2′4 [B] 86-511
ISBN 0-306-80262-7

This Da Capo Press paperback edition of *Louis Pasteur: Free Lance of Science* is an unabridged republication of the edition published in New York in 1960. It is reprinted by arrangement with Charles Scribner's Sons.

Copyright © 1950, 1960 by René Dubos

Published by Da Capo Press, Inc.
A Subsidiary of Plenum Publishing Corporation
233 Spring Street, New York, N.Y. 10013

All Rights Reserved

Manufactured in the United States of America

To Jean

Contents

	A Portfolio of Pictures	xi
	Introduction: The Complexities of Genius	xxvii
I	The Wonderful Century	3
II	The Legend of Pasteur	21
III	Pasteur in Action	58
IV	From Crystals to Life	90
V	The Domestication of Microbial Life	116
VI	Spontaneous Generation and the Role of Germs in the Economy of Nature	159
VII	The Biochemical Unity of Life	188
VIII	The Diseases of Silkworms	209
IX	The Germ Theory of Disease	233
X	Mechanisms of Contagion and Disease	267
XI	Medicine, Public Health and the Germ Theory	292
XII	Immunity and Vaccination	317
XIII	Mechanisms of Discoveries	359
XIV	Beyond Experimental Science	385
	Appendix: Chronology of Events in Pasteur's Life	403
	Reference Notes to the Introduction	407
	Bibliography	409
	Index	413

A PORTFOLIO OF PICTURES

Louis Pasteur at the age of thirty, when he was a professor at the University of Strasbourg

Pasteur's father, Jean Joseph Pasteur

Pasteur's mother, Madame Jean Joseph Pasteur

(The portraits on these two pages are pastel drawings made by Pasteur when he was fifteen)

Pasteur as a student at the Ecole Normale Supérieure in Paris

(*drawing by Labayle from a daguerreotype*)

Charles Chappuis, schoolmate and lifelong confidant of Pasteur

(*drawing by Pasteur*)

Jean Baptiste Dumas, chemist, professor, and statesman, whose lectures were an inspiration to Pasteur in his student days

Antoine Jérôme Balard, professor of chemistry under whom Pasteur studied at the Ecole Normale

Jean Baptiste Biot, famous physicist and patron of the young Pasteur

Pasteur with his long-time friend the physicist Pierre Auguste Bertin-Mourot, about 1865; this is the only known picture that shows Pasteur smiling

Marie Laurent some years before her marriage to Pasteur

Pasteur at about forty-five, dictating a scientific paper to his wife at the silkworm laboratory at Pont Gisquet near Alais (now Alès) in southeastern France

Pasteur in 1868

Madame Pasteur and the Pasteurs' daughter Marie-Louise, about 1877

Pasteur in the animal rooms of his laboratory at the Ecole Normale

(*reproduced from* Le Journal Illustré, *Paris, March 30, 1884*)

Pasteur at about sixty-seven

INTRODUCTION TO THE 1976 EDITION

The Complexities of Genius

The Price of Success

A BIOGRAPHY of Pasteur focused on his professional achievements gives the impression that he led an enchanted life. His contributions to science, technology, and medicine were prodigious and continued without interruption from his early twenties to his mid-sixties. His skill in public debates and his flair for dramatic demonstrations enabled him to triumph over his opponents. His discoveries had practical applications that immediately contributed to the health and wealth of humankind. His worldwide fame made him a legendary character during his lifetime; he was then and remains now the white knight of science.

While writing Pasteur's biography a quarter of a century ago, I could readily document the fact that his extraordinary successes had been achieved at the cost of immense labor and against tremendous odds — including the stroke that paralyzed him on the left side at the age of forty-six. However, I felt I could also read between the lines of his public statements the frequent expressions of a melancholy mood, an intellectual and emotional regret at having sacrificed great theoretical problems to the pursuit of practical applications. Writing as if he had not been complete master of his own life, Pasteur stated time and time again that he had been "enchained" (*enchaîné*) by the inescapable logic of his discoveries; he had thus been compelled to move from the study of crystals to fermentation, then to spontaneous generation, on to infection and vaccination.

One can indeed recognize a majestic ordering in Pasteur's scientific career. Yet, as I indicated in Chapter XIII, Mechanisms of Discovery, the logic that governed the succession of his achievements was not as inescapable as he stated. At almost any point in the evolution of his scientific career, he could have followed, just as logically, other lines of work that would have led him to discoveries in fields other than fermentation and vaccination. Some of his casual remarks quoted in Chapter XIII indicate that he was aware of the potentialities he had left undeveloped. He must have often wondered whether "the road not taken" might not have been the better road.

My interest in these inner conflicts of Pasteur's intellectual life was reawakened by a conversation I had some twenty years ago with his grandson, Professor Pasteur Vallery-Radot who told me of an incident that occurred around 1891. Pasteur was then living at the Pasteur Institute in an apartment that had been reserved for his family and that has now become a museum dedicated to mementos of his life. Because of his poor health at that time, he was no longer able to engage in laboratory work and was given to daydreaming about science. Pasteur Vallery-Radot subsequently described the same incident at a symposium organized in 1957 by an industrial firm to celebrate the one-hundredth anniversary of Pasteur's first paper on lactic acid fermentation:

"I see again that face, that appeared to be carved from a block of granite — that high and large forehead, those grayish-green eyes, with such a deep and kind look. . . .

"He seemed to me serious and sad. He was probably sad because of all the things he had dreamed of but not realized.

"I remember one evening, at the Pasteur Institute. He was writing quietly at his desk, his head bent on his right hand, in a familiar pose. I was at the corner of the table, not moving or speaking. I had been taught to respect his silences. He stood up and, feeling the need to express his thoughts to the nearest person, even a child, he told me: 'Ah, my boy, I wish I had a new life before me! With how much joy I should like to undertake again my studies on crystals!' To have given up his research on crystals was the eternal sorrow of his life." [1]

Although Pasteur had abandoned the field of crystallography in which he had won his first laurels, he had kept informed about its developments and moreover fervently maintained the romantic belief of his youth that crystalline assymetry would prove the key to the chemical understanding of life. This, however, was only one of the many reasons that he was "sad because of all the things that he had dreamed of but not realized." Many other aspects of his early scientific work continued to occupy his mind throughout his life and frequently surfaced in the form of casual remarks, suggestions for new lines of experiments, and prophetic views on the direction science should take.

The effect of environmental factors on the characteristics and activities of living things was a particular theme which he did not develop in his experimental work but which continually emerged in his writings. Here again one of his statements betrays regret at not having followed his early hunches. He had entered the field of pathology almost by accident through his work on the diseases of silkworms. His first hypothesis had been that these diseases were nutritional and physiological in nature, but he eventually discovered that they could be controlled by protecting the worms against microbial contamination. However, despite the outstanding success of this control technique, he continued to believe that the resistance of the worms could be increased by measures that would improve their physiological state. In *Etudes sur la maladie des vers à soie,* he went as far as to state: "If I were to undertake new studies on the silkworm diseases, I would direct my effort to the environmental conditions that increase their vigor and resistance." (*"Si j'étais amené, par des circonstances imprévues, à de nouvelles études sur les vers à soie, c'est des conditions propres à accroître leur vigueur que j'aimerais à m'occuper. . . . J'ai la conviction qu'il serait possible de découvrir des moyens propres à donner aux vers un surcroît de vigueur qui les mettraient davan-*

tage à l'abri des maladies accidentelles.") [2] This phrase clearly reveals an aspect of his thought that greatly intrigued him but that he did not have the time to convert into experimental work.

Even though Pasteur's name is identified with the "germ theory" of fermentation and disease — namely, the view that many types of chemical alterations and of pathological processes are caused by specific types of microbes, it is certain that his concern was not limited to the causative role of microbes. He was intensely interested in what he called the "terrain," a word he used to include the environmental factors that affect the course of fermentation and of disease. I now see more clearly than I did when writing Pasteur's biography that the magnitude of his theoretical and practical achievements derives in large part from the fact that his conceptual view of life was fundamentally ecological.

Pasteur as Ecologist

FROM the very beginning of his biological investigations, Pasteur became aware of the fact that the chemical activities of microbes are profoundly influenced by environmental factors. Furthermore, he developed very early a sweeping ecological concept of the role played by microbial life in the cycles of matter. During the 1860's he wrote letters to important French officials to advocate support of microbiological sciences on the grounds that the whole economy of nature, and therefore man's welfare, depended upon the beneficial activities of microorganisms (see pages 160–162 of text). He boldly postulated that microbial life is responsible for the constant recycling of chemical substances under natural conditions — from complex organic matter to simple molecules and back into living substance. In a language that was more visionary than scientific, he asserted that each of the various microbial types plays a specialized part in the orderly succession of changes that are essential for the continuation of life on earth. Long before the word ecology had been introduced into the scientific literature, he thus achieved an intuitive understanding of the interplay between biological and chemical processes that brings about the finely orchestrated manifestations of life and of transformations of matter in natural phenomena.

Pasteur's ecological attitude can also be recognized in his repeated emphasis — to the point of obsession — on the fact that the morphology and chemical activities of any particular microbial species are conditioned by the physicochemical characteristics of the environment. He pointed out, for example, that molds can be filamentous or yeast-like in shape, depending upon the oxygen tension of the medium in which they grow. He demonstrated also that the gaseous environment determines the relative proportions of alcohol, organic acids, carbon dioxide, and protoplasmic material produced by a particular microbial species from a particular substrate. Observations of this type give to the book in which he assembled his studies on beer (*Etudes sur la bière*, published in 1876) an importance that far transcends the practice of beer making. In that book he approached the problem of fermentation from an ecological point of view. By demonstrating that "fermentation is life without oxygen" ("*La fermentation est la conséquence de la vie sans gaz oxygène libre*"), he introduced the first sophisticated evidence of biochemical mechanisms in an ecological relationship.[3]

The ecological attitude in Pasteur's laboratory certainly helped his associate Emile Duclaux, who eventually became director of the Pasteur Institute, to recognize that the enzymatic equipment of microbes can be modified at will by altering the composition of the culture medium. This was the first demonstration of a phenomenon that opened the way for modern discoveries on enzyme induction, and thus constitutes another fundamental link in the understanding of the ecological relation between environmental factors and biological characteristics.

Pasteur's recognition of the effects that environmental factors exert on metabolic activities is now incorporated into theoretical microbiology and technological applications. In contrast, his forceful statements concerning the importance of the terrain in infectious diseases have been overlooked, in part because he did not have time to support his intuitive views by systematic laboratory investigations, and perhaps even more because medical scientists continue to neglect this field, except with regard to the special approach that Pasteur himself opened — immunological protection. Yet he had a sophisticated ecological concept of infectious processes based on an awareness of the genetic and environmental parameters that condition evolutionary and phenotypic adaptations. This aspect of his biological philosophy can be illustrated with statements paraphrased from his writings.

Early in his work on disease, Pasteur recognized that it was a biological necessity for living things to be endowed with natural resistance to the agents of destruction ubiquitous in their environment. (*"Le propre de la vie chez tous les êtres est de résister aux causes de destruction dont ils sont naturellement entourés."* [4])

As he saw it, populations — be they of microbes or of men — usually achieve some sort of evolutionary adaptation to their environment which renders them better able to resist the causes of disease with which they frequently come into contact (*"Le microbe du choléra des poules, par sa culture et son séjour dans le corps de la poule, reprend de la virulence, c'est-à-dire qu'ayant vécu et ayant formé un grand nombre de générations successives dans ce milieu, il devient plus apte à se propager et à vaincre la résistance vitale de l'animal, à peu près comme on voit une race d'hommes ou d'animaux s'acclimater peu à peu dans un pays nouveau, y prospérer et résister peu à peu à des causes naturelles de maladie et destruction."* [5])

Furthermore, he took it for granted that the body in a state of normal physiological health exhibits a striking resistance to many types of microbial agents. (*"Dans l'état de santé, notre corps oppose naturellement une résistance au développement et à la vie des infiniments petits. Dans les conditions physiologiques normales principalement et dans une foule de circonstances, la vie arrête la vie qui lui est étrangère."* [6])

As he pointed out, the various body surfaces harbor various microorganisms which can cause damage only when the body is weakened. (*"L'homme porte sur lui ou dans son canal intestinal sans grand dommage les germes de certains microbes prêts à devenir dangereux lorsque, dans des corps affaiblis ou autrement, leur virulence se trouve progressivement renforcée."* [7])

In contrast, infection often fails to take hold even when antiseptic measures are neglected in the course of surgery. Indeed, human nature possesses a remarkable ability to overcome foci of infection. ("*. . . si toute amputation, toute plaie n'entraîne pas nécessairement la mort lorsqu'on s'affranchit des précautions antiseptiques inspirées par les résultats de mes travaux de ces vingt et une dernières années, cela est dû principalement à la vie, à la résistance vitale. . . . La facilité avec laquelle la nature, prenant le dessus, se débarrasse de foyers purulents . . .*" [8])

Pasteur's attitude regarding the importance of physiological well-being in resistance to infection had developed during his studies with silkworms. He had soon recognized profound differences in the pathogenesis of two diseases in these insects. In one, pébrine, the presence of the specific protozoan was a sufficient cause of the disease, provided the infective dose was large enough. (*"Dans le pébrine, c'est la présence des corpuscules qui fait tout le mal, encore faut-il qu'ils soient abondants."* [9]) In the other, flacherie, the resistance of the worms to infection was profoundly influenced by environmental factors. Among these, Pasteur considered that excessive heat and humidity, inadequate aeration, stormy weather, and poor food were inimical to the general physiological health of the insects. (*"Par ces mots, vigueur des vers, j'entends la résistance plus ou moins grande qu'ils offrent aux maladies accidentelles." "La contagion aurait d'autant moins de prise et de rapidité dans ses effets que les vers seraient plus robustes. La flacherie héréditaire serait la conséquence d'un affaiblissement de la graine ou des vers."* [10])

As he put it, the proliferation of microorganisms in the intestinal tract of worms suffering from flacherie was more an effect than a cause of the disease. (*"Bien que la fermentation de la feuille dans le canal intestinal soit accompagnée de la présence de diverse organismes, par exemple de vibrions, ces organismes paraissent être un effet plutôt qu'une cause."* [11]) Here Pasteur was anticipating George Bernard Shaw's remark in the preface to *The Doctor's Dilemma* (1906): "The characteristic microbe of a disease might be a symptom instead of a cause."

Pasteur did not hesitate to extend these views to the most important human diseases. He accepted that resistance to tuberculosis was on the one hand an expression of hereditary endowment and on the other hand was influenced by the state of nutrition and by certain factors of the environment, including the climate. ("*. . . si vous placez cet enfant dans des conditions de nourriture et dans des conditions climatériques convenables, très souvent vous le sauverez, et il ne mourra pas phtisique. . . . Il y a donc, je le répète, une différence essentielle entre une maladie avec ses caractères, c'est-à-dire la maladie prise en soi, et les causes prédisposantes, les occasions qui peuvent lui donner naissance. . . . Il y a peut-être plus de rapport qu'on ne saurait le dire entre tous les caractères relatifs à la phtisie pulmonaire et les caractères relatifs à l'affaiblissement qui détermine, pour ainsi dire forcément, la flacherie chez les vers."* [12])

Even more boldly, he suggested that mental states affected resistance to infection. ("*. . . combien de fois la constitution du blessé, son affaiblissement, son état moral . . . n'opposent qu'une barrière insuffisante à l'envahissement des infiniment petits."* [13])

This point of view naturally led Pasteur to conclude that resistance to infection could probably be increased by improving the physiological state of the infected individual. He urged his collaborator Emile Duclaux to look for procedures that would increase the general resistance of silkworms. ("*Peut-être M. Duclaux trouvera-t-il un moyen de fortifier les jeunes vers, de façon à les mettre davantage à l'abri des maladies accidentelles."* [14])

And he expressed the opinion that in man also successful therapy often depends upon the ability of the physician to restore physiological conditions favorable to natural resistance. ("*C'est un principe qui doit être sans cesse présent à l'esprit du médecin et du chirurgien, parce qu'il peut devenir souvent un des fondements de l'art de guérir, comme il peut constituer d'autres fois un des plus grands dangers dans le développement des maladies."* [15])

Although Pasteur thus had a clear view of the influence that the physiological state and environmental factors exert on resistance to infection, he did not carry out any significant experimental work in this area. (See, however, his dramatic experiments illustrating the effects of temperature on susceptibility to infection, pages 284–287 of text.) He probably felt that in the state of scientific knowledge of his time the more urgent task was to determine the specific causes of infection and to search for specific methods of protection. It is indeed certain that biological sciences in general and microbiological sciences in particular could not have gone far without the precise knowledge and the intellectual discipline provided by the concept of specificity. The time has come, however, when it would be profitable to follow more actively the other approaches that Pasteur visualized but did not follow — the physiological and ecological study of microorganisms in natural systems and in pathological processes.

Doer and Seer

PASTEUR's ecological philosophy had little influence on the practical policies he advocated for controlling the phenomena of fermentation and infection. When he discussed large theoretical problems in the light of ecological concepts, he professed that the activities of microbes are essential for the continuation of life on earth; he also suggested that microbes might safely coexist with animals and human beings if the infectious process took place under the proper environmental and physiological conditions. In practice, however, he devoted most of his laboratory work to the development of practical techniques for the domestication or destruction of microbes. This dichotomy between conceptual theory and scientific practice can be explained by the climate of scientific and public opinion in the nineteenth century, and also by some peculiarities of his complex genius.

The germ theory was formulated at a time when many biologists and social philosophers believed that one of the fundamental laws of life is competition, a belief symbolized by phrases such as "nature red in tooth and claw" and "survival of the fittest." The ability of an organism to destroy or at least to master its enemies or competitors was then deemed an essential condition of biological success. In the light of this theory, microbes were to be destroyed, unless they could be used for some human purpose, as in desirable fermentations. Aggressive warfare against microbes was particularly the battle cry of medical microbiology, and is still reflected in the language of this science. The microbe is said to be an "aggressor" which "invades" the tissues; the body "mobilizes" its defenses; the physician or the scientist is a disease "fighter" whose goal is to achieve the "conquest" of this or that infection.

As we have seen, Pasteur did not share the simple-minded view that killing and being killed are the only alternatives in biological relationships; indeed he had perceived the ecological possibilities and advantages of peaceful coexistence. But he lived in a period when knowledge meant power used for the conquest of nature. It was during the nineteenth century that the findings of experimental science were for the first time converted into large-scale technological applications. Like his contemporaries, Pasteur identified progress with the use of science for achieving mastery over natural forces. As he was very much a man of his time, he focused most of his effort on the kind of scientific problems most likely to yield results of practical significance — for example, by helping in the "control" of fermentation and in the "conquest" of disease. For his public life, scientific progress meant the development of techniques such as sterilization, pasteurization, and vaccination, even though these practical lines of work prevented him from pursuing other questions that he considered of larger theoretical significance.

Some peculiarities of Pasteur's genius also contributed to the seeming lack of unity in his scientific life. Both in temperament and in social attitudes, he exhibited a combination of contradictory qualities.

On the one hand, Pasteur had the temperament of a dreamer and of a romantic artist; he allowed himself to be guided and inspired by the mirage of an imagination that ranged far beyond the horizon of established knowledge, at times even beyond common sense. On the other hand, he had a compulsive urge to observe and investigate concrete situations, as well as to exert a puritanical control over himself and the external world.

He had a great sense of pride; his confidence in his scientific work and in his opinions often amounted to intellectual arrogance and resulted in expressions of obvious contempt for people who did not share his views. In contrast, he was profoundly respectful of his teachers, of academic and social institutions, of the established order and conventional values.

These seemingly incompatible aspects of his temperament were reflected in his private life. In Chapter XIV I made frequent references to his allegiance to the Catholic faith. These were based on the statement in the semiofficial biography, *La Vie de Pasteur*, by his son-in-law René Vallery-Radot: "Absolute faith in God and in Eternity . . . were feelings which pervaded his whole life; the virtues of the Gospel had ever been present to him. Full of respect for the form of religion which had been that of his forefathers, he came to it simply and naturally for spiritual help in the last stages of his life." [16] Since writing this biography, however, I have found convincing evidence that Pasteur did not attend church services and was in fact completely indifferent to traditional religion. The most that can be said is that he always took a vigorous stand against materialism and that he began once more to receive religious sacraments six months before his death.[17] But these facts do not change significantly the picture I had formed of his character.

On many occasions, also mentioned in Chapter XIV, Pasteur expressed his belief in the validity and importance of spiritual and religious values that transcend scientific knowledge. The passionate manner in which he expressed this belief, even on such an official occasion as his reception into the Académie Française (Academy of Letters), leaves no doubt that it corresponded to a deep concern of his intellectual and emotional life. He felt that "the mystery of the infinite . . . establishes deep in our hearts a belief in the supernatural." The point of interest here is that he stated such views in broad philosophical terms applicable to almost any form of religion ". . . whether God be called Brahma, Allah, Jehovah or Jesus." (*"Tant que le mystère de l'infini pèsera sur la pensée humaine, des temples seront élevés au culte de l'infini, que le Dieu s'appelle Brahma, Allah, Jéhova ou Jésus."*) [18]

Pasteur's references to God and the infinite were manifestations of his romantic visionary temperament. Another aspect of his character was his bourgeois acceptance of conventional values. The facts that he lived in a predominantly Catholic country and that his wife was a devout Catholic probably made him accept as a matter of course practices to which he did not attach importance. He displayed a double standard in religious matters just as he did in matters pertaining to ecological equilibriums and control of microbial life. While he accepted the conventions of his time and society, he also had in mind larger values and preoccupations. He searched for a meaning in life which not only went deeper than that provided by experimental science but also reached beyond the official nineteenth-century religion.

Scientists, like artists, unavoidably reflect the characteristics of the civilization and the time in which they live. In this sense, they are "enchained," as Pasteur complained he had been, by the inexorable logic of their time and their work. A few of the greater ones, however, have visions that appear to be without roots in their cultural past and that are not readily explained by direct environmental influences. These visionaries appear indeed almost as eruptive phenomena, seemingly unpredictable from their environment. Yet even they are not freaks in the natural sequence of cultural events. They constitute mentalities through which emerge and become manifest social undercurrents that remain hidden to less perceptive minds. Some of these visionaries succeed in converting their preoccupations — which are signs from the cultural subconscious — into messages and products of immediate value to their fellow men; they become the heroes of their societies. Others perceive the hopes and the tasks of the distant future, but without providing definite answers or practical solutions; they give warnings of the questions and problems to come, but their anticipations are usually not understood by their contemporaries.

Pasteur belongs in both classes. As a representative of nineteenth-century bourgeois civilization, he focused much of his scientific life on the practical problems of his time. But as a visionary he saw beyond the needs and concerns of his contemporaries; he formulated scientific and philosophical problems that were not yet ripe for solution.

His immense practical skill in converting theoretical knowledge into technological processes made him one of the most effective men of his century; he synthesized the known facts of biology and chemistry into original concepts of fermentation and disease and thus created a new science which dealt with urgent needs of his social environment. The other side of his genius, although less obvious, is more original and perhaps more important in the long run. His emphasis on the essential role played by microorganisms in the economy of nature, and on the interplay between living things and environment, made him perceive an area of science that is only now beginning to develop; he contributed to scientific philosophy by perceiving that all forms of life are integrated components of a global ecological system.

LOUIS PASTEUR

FREE LANCE OF SCIENCE

CHAPTER I

The Wonderful Century

> Although the roads to human power and to human knowledge lie close together, and are nearly the same, nevertheless . . . it is safer to begin and raise the sciences from those foundations which have relation to practice, and to let the active part itself be as the seal which prints and determines the contemplative counterpart.
>
> — FRANCIS BACON

LOUIS PASTEUR was born on December 27, 1822, at Dôle in the eastern part of France, where his father owned and managed a small tannery. When he died on September 28, 1895, at Villeneuve l'Etang near Paris, his name had already become legendary as that of the hero who had used science to master nature for the benefit of mankind. Many fields had been opened or enriched by his labors: the structure of the chemical molecule; the mechanism of fermentation; the role played by microorganisms — in the economy of matter, in technology, in disease; the theory and practice of immunization; the policy of public hygiene. But the importance of his discoveries is not in itself sufficient to account for his immense fame. Among Pasteur's contemporaries, several equaled and a few surpassed him in scientific achievement, yet of him only was it said that "he was the most perfect man who has ever entered the kingdom of science." For Pasteur's name evokes not only the memory of a great scientist, but also that of a crusader who devoted his life to the welfare of man.

There were many traits in Pasteur's personality which enormously magnified the importance of his scientific contributions to society. His intense awareness of the problems of his environment, his eagerness to participate in their solution, his passionate desire to convince his opponents, his indefatigable vigor and skill in controversy — all these characteristics were as important as his experimental genius in making him not only the arm but also the voice, and finally the symbol, of triumphant science.

In reality, Pasteur achieved this great popular success by the sacrifice of higher ambitions. As a young man, he had planned to devote his life to the study of lofty theoretical problems: the fundamental structure of matter and the origin of life; but instead he soon began to devote more and more time to practical matters — asking of nature questions relevant to the immediate preoccupations of his time.

Although he was unquestionably one of the greatest experimenters who ever lived, he did not create a new scientific philosophy — as had Galileo, Newton, Lavoisier and the other men of genius that he so desired to emulate. Nevertheless, Pasteur kept to the end his youthful hope of gaining, through science, an insight into the problems of natural philosophy — and in most of his writings, broad chemical and physiological theories are propounded side by side with details for the practical application of his discoveries. Nurtured in the classical tradition of the French Enlightenment, he worshiped the experimental method as the oracle which would reveal to man the universal laws of the physical world; as a child of the nineteenth century, on the other hand, he responded to the impact of the astonishing power displayed by the exact sciences in solving the technical problems of industrial civilization. Indeed, he symbolizes the position reached by science in 1850, when experimental technology was replacing natural philosophy in the preoccupations of most scientific men. Theory and practice fought to rule Pasteur's life, as they did to control his times.

Until the nineteenth century, society had demanded little from the man of science — less than from the artist, who, according to the mood of the time, was expected to illustrate the Holy Scriptures, or to depict the sumptuous life of Pompeii or of Venice, or the bourgeois atmosphere of Flanders, or the pomp of Louis XIV. From the earliest times, true enough, mathematicians and physicists had served governments and princes as architects — had built for them tombs and palaces, ramparts and catapults, harbors, ships, canals and roads — while most naturalists, alchemists and chemists had been physicians, devoting some of their talents to the art of healing or to compounding poisons.

It had sufficed the man of science that his activities matched in general the preoccupations of his day; he might search for gold or for the elixir of life; he might investigate natural phenomena in order to make manifest the glory of God or satisfy the curiosity of man; or, at the most, he might devise a few instruments and techniques to make life easier and more entertaining. Yet science was predominantly the concern of the philosophical mind, more eager to penetrate the mysteries of the universe than to control nature.

This point of view had dictated the attitude even of those engaged in studies of immense practical importance. For example, Harvey, whose physiological discoveries were the beginning of scientific medicine, bequeathed his estate to the Royal College of Physicians with the stipulation that the proceeds be used "to search out and study the secrets of nature"; he did not voice much interest in the practical consequences of this search.

The men of genius of the seventeenth century had discovered many of the fundamental laws of the physical world. During the following century, the scientists of the Enlightenment exploited the philosophical consequences of these laws in the faith that they had arrived at a rational concept of the relation of man to the universe. Whether or not they erred in their premature conclusions, this striving after aims which transcend the preoccupations of everyday life justifies their claim to be recognized as "natural philosophers." That expression survived into the early nineteenth century, when Geoffroy Saint-Hilaire entitled his great work on the analogies of living creatures *Philosophie anatomique*. Even Faraday, on the eve of the profound industrial revolution which was to result from his electrical and chemical discoveries, preferred to be called a "philosopher," rather than a "scientist." It was perhaps as a silent protest against the encroachment of society into the activities of natural philosophers that, while still in full scientific productivity, he withdrew from all his consulting and industrial connections into the sanctuary of the Royal Institution.

The integration of science and social economy, nevertheless, had had several isolated sponsors before Pasteur's time. Francis Bacon had pictured, in *The New Atlantis,* a society of scholars organized for the acquisition of a knowledge that would permit man to achieve mastery over nature. "The end of our Foundation," he wrote, "is the knowledge of causes and secret motions of things; and the enlarging of the bounds of human empire, to the effecting of all things possible." In 1666 Colbert, that prototype of American efficiency who conducted the business of France under Louis XIV, had created the French Academy of Sciences and had supplied it with funds for the support of academicians, and of their instruments and experiments. As early as 1671, he organized a cooperative project for the survey of the kingdom and its dependencies. Thus, under that Most Christian Monarch, King of France by Divine Right, was born a tradition which the leaders of Communist Russia were to follow systematically two hundred and fifty years later. In England, the Royal Society and the Royal Institution were founded for the cultivation of "such knowledge as had a tendency to use" and "to make science useful as well as attractive." When, in 1751, the French Encyclopedists, under the leadership of Diderot and d'Alembert, undertook the publication of a universal dictionary of arts, sciences, trades and manufactures, they devoted much of their attention to technical processes as carried out in workshops. "Should not," they asked, "the inventors of the spring, the chain, and the repeating parts of a watch be equally esteemed with those who have successfully studied to perfect algebra?" The Paris Academy of Sciences followed this lead and published, between 1761 and 1781, twenty volumes of illustrated accounts of arts and crafts. The activity of scientists everywhere then began to embrace practical applications along with philosophical inquiries.

An example was the study of power. The primitive steam engine was invented by Newcomen in 1705 and had been much improved by James Watt in 1765. As the use of the Watt engine spread, the need for evaluating the yield of energy per unit of fuel consumed, as a basis for improving the efficiency of the machine, led the young French physicist Sadi Carnot to investigate the theoretical relation of heat to power. Study of this relation continued to occupy the minds of physicists. Joule, Meyer, Kelvin and Helmholtz finally supplied the theoretical information from which the modern world learned to harness steam power for transportation and industry. Railroads, steamships, power plants of large factories soon emerged from the calculations and experiments of these scientists.

The passage of electricity from the cabinet of the natural philosopher to workshops and homes was an even greater miracle to the man of the nineteenth century. In 1819, Oersted of Copenhagen found that an electric current tended to twist a magnetic pole around it; and, shortly thereafter, the theory of the interaction between currents and magnets was developed by Ampère, who also pointed out that the deflection of magnets by currents could be used for telegraphic transmission. It was not long before Morse and Wheatstone had made a practical reality of the electric telegraph. In 1823, Faraday showed that a wire carrying a current could be made to rotate around the pole of a magnet, and thus created the first electric motor. The electromagnet and the commutator were invented by Sturgeon during the next few years and, about 1830, the work of Joseph Henry in America and of Michael Faraday in England led to the discovery of electromagnetic induction. The scientific armamentarium which made possible the dynamo and other electromagnetic machines was thus complete.

Although the practical achievements of science during the early nineteenth century were most spectacular in the production and distribution of power, other scientific pursuits also helped to transform everyday life. For instance, when Daguerre and Niepce invented photography in 1835, "daguerreotypes" became overnight a popular fad, and frequently reached such a high level of technical perfection as to give them great documentary and artistic interest. Photography, it then appeared, was to do for the recording of the external forms of nature what printing had done for the recording of thought.

Chemistry was abandoning the romantic den in which the alchemist had pursued the elixir of life and the dream of gold. Lavoisier, who initiated the modern era of theoretical chemistry, started his scientific life by collaborating in the preparation of an atlas of the mineralogical resources of France. Elected a member of the Royal Academy at the age of twenty-five, he prepared reports on a variety of technical problems. This made him familiar with the operations of most of the national industries: mines, iron and bleaching works, starch and soap factories, and others. He also improved the manufacture of saltpeter and gunpowder. It was in part his work on the Paris water supply and his interest in mineral waters that led him to investigate the chemistry of water, and his classical studies on the composition of air originated from his efforts to design lanterns for the lighting of Paris.

During the eighteenth and early nineteenth centuries France was leading Europe in theoretical and industrial chemistry, and her self-sufficiency during the Revolutionary and the Napoleonic Wars was in no small part the result of her scientific superiority.

The place of chemistry in the economy of the rest of the world continued to expand after the Napoleonic Wars. Disasters like the mine explosion of 1812 near Gateshead-on-Tyne led Humphry Davy to study the behavior of firedamp, and to demonstrate that explosion would not pass through fine gauze. In 1816 he devised the safety lamp, which decreased the hazards in coal mining and thus contributed to the industrial supremacy of England. The synthesis of urea by Wöhler in 1828 opened the way for the synthesis of medicaments and dyestuffs. Even the technology of food was influenced by the new knowledge. Marggraf applied chemistry to the production of sugar from beetroot; the polariscope permitted direct measurement of the concentration of sugar in crude extracts of the root; soon fields of sugar beets covered vast areas of northern Europe.

Justus von Liebig organized in Giessen the first and most famous laboratory of biochemistry. Stimulated by the desire to correct the poverty of the surrounding land, he undertook studies which elucidated the principles of soil fertility and led to the rational utilization of fertilizers. Scientific agriculture had begun. More than anyone else Liebig made the world conscious of the fact that even living processes would someday become amenable to chemical control.

By the early part of the nineteenth century scientific knowledge was no longer the peculiar diversion of a few philosophers and curious minds. Whereas the technical advances of the eighteenth century – in textiles, in the metallurgy of iron and steel, in power – had been inventions, made by practical men, and were not based on the discoveries of experimental scientists, this relationship was obviously changing. More and more frequently, during the nineteenth century, research in the laboratory was preceding industrial applications. Scientific knowledge was becoming a source of wealth.

Science had also become essential to the security of the state. True enough, the Committee of Public Safety had sent Lavoisier to the guillotine in 1794 with the statement that "The Republic is in no need of chemists," but soon the statesmen responsible for the conduct of the French Revolutionary Wars had discovered their need for such scientists in time of emergency. "Everything," writes Maury in his history of the French Academy of Sciences, "was wanting for the defense of the country – powder, cannons, provisions. The arsenals were empty, steel was no longer imported from abroad, saltpeter no longer came from India. It was precisely those men whose labours had been proscribed who could give to France what she wanted. On the basis of investigations begun by Lavoisier, Fourcroy taught the methods of extracting and refining saltpeter; Guyton de Morveau and Berthollet made known a new method of manufacturing gunpowder and studied the making of iron and steel; Monge explained the art of casting and boring cannons of brass for land use, and cast-iron cannons for the navy."

In the space of a few years, science had become a necessity to society. Bacon's dictum had come true: knowledge was power.

Thus was born the tradition of mobilizing scientists to perfect the instrumentalities of war, and the importance of the military aspects of science has ever since grown in magnitude with each new conflict. During the Civil War in the United States, Joseph Henry became the chief adviser to the government on scientific military inventions, publishing several hundred reports, based on much experimentation. Out of this activity arose the National Academy of Sciences. Such was also the ancestry of the National Research Council and of the Office of Scientific Research and Development, organized in the United States during the First and Second World Wars respectively. Similar associations of scientists were created in the other belligerent countries, not only to devise weapons of offense and defense, but also to adapt the national economy to shortages of food and other supplies.

The English blockade during the Napoleonic Wars greatly stimulated the development of practical chemistry in France. In order to foster the search for home products to replace colonial and foreign goods, encouragement of all sorts was given to investigators; technical schools and colleges were established; exhibitions were promoted. Because France had been cut off from her usual supply of crude soda, the Paris Academy of Sciences offered a prize which stimulated Leblanc's discovery of a method to make carbonate of soda from salt. This in turn led, somewhat later, to the enormous development of the sulfuric acid industry in England and on the Continent.

Just as the absence of cane sugar had encouraged the cultivation of the sugar beet in the plains of northern France, it was to answer a state need that, stimulated by a prize offered by Napoleon, Appert invented a method for the preservation of perishable food. A few decades later, this method was improved by a

Scottish firm — Donkin, Hall & Gamble — which sold preserved food and meat to the English Navy, to the East India Company, and to the British and French governments during the Crimean War.

As scientists came to occupy an increasingly important place in the affairs of the modern state, concern with scientific matters spread to broader areas of the population.

Interest in science on the part of some laymen was not entirely new, of course: the fashionable salons had long accepted scientific discussions as a worthy subject for their intellectual if often artificial commerce. In most elegant terms, Fontenelle had written in 1686 his *Entretiens sur la pluralité des mondes* for refined and powdered marchionesses, but although his writings were accurate and skillful, literary predominated over scientific interest in his discussion of astronomy. Buffon and Voltaire had given to science a more philosophical tinge; and the Encyclopedists wrote informatively about it to educate the public. But with the Revolution came the descriptive, utilitarian and economic aspects of science — soon displacing all others. It is interesting to recognize, in the proceedings of scientific academies of the time, some of the most notorious names of Revolutionary France. Marat lectured at Rouen and at Lyon on electricity and optics; Danton and Bonaparte competed for the *Prix Raynal* at Lyon, and Robespierre's name is connected with the Academy of Arras.

Napoleon I professed a great interest in theoretical science. He discussed problems of celestial mechanics with Laplace, and took a large number of scientists with him during his Egyptian campaign. In 1807, he made a special court performance of the presentation of a report on the progress of sciences. Following the discovery of the electric current, he invited Volta to demonstrate his battery in Paris, where it aroused an enormous interest. He founded a medal with a prize of three thousand francs for the best experiment on "the galvanic fluid," and — despite the fact that France and England were then at war — the medal was awarded to Humphry Davy in 1807. Again in the course of the war, in 1813, he granted Davy permission to visit the volcanoes of Auvergne and the English party was honored and entertained by the French chemists and by the court, despite Davy's rudeness and arrogance. This trip, be it said in passing, was of considerable moment for the history of chemistry, since on that occasion Faraday began his apprenticeship with Davy, and the latter received from conversations with Ampère information that led him to the discovery of iodine.

The welcome granted by France to Humphry Davy in the midst of the war with England was a striking manifestation of that respect for culture and knowledge which transcended national rivalries during the early nineteenth century. It reflects also the glamour of the English chemist, who had achieved fame not only by his spectacular discoveries — the electrolysis of water, the preparation of sodium and potassium, the chemistry of nitrous oxide and the anesthetic effect of this gas — but also by his genius as an exponent of science. In 1802 Davy had become professor of chemistry in the Royal Institution. It had been founded in 1799 with the object of "diffusing knowledge and facilitating the general and speedy introduction of new and useful mechanical inventions and improvements; and also for teaching, by regular courses of philosophical lectures and experiments, the applications of the new discoveries in science to the improvement of arts and manufactures." Although Davy devoted much care to the preparation of his lectures and demonstrations, he composed them only a few hours in advance, thus achieving in his presentation the immediacy of journalism. His rapidity of comprehension and performance appeared to the public as pure intuition and conformed to the popular idea of genius. The success of his lectures increased from year to year, and soon established him in the fashionable life of London. His audience at the theater of the Royal Institution was close to one thousand, and included many of the celebrities of the time, among them Coleridge, who attended regularly in the hope of increasing his stock of literary metaphors.

When Davy was asked to lecture on chemistry and geology in Dublin in 1810, the rooms which had been arranged to hold five hundred and fifty persons proved much too small for the enthusiastic audience. The charge for admission was two guineas but, when the supply of tickets had been exhausted, ten to twenty guineas were offered by those eager to attend.

Faraday's lectures at the Royal Institution were no less successful than those of his celebrated predecessor. Despite his scorn of social life, Faraday was well aware of the significance of science for the public of his time and prepared in writing a careful analysis of the art of lecturing. When in 1861 he gave a course of lectures on "The Chemical History of the Candle," large audiences of school children gave up their Christmas holidays to hear him.

John Tyndall followed Faraday at the Royal Institution and continued the great tradition of popular scientific lectures. In his published *Fragments of Science* he covered all fields of inquiry from the theory of color to the origin of tuberculosis. So great was his fame that the lectures on light which he delivered in the United States during the winter 1872–1873 gained him thirteen thousand dollars; even the rigorous winter weather of the Atlantic Coast could not discourage his large audiences in Boston, New York and Philadelphia.

Throughout Western civilization, in the nineteenth century, the men of science established contact with a large and responsive public – by means of lectures, books and pamphlets. Interest in von Humboldt's writings on cosmography proved an impetus to scientific explorations; Liebig published his famous *Familiar Letters on Chemistry;* Helmholtz brought to international audiences, beyond the German university towns where he lectured, his brilliant views on the union of physics, physiology, psychology and aesthetics.

Needless to say, biological problems also loomed large in the intellectual preoccupations of the day. During the early part of the century, the anatomist and paleontologist Cuvier became the eloquent voice of official French science. On February 15, 1830, his friend and scientific opponent Saint-Hilaire expounded before the Royal Academy of Sciences in Paris the doctrine of the unity of organic composition which, because it implied some form of transformism from a universal animal ancestor, was in conflict with Cuvier's belief in the fixity of species. Thus was launched a debate that lasted several months and that attracted wide notice. When either of the two champions was to speak, the visitors' seats were crowded and the grave academic hush was replaced by tense and eager excitement.

Thirty years later, the publication of Darwin's *Origin of Species* was to open another debate which spread even more widely through the occidental conscience. In 1860, at Oxford, more than one thousand persons attended the historic meeting during which Huxley convinced his audience, at the expense of Bishop Wilberforce, that theories of science must be judged on the basis of fact and reason, not by the authority of dogma. The theory of evolution in this way became a part of social philosophy; the new scientific faith, Darwinism, tore Europe asunder as had the Reformation two centuries before.

The first edition of 1250 copies of *Origin of Species* was sold out on the day of its publication (November 24, 1859). The second edition of 3000 copies was also snatched when it appeared six weeks later. Similarly, 8500 copies of *Formation of Vegetable Mould through the Action of Worms,* also by Darwin, were sold within three years. These numbers acquire greater significance when it is realized that neither of the two books had been written for the general public, and that a popular novel of the day would sell at the most 30,000 to 40,000 copies. In order to publish his *First Principles* Spencer issued a prospectus outlining the work and asking for subscribers, and arranged for publication in periodicals, as Dickens and Thackeray had published their writings. With his article *Problème de la physiologie générale,* Claude Bernard brought the spirit of modern physiology to the lay readers of the *Revue des Deux Mondes,* and when, in 1865, he published his *Introduction à l'étude de la médecine expérimentale* a large, educated public shared with the professional scientists an intimate understanding of the experimental method.

Even the purely technical aspects of nineteenth-century science excited interest among laymen. Thus, in 1819, Chateaubriand found it worth while to mention the invention of the stethoscope by Laënnec, and predicted a great future for the instrument in the study of cardiac and respiratory diseases. The first international loan exhibition of scientific apparatus was organized in London in 1876. In a single day, 11,969 persons visited the exhibits, to which many columns were devoted in *The Times.* Under the guidance of James Clerk Maxwell, Queen Victoria herself considered it proper to display great interest in the show, and listened with dignified attention to the description of the air pump and the Magdeburg hemispheres.

Pasteur, as we shall see, was also to become involved in many public debates and in demonstrations of technical problems to laymen and artisans. When, in 1861, he delivered in the Sorbonne his famous lecture on spontaneous generation, one could recognize in the audience such celebrities as Victor Duruy, Alexandre Dumas senior, George Sand, Princess Mathilde. A few years later, a farm at Pouilly le Fort became a center of international interest – when journalists and scientists, as well as farmers, assembled there to witness the demonstration that sheep could be immunized against anthrax.

Medical science had become front-page news.

As one reads the accounts of these great scientific performances – the magnitude of the problems which were raised, the intellectual majesty of the scientists who were the main performers, the brilliance and responsiveness of the audience – one returns with a sense of frustration to the dull scene where science and public come into contact in the present world. Yet – the subject of the drama has remained no less exciting. Science is still the versatile, unpredictable hero of the play, creating endless new situations, opening romantic vistas and challenging accepted concepts. But the great actors no longer perform for the public, and the audience has lost its glamour. Gone are the days when such men as Davy, Faraday, Tyndall, Huxley, Helmholtz, Cuvier, Saint-Hilaire, Arago, Bernard and Pasteur introduced – in simple and elegant but also accurate terms – the true concepts and achievements of science and the mental processes of scientists to appreciative audiences of children and adults, artisans and artists, earnest scholars and fashionable ladies. The great pageant of science is still unfolding; but now, hidden behind drawn curtains, it is without audience and understandable only to the players. At the stage door, a few talkative and misinformed charlatans sell to the public crude imitations of the great rites. The world is promised cheap miracles, but no longer participates in the glorious mysteries.

As a token of its respect for science, the nineteenth century bestowed upon many scientists honors and privileges as great as those which are today the monopoly of soldiers, politicians and businessmen. During Davy's illness in 1807, bulletins on the state of his health were issued similar to those published for royalty; eminent medical specialists refused to accept fees for their services. His convalescence stimulated public subscriptions which yielded sufficient funds for the construction of large voltaic batteries to be used in the furtherance of his work. Despite the early conflict between the doctrine of evolution and Christian dogma, Darwin, loaded with awards and honors during his lifetime, was buried with High Church ceremony in Westminster Abbey. In France, Cuvier remained one of the important personages of the state under Napoleon I, Louis XVIII and Charles X. The chemist Jean Baptiste Dumas and the physiologist Paul Bert passed from their chairs of the Sorbonne to the highest seats of government during the Second Empire and the Third Republic. Claude Bernard, Olympic in his aloofness from practical medicine, was made a senator without his asking; his funeral, like that of Darwin, was a national event attended by the highest officials of the state. Napoleon III entertained the famous men of science at court in Paris and Rambouillet. There Pasteur, even before the studies on the germ theory of disease had made his name an object of veneration, was invited to demonstrate with his microscope the teeming "world of the infinitely small" before the Emperor and his court in crinoline.

During the days when an attack of hemiplegia threatened Pasteur's life in 1868, the Emperor daily sent a personal courier to the house of the patient. Pasteur himself recounts with obvious pride that, at the International Congress of Physiology in Copenhagen in 1884, the Queen of Denmark and the Queen of Greece, breaking all social etiquette, walked to him to greet him. In 1892 France, now a republic, delegated her President to the Pasteur Jubilee, which took place at the Sorbonne and was attended by representatives from the whole world. In 1895, the national funeral of the great scientist was celebrated with a pomp that was to be equaled only by that of Victor Hugo, the hero of literary France.

The pageant of discoveries which thus revolutionized life during those exciting years, and the hope that man would soon complete his mastery over nature, created in the Western world an atmosphere of faith in science and an enthusiasm which was to find a somewhat naïve expression in many books.

In 1899, A. R. Wallace, who had proposed the theory of evolution simultaneously with Darwin, published under the title *The Wonderful Century* an enthusiastic account of the achievements of his age. To the nineteenth century he credited twenty-four fundamental advances, as against only fifteen for all the rest of recorded history. Many of the great inventions and scientific theories listed by Wallace matured only during the second half of the century; but even while Pasteur was still a schoolboy, science was influencing habits, thought and language.

In the sheltered atmosphere of the College of Besançon and of the Ecole Normale, Pasteur may not have felt the full impact of the social forces urging every scholar to devote his talent, knowledge and energy to the solution of practical problems. But when he became professor of chemistry and dean of the newly created

Faculty of Sciences at Lille in 1854, the impact reached him through official channels. The decree organizing the new Science Faculties throughout France was very explicit; their role was to encourage the applications of science to the local industries. In a letter written during March 1855, the Minister of Public Education followed his appreciation of the success of Pasteur in his new functions by the following warning: "Let M. Pasteur be careful, however, not to be guided exclusively by his love for science. He should not lose sight of the fact that to produce useful results and extend its favorable influence, the teaching in the faculties, while remaining at the highest level of scientific theory, should nevertheless adapt itself, by as many applications as possible, to the practical needs of the country." M. Fortoul, Minister of Public Education in the conservative government of Napoleon III, would have been much surprised and disturbed had he recognized, in his recommendation to Pasteur, the echo of another statement made almost simultaneously by Karl Marx: "Hitherto, philosophers have sought to understand the world, henceforth they must seek to change it."

In his letters from Paris and Strasbourg to his friend Chappuis, Pasteur talked of crystals as a lover of pure science, without ever referring to the possible role of his work in modifying the life of man. In response to his new responsibilities in Lille he soon became acutely conscious of wider social duties, emphasizing in his lectures the role of science in the practical life of the citizen and of the nation. Said he: "Where will you find a young man whose curiosity and interest will not immediately be awakened when you put into his hands a potato, when with that potato he produces sugar, with that sugar, alcohol, with that alcohol ether and vinegar? Where is he that will not be happy to tell his family in the evening that he has just been working out an electric telegraph . . . ?

"Do you know when it first saw the light, this electric telegraph, one of the most marvelous applications of modern science? It was in that memorable year, 1822: Oersted, a Danish physicist, held in his hands a piece of copper wire, joined by its extremities to the two poles of a Volta pile. On his table was a magnetized needle on its pivot, and he suddenly saw, by chance you will say, but chance only favours the prepared mind, the needle move and take up a position quite different from the one assigned to it by terrestrial magnetism. A wire carrying an electric current deviates a magnetized needle from its position. That, gentlemen, was the birth of the modern telegraph . . ."

He warned that there are not two forms of science — pure and applied — but only science, and the application of science. "Without theory, practice is but routine born of habit. Theory alone can bring forth and develop the spirit of invention." From then on, the applications of science were to loom large in his activities; for he had tasted the intoxicating atmosphere which society provided for those who moved from the cabinet of the philosopher to the busy market place. His life was henceforth to be divided between the serene peace of the laboratory and the full-blooded excitement which surrounds the application of science to practical problems.

Pasteur was not alone in dedicating his genius to the service of a society intent on mastering the physical world. For example, William Thompson had also started by concerning himself with abstract scientific problems but soon felt impelled to facilitate the social desires of his times. First distinguished in theoretical physics and mathematics, he later was willing to devote more and more of his energy to the production of wealth. He it was who first organized a laboratory specially adapted to industrial research. A few decades later, Edison abandoned any pretense of interest in theoretical inquiries for their own worth, selecting his research problems only on the basis of the demands of the industrial markets around him.

Thus, within a few generations, the scientist had evolved from natural philosopher to technologist. Were Michael Faraday and Claude Bernard, men who refused to become involved in the practical applications of their sciences, the greater for obeying the spiritual urge to pursue their theoretical inspiration to the very end, and for leaving to more limited minds the conversion of their findings into social commodities? It is too early to judge. The history of experimental science is far too short to permit an adequate perspective of its true relation to human welfare and to the understanding of the universe.

But whatever the ultimate judgment of history, Wallace was right: the nineteenth was a wonderful century. Its scientists were masterful practitioners of the experimental method and, at the same time, they knew how to integrate their efforts into the heritage of classical ages. Faithful to the tradition of the Enlightenment, they never forgot, even while solving the technological problems of industrial civilization, that science *is* natural philosophy. In their hands, science was not only a servant of society, an instrument for the control of the physical world, but also an adornment of our Western culture.

CHAPTER II

The Legend of Pasteur

> I learned this, at least by my experiment, that if one advances confidently in the direction of his dreams, and endeavours to live the life which he has imagined, he will meet with a success unexpected in common hours. In proportion as he simplifies his life the laws of the Universe will appear less complex, and solitude will not be solitude, nor poverty poverty, nor weakness weakness.
>
> — HENRY D. THOREAU

FEW LIVES have been more completely recorded than that of Louis Pasteur. His son-in-law René Vallery-Radot has presented, in *La Vie de Pasteur,* a chronological account of the master's origins, family life, labors, struggles, trials and triumphs. His grandson, Professor Pasteur Vallery-Radot, has reverently collected and published all his scientific and other writings, as well as his correspondence. The portraits painted by the young Louis in his home town at Arbois and at school in Besançon are readily available in private collections, and in the form of excellent reproductions. Emile Duclaux, one of Pasteur's students and early collaborators, his intimate associate to the end, has described and analyzed, in *Pasteur: l'Histoire d'un Esprit,* the evolution of the master's scientific mind and discoveries. The dwelling in which Pasteur was born, those in which he lived, toiled and died, are carefully maintained in their original condition as national shrines, helping us to recapture the atmosphere in which the son of a modest tanner moved from a quiet French province to become a legendary hero of the modern world. Numerous photographs, statues, paintings and medals reveal the evolution from the young thoughtful schoolboy, through the stern professor and eager experimenter of early adulthood, the passionate fighter and apostle of maturity, to the tired warrior dreaming in his old age.

Because Pasteur touched on so many problems and influenced so many lives during his tempestuous career, the different aspects of his personality are reflected — as by a multifaceted mirror — in the reaction to his performance of men at all levels of society and in all walks of life. There are many records of the admiration of his colleagues for his scientific discoveries, but also of impatience for his intolerence and overbearing attitude when he knew — or believed — that truth was on his side. Other philosophers and scientists shared his faith that the exact sciences constituted — outside of revealed religion — the only avenue to wisdom and to power open to man; but there were also those who sneered at that naïve philosophy, certain as they were that nature and truth would not be conquered by such primitive means. Countless human beings have worshiped him as the savior of their children or of their humble trades; but he had also to face the opposition of those who questioned the practical value of his discoveries — sometimes on the basis of healthy and informed criticism, too often because man is blind and deaf to the new, or resents any changes to the old order of things.

Soon, however, worship triumphed over criticism; legend captured Pasteur from history. France took him as the symbol of her genius for logic and of her romantic impulses. His name now calls forth in French hearts poetical and haunting associations: the small towns of Dôle and Arbois where he was born and raised, along graceful rivers called the Doubs and the Cuisance; Paris — its great schools, the atmosphere of meditative scholarship and of feverish participation in the affairs of the world; a revered old man, exhausted by years of endless toil in the service of humanity, recalling under the huge trees of the park of Saint-Cloud the dreams of the idealistic student who — fifty years earlier — had planned to consecrate himself to the solution of some of the eternal problems of life. Across half a century, his voice still resounds with this message from a romantic age: "The Greeks have given us one of the most beautiful words of our language, the word 'enthusiasm'—a God within. The grandeur of the acts of men is measured by the inspiration from which they spring. Happy is he who bears a God within!"

It is not in France alone that Pasteur has become a legendary hero. Scientific institutes, broad avenues, even provinces and villages, carry his name all over the world. From monuments and statues, he supervises students entering halls of learning and watches over children playing in public squares. Even during his lifetime, "pasteurization" became a household word connoting healthy food and beverages. Had Pasteur lived in the thirteenth century, his silhouette would adorn the stained glass windows of our cathedrals; we would know him in the monastic garb of an abbot—the founder of some new religious order—or in the armor of a knight fighting a holy war. For, as much as a scientist, he was the priest of an idea, an apostle and a crusader. It is the champion of a cause, rather than the intellectual giant, that mankind remembers under his name, and that an anonymous writer in the London *Spectator* of 1910 evoked in the following lines:

> There are more than sixty Pasteur Institutes: but I am thinking of the Paris Institute. At the end of one of its long corridors, down a few steps, is the little chapel where Pasteur lies. . . . From the work of the place, done in the spirit of the Master, and to his honour, you go straight to him. Where he worked, there he rests.
>
> Walls, pavement, and low-vaulted roof, this little chapel, every inch of it, is beautiful: to see its equal you must visit Rome or Ravenna. On its walls of rare marbles are the names of his great discoveries—*Dyssymétrie Moléculaire. Fermentations . . . Générations dites Spontanées . . . Etudes sur le Vin . . . Maladies des Vers à Soie . . . Etudes sur la Bière . . . Maladies Virulentes . . . Virus Vaccins . . . Prophylaxie de la Rage*. . . . In the mosaics, of gold and of all colours, you read them again; in the wreathed pattern of hops, vines and mulberry leaves, and in the figures of cattle, sheep, dogs, and poultry. In the vault over his grave are four great white angels, Faith,

> Hope, Charity, and Science. From time to time Mass is said in the chapel: the altar is of white marble. Twice a year, on the day of the master's birth and the day of his death, the workers at the Institute, the "Pasteurians," come to the chapel, some of them bringing flowers in memory of him, and afterwards pay a visit of ceremony to Madame Pasteur, whose apartments are on the second floor of the Institute, above the chapel. . . .
>
> Yet, to me, who remember him, saw him, heard him talk, shook hands with him, all the adornments round his grave were not sufficient, and the half was not told me. For he was, it seems to me, the most perfect man who has ever entered the kingdom of Science. . . . Here was a life, within the limits of humanity, well-nigh perfect. He worked incessantly: he went through poverty, bereavement, ill-health, opposition: he lived to see his doctrines current over all the world, his facts enthroned, his methods applied to a thousand affairs of manufacture and agriculture, his science put in practice by all doctors and surgeons, his name praised and blessed by mankind: and the very animals, if they could speak, would say the same. Genius: that is the only word. When genius does come to earth, which is not so often as some clever people think, it chooses now and again strange tabernacles: but here was a man whose spiritual life was no less admirable than his scientific life. In brief, nothing is too good to say of him: and the decorations of his grave, once you know his work, are poor, when you think what he was and what he did. Still, it is well that he should lie close to the work of the Institute, close to the heart of Paris, with Faith, Hope, Love and Science watching over him.[1]

Thus the son of a former sergeant in Napoleon's army had found his place in the golden legend of the modern world.

After the collapse of the Emperor, Sergeant Jean Joseph Pasteur had taken refuge in the humble profession of tanning – first at Dôle, then at Arbois in eastern France. A crude painting which he made, of a man in a soldier's uniform, leaning on the plow while gazing into a distant dreamland, suggests that the peaceful

[1] We have been informed by the editor of the *Spectator* that this article was written by Stephen Paget (1855–1926) F.R.C.S., Vice Chairman Research Defence Society, and author of several medical and historical books.

citizen had not forgotten the intoxicating dreams of the imperial epic; and yet, perhaps because he was tired from having seen too many social and military upheavals, all he desired for his son was that he should rise above the status of the small businessman into the safe if obscure dignity of a teaching appointment in the provincial secondary school. In the melancholy eyes of the portrait of his father painted by the young Louis Pasteur in 1837, one recognizes the resigned wisdom of so many sensitive and reflective citizens of the old European communities, to whom history has given the vicarious excitement of adventure and of political progress, but who also know that society exacts a painful toll of those who want to rise above its norm. In his many letters to his son away at school – first in Besançon, then in Paris – the old soldier expressed a homely philosophy, seeing in excessive social or intellectual ambition a danger far greater than those lurking in wicked Paris. "There is more wisdom in these hundred liters of wine," he would assure his overeager son, "than in all the books of philosophy in the world."

But despite this counsel of resignation, Jean Joseph Pasteur devoted his own evenings, after the hard days of labor in the tannery, to reading in history books some accounts of the past glories of France, and to acquiring the education which appeared to him the symbol of greater human dignity. How much yearning for a broader life appears in the efforts of the old soldier, attempting to understand in later life the scientific achievements of his son, and to educate himself in order to become the advocate of learning for his turbulent daughters!

Louis Pasteur's mother forms the silent and poetical background of this delightful family picture. We see her draped in a lovely shawl, with all the dignity of a provincial housewife, in a masterly pastel made by Louis at the age of fifteen. And behind the charm of her disciplined face, one can read all the emotional intensity which inspired her to write to her son on January 1, 1848, shortly before her death: "Whatever happens, do not become unhappy, life is only an illusion."

Nothing obvious in the home atmosphere of the young Louis

Pasteur appeared designed to prepare him for the exciting role which he was to play in science and society. His peaceful and humble family, the gentle, comfortable and settled country in which he was born and raised, his teachers' disciplined acceptance of a limited environment – all invited him to a quiet life, adorned but not monopolized by study. Outwardly, he appears as a sentimental, hard-working boy, serious-minded, dutiful, eager to assimilate from the well integrated atmosphere of his environment the classical culture of France, the knowledge of the glorious role that his country had played in the history of Western civilization. When he left his native province for the great centers of learning in Paris, it was not to find the answer to some soul-searching query, not for the sake of intellectual adventure, not with the ambition of the social conqueror. It was merely as an earnest student, going where teaching was most enlightened in order to prepare himself as best he could for a worth-while place in his community. He had not yet dreamed that fate had selected him for a historical role to be played beyond his native province and even beyond France – a legend in the annals of humanity.

At least, there is nothing to reveal that the magic wand had yet tapped him when he entered the great Ecole Normale Supérieure in Paris. Only the fact that he had engaged in portrait painting between the ages of thirteen and eighteen differentiated him slightly from the ordinary good student. However skillful, these portraitures were no more than the conscientious expression of his immediate surroundings – his father, his mother, the town officials and notables, a picturesque old nun and his school friends – all witnesses of the vigorous but settled life of his town and school.[2] But who knows what strivings and urges

[2] The Finnish artist Albert Edelfeldt, who painted a famous portrait of Pasteur in his laboratory in 1887, expressed in a letter to one of his friends the following judgment on Pasteur as a painter: "Outside of science, painting is one of the few things that interest him. At the age of 16, he had intended to become a painter and amused himself making pastel drawings of his parents and of other citizens of Arbois; some of these pastels are in his home at the Institute and I have looked at them very often. They are extremely good and drawn with energy, full of character, a little dry in color, but far

such humble efforts conceal? The travels of the explorer into the dangerous and unknown, the literary and artistic projections of imaginative wanderings into unusual or unreal worlds, the visions of wild dreams, certainly are not the only manifestations of the restless mind. The mere copying of one's environment may at times be a naïve effort to dominate the world by an act of re-creation. And so Pasteur may have begun, in these youthful portraits, an attempt at the intellectual mastery and control of his environment.

Like his schooling, his early letters and writings fail to give an obvious omen of the adventurous life he was to live. To his parents, he faithfully recorded conscientious scholastic efforts; to his mother, he recommended that she not interfere with his sisters' schoolwork by too many small household chores; to his sisters, he advised good behavior and diligent study.

"Work, love one another. Work . . . may at first cause disgust and boredom; but one who has become used to work can no longer live without it . . . with knowledge one is happy, with knowledge one rises above others."

". . . Action, and work, always follow will, and work is almost always accompanied by success. These three things, will, work and success, divide between themselves all human existence; will opens the door to brilliant and happy careers; work allows one to walk through these doors, and once arrived at the end of the journey, success comes to crown one's efforts."

This rigid sense of discipline was softened by a great sentimentality and a profound devotion to his family, friends and country. He read edifying books and attempted to mold his life, and that of others, according to their teachings. So strong was

---

superior to the usual work of young people who destine themselves to an artistic career. There is something of the great analyst in these portraits: they express absolute truth and uncommon will power. I am certain that had M. Pasteur selected art instead of science, France would count today one more able painter. . . ."

his attachment to the home atmosphere that when he first went to school in Paris he could not conceal in his letters many a pathetic expression of loneliness. "Oh! what would I not give for a whiff of the old tannery!" — and he returned to Arbois for a year before again gathering enough courage to go and meet his destiny in the capital.

His father, mother and sisters, and later his wife and children, constituted his emotional universe, supplemented by a very few friends, and by some of his masters upon whom he bestowed unbounded devotion. Chappuis and Bertin, comrades of his youth, remained his confidants to the end. And to the old family home in Arbois he returned every summer, and in periods of familial tragedies, there to recover physical and moral strength. From his father, and from his schoolbooks, he learned to identify his life with that of France, and he maintained unaltered until his death a deep loyalty to family, friends and country.

Not until he was twenty-five do his writings express a philosophical query, an overwhelming question; they state, rather, with a force born of good upbringing, only the moral standards of his environment and his determination to live according to them. Is it not possible, however, that even this homely philosophy may at times be the product of an intense pressure to escape from oneself and one's environment? Most adolescents experience the urge — often obsessing — to grow above and beyond their physical needs and comforts, long before an ideal or an objective has been recognized toward which to proceed. Perhaps, in the life of many, the direction in which one goes, the special nature of the outlet, is of far less importance than the opportunity to move, to transcend oneself, to emerge from plant and animal life into these immensely varied areas which are the reserved hunting grounds of the human mind.

Pasteur did not early find the formula of his life, except that it should be devoted to work and to some worth-while cause. His immediate environment did not suggest any field in which he could expand, any channel in which he could direct his energy. There was seemingly no overpowering interest, no philosophical question or scheme to harness his mind, no passionate urge to monopolize his thoughts. In haphazard manner he responded, at first, to any voice which pleaded before him with a respectable argument. As a student, in Besançon or in Paris, his dominating preoccupation was to reach the top of his class in mathematics, physics and chemistry; and he often succeeded, through application and industry. Admitted in 1842 to the scientific section of the Ecole Normale Supérieure, he refused to enter the school because he had been received as only the sixteenth in his class; he competed again the following year, to be readmitted as fifth in rank. Training himself to become a professor, he begged the famous chemist, Jean Baptiste Dumas, to accept him as a teaching assistant—not, he assured him, that he wanted the job for the sake of money or for the purpose of forming a closer acquaintance with an important man, but because he had "the ambition to become a distinguished professor"—"My chief desire is . . . to secure the opportunity to perfect myself in the art of teaching." To his friend Chappuis, he wrote with pride that he had been highly successful in his practical classroom test as a teacher of physics and chemistry. "M. Masson told me that . . . my lesson in physics was good, the one in chemistry was perfect. . . . Those of us who are to become professors must make the art of teaching our chief concern." And, in fact, it is at this level that his instructors judged him. "Will make an excellent professor," was the laconic and uninspired comment which ushered him into the world from the Ecole Normale.

Within the walls of the old school, however, Pasteur had already received, unknown to his schoolmates and to most of his instructors, the visitation of the Muse of Science. The dutiful student was no longer satisfied with being a passive recipient of knowledge, or even with the prospect of merely passing it on to others. He had tasted the excitement of discovery. The passion—the almost insane urge—to move on into the unchartered lands of nature had taken hold of him.

The investigator was beginning to claim precedence over the

professor. It was while repeating some classical experiments on the formation and properties of crystals that he became aware of the world of mystery hidden behind the polished teaching of textbooks and professors. From then on, the torment of the unknown became a dominating component of his life. But even before this revelation, he had received from Dumas the spark which had fired his eagerness to understand the chemical laws governing the world of matter, and his awareness of the power that chemistry could exert in the affairs of man.

Like many chemists and physiologists of the nineteenth century — Liebig and Claude Bernard for example — Dumas had begun his scientific career as apprentice to an apothecary, at a time when pharmacy had not yet degenerated into the distribution of ready-made, highly advertised packages. From Alais in the South of France, where he was born in 1800, Dumas had gone to study in Geneva. It was as a young pharmacist that he had signed the studies on iodine, blood, muscle contraction and plant physiology which first made his name familiar to European scientists. In Paris, he soon became one of the scientific leaders and one of the founders of organic chemistry. He formulated in particular the theory of substitution of chemical radicals, then the theory of alcohols and of fatty acids, and finally devoted himself, with his friend Boussingault, to the study of the chemical changes associated with living processes. These strenuous studies did not suffice to satiate his creative vigor, for he retained the exuberance, generosity and communicativeness of the sunny land of his birth. He was not only a leader in science but even more a leader of men, and soon found himself engaged in the reorganization of higher education. An influential senator and minister during the Second Empire and the Third Republic, he sat in all committees concerned with the relation between science and society. He loved authority, not for the mere sake of exercising power, but because he had a physiological need to operate on a large scale, to spend his varied and great talents on matters of national interest. He liked to recognize and support ability and genius. He was one of the first who guessed how much Pasteur would contribute to science and to France, and he never spared his influence and his wisdom to encourage and guide the younger man who was his student, then his colleague, and always his friend and admirer. With similar vision and generosity, Dumas had protected Daguerre during the fifteen years when the inventor of photography had to struggle against technical difficulties, and against the ridicule – surprising to us as it is – that his contemporaries first poured upon him.

Dumas was a great teacher. He brought to the lecture room the authority of his name, an immense sense of the dignity of his calling, and an eloquence made of the thorough preparation of his delivery and of the warmth of his meridional accent. On the days when he taught his course at the Sorbonne, the eight hundred seats of the amphitheater were filled with a varied audience attracted by the great manner of the chemist, as much as by the subject which he taught. Fortunately, the first row was reserved for the students of the Ecole Normale, and the enthusiastic Pasteur came out of each lecture intoxicated with vast projects. Pasteur retained for Dumas a veneration which he never tired of expressing, and he often spoke of the unforgettable days when his mind and his heart had been opened by the great teachers whom he called *allumeurs d'âme.* Those had been his greatest emotions, and at the end of his own glorious life, he liked to refer to himself as the disciple of the enthusiasms that Dumas had inspired.

While still a student, Pasteur had attracted the attention of another celebrated chemist, Antoine Jérôme Balard, who was then professor at the Ecole Normale. Like Dumas, Balard was a Southerner and had been a druggist's apprentice. He had discovered bromine at the age of twenty-four and had increased his fame in Parisian scientific circles by a delightful contempt for the conventions of social life. Even after becoming a member of the Institut de France, he continued to live in a primitive student room furnished with two old shaky armchairs painted with his own hands in a peculiar red color, under the illusion that he was imitating mahogany. When traveling, his total luggage consisted of a shirt and a pair of socks, wrapped in a newspaper, which he would slip into his large pocket. He had adopted in his work the same simplicity which ruled his daily life. Having read in the writings of Benjamin Franklin that a good workman should know how to saw with a file and to file with a saw, he liked his students to work without equipment. He rejoiced at seeing Pasteur compelled to build with his own hands the goniometer and polarimeter needed for crystallographic studies, as well as the incubator in which were carried out the classical experiments on fermentation and spontaneous generation. As is often – indeed, usually – the case, Balard made most of his discoveries while working without means, on a corner of his apothecary's bench. When he became professor of chemistry at the Ecole Normale, now occupying new quarters in the Rue d'Ulm, he cheated the administration out of a few rooms by pretending that they were to be used for the display of collections; he transformed them into research laboratories. There also he put a bed, so as to become even more independent of conventional life. It was in these humble quarters that Balard took on young Pasteur as his assistant. By that time, however, he had become so much more interested in the work of others than in his own that he let the young student go his own way and merely encouraged him with his jovial optimism.

Balard, fanciful in his habits and picturesque in the vehemence of his speech and gestures, was also a man of strong convictions. When he heard that Pasteur was to be sent to a small secondary school far away from Paris by administrative order from the Ministry of Education, he unleashed a one-man campaign against the decision, and the Ministry had to yield under the barrage of a torrent of words. Pasteur was allowed to spend an additional year at the Ecole Normale, and remained always grateful to his master for this timely help. With ever-increasing industry, he now devoted all his spare time to chemical experiments in Balard's laboratory.

Delafosse, one of the chemistry instructors, had published a conscientious study dealing with the geometrical, physical and chemical properties of crystals. The elegance and precision of this field of research appealed to the neat and orderly Pasteur. Moreover, it soon provided him with a specific question worthy of his industry and imagination. He had read in the school library a recent note in which the celebrated German crystallographer and chemist Mitscherlich had stated that the salts of tartaric and paratartaric acids, although identical in chemical composition and properties, differed in their ability to rotate the plane of polarized light. This anomaly had remained in Pasteur's mind as an obsessing question, and it was to clarify it that he undertook the study which led him to recognize that paratartaric acid was, in reality, a mixture of two different tartaric acids possessing equal optical activity, except for the fact that one (the right or dextro form) rotated a polarized beam of light to the right, whereas the other (the left or levo form) rotated light to the left. The genesis and significance of this discovery will be discussed in succeeding chapters. Suffice it to point out here that Pasteur had demonstrated, with one stroke, independence of mind in questioning the statement of a world-famous scientist, imagination in recognizing the existence of an important problem, and experimental genius in dealing with it. He had exhibited extraordinary power of detailed observation, a superb competence in planning the strategy and tactics of his experimental attack, tireless energy and meticulous care in its execution.

Pasteur had become interested in crystal structure before realizing that this study would lead him into questions of immense theoretical significance, but the implications of his findings soon became apparent to him. That he found the problem worthy of his metal is obvious from the enthusiasm displayed in a letter to his friend Chappuis: "How many times I have regretted that we did not both undertake the same studies, that of physical sciences! We who so often used to speak of the future, how little we understood! What beautiful problems we would have undertaken, we would undertake today, and what could we have not solved, united in the same ideas, the same love of science, the same ambition? I wish that we were again twenty and that the three years of the School were to start under these conditions." Again — a few years later — he wrote to his friend: "That you were professor of physics or chemistry! We would work together and within ten years, we would revolutionize chemistry. There are marvels hidden behind the phenomenon of crystallization, and its study will reveal some day the intimate structure of matter. If you come to Strasbourg, you will have to become a chemist despite yourself. I shall speak to you of nothing but crystals."

Balard took great personal pride in the work done in his laboratory by the young Pasteur and, with his customary exuberance and loud voice, soon undertook to promote it during conversations at the meetings of the Paris Academy of Sciences. Among his listeners none was more interested — even though somewhat skeptical — than the veteran physicist, Jean Baptiste Biot.

Biot was then seventy-four. Aloof from the world, he maintained a haughty independence, based on immense scientific and literary culture and on the most exacting ideals. He denounced sham and pretense wherever he found them, irrespective of the consequences of his actions, undisturbed by the enmity that he caused; when later he became convinced that the influential Balard no longer took an active part in research, he fought alone against his appointment to the chair of chemistry at the Sorbonne. Speaking of his scientific colleagues who affected to disdain letters and who were careless in their use of the noble French language, he publicly said with scorn: "I do not see that the quality of their science becomes the more obvious for their lack of literary culture." Among his many scientific achievements, Biot counted some of the pioneer work on the ability of organic compounds to change the direction of polarized light (optical activity), and he immediately perceived, therefore, the importance of the separation of paratartaric acid into two opposite forms of tartaric acid. Unconvinced by Balard's heated reports, however, he demanded to see the evidence which justified these extraordinary claims.

To a letter from Pasteur asking for an interview, Biot replied with his usual dignity: "I shall be pleased to verify your results if you will communicate them confidentially to me. Please believe in the feelings of interest inspired in me by all young men who work with accuracy and perseverance."

An appointment was made at the Collège de France where Biot lived, and there the young Pasteur demonstrated the validity of his claims to the distinguished master. From that day on began, between Pasteur and Biot, one of the most exquisite relationships in the annals of science, made up of filial affection, of common ideals and interests, of respect and admiration.

The warm and sensitive heart which Biot hid beneath his austerity and skepticism becomes manifest in a note from him to Pasteur's father: "Sir, my wife and I appreciate very much the kind expressions in the letter you have done me the honour of writing me. Our welcome to you was indeed as hearty as it was sincere, for I assure you that we could not see without the deepest interest such a good and honorable father sitting at our modest table with so good and distinguished a son. I have never had occasion to show that excellent young man any feelings but those of esteem founded on his merit, and an affection inspired by his personality. It is the greatest pleasure that I can experience in my old age, to see young men of talent working industriously and trying to progress in a scientific career by means of steady and persevering labour, and not by wretched intriguing." To Pasteur himself he wrote, following this visit: "We highly appreciated your father, the rectitude of his judgment, his firm, calm, simple reason, and the enlightened love he bears you." And shortly before his death he gave his photograph to Pasteur, with these words as a further symbol of his affection: "If you place this portrait near that of your father, you will unite the pictures of two men who have loved you very much in the same way."

Despite the vigorous protests of Dumas, Balard and Biot, and of other eminent members of the Academy, Pasteur could no longer escape the decision of the Ministry of Education to send him off – as was the custom – to a teaching appointment away from Paris. In 1847, he took up his new post in Dijon where he taught elementary physics with his usual thoroughness, while lamenting the lack of time and facilities for his investigations. Soon, however, his sponsors obtained for him a better appointment at the University of Strasbourg, where – in January 1848 – he became acting professor of chemistry. In Strasbourg began one of the richest and happiest periods of his life.

He took quarters in the house of Pierre A. Bertin-Mourot, professor of physics on the faculty, whom he had known while at school in Besançon and at the Ecole Normale. A conscientious and able teacher, equally devoted to his students and to his friends, Bertin brought into Pasteur's life the smiling help of a benevolent philosophy appreciative of wine, beer and all the simple pleasures of a normal existence. To Pasteur's intensity, impetuosity and lack of humor, he opposed an amiable skepticism, a robust heartiness, balanced by great common sense and an exacting conception of duty. This excellent man wanted his efforts to remain unknown – be they concerned with the pains he took in preparing his lectures or with the help that he so generously gave to others – because, as he put it, "they are my own business." He remained Pasteur's close friend throughout life, and when later he became assistant director of the Ecole Normale in replacement of Pasteur, his jovial and generous attitude once more helped to ease the tension of the stormy life of his famous colleague.

A letter from his father reveals that Pasteur made plans to arrange his life in a more permanent manner as soon as he arrived in Strasbourg: "You say that you will not marry for a long time, that you will ask one of your sisters to live with you. I would like it for you and for them, for neither of them wishes for a greater happiness. Both desire nothing better than to look after your comfort; you are absolutely everything to them."

These plans were soon to be modified for, in the meantime, Pasteur had been introduced to the home of the University Rector, M. Laurent. He wrote him the following letter on February 10, 1849 – to ask for his daughter, Marie Laurent, in marriage:

SIR:

An offer of the greatest importance to me and to your family is about to be made to you on my behalf; and I feel it my duty to put you in possession of the following facts, which may have some weight in determining your acceptance or refusal.

My father is a tanner in the small town of Arbois in the Jura. I have three sisters. The youngest suffered at the age of three from a cerebral fever which completely interrupted the development of her intelligence. She is mentally a child, although adult in body. We expect to place her shortly in a convent where she probably will spend the rest of her life. My two other sisters keep house for my father, and assist him with his books, taking the place of my mother whom we had the misfortune to lose May, last.

My family is in easy circumstances, but with no fortune; I do not value what we possess at more than fifty thousand francs, and I have long ago decided to hand over my share to my sisters. I have therefore absolutely no fortune. My only means are good health, some courage, and my position in the University.

I left the Ecole Normale two years ago, an *agrégé* in physical science. I have held a doctor's degree eighteen months, and I have presented to the Academy of Sciences a few works which have been very well received, especially the last one, and upon which there is a report which I have the honor to enclose.

This, Sir, is all my present position. As to the future, unless my tastes should completely change, I shall devote myself entirely to chemical research. I hope to return to Paris when I have acquired some reputation through my scientific studies. M. Biot has often told me to think seriously about the Academy; I may do so in ten or fifteen years' time, and after assiduous work; but this is only a dream, and not the motive which makes me love science for science's sake.

My father will himself come to Strasbourg to make the proposal of marriage. No one here knows of the project which I have formed and I feel certain, Sir, that if you refuse my request, your refusal will not be known to anyone. . . .

P.S. I was twenty-six on December 27.

Thus, Pasteur took this most important step of his personal life a few weeks after having first met Marie Laurent, with the same impetuosity that led him to rapid and at times instantaneous decisions in his scientific career. His published letters to Marie Laurent give some measure of the intensity of his emotion: "I have not cried so much since the death of my dear mother. I woke up suddenly with the thought that you did not love me and immediately started to cry. . . ." "My work no longer means anything to me. I, who so much loved my crystals, I who always used to wish in the evening that the night be shorter to come back the sooner to my studies." But the disturbance caused in his life of labor by this sentimental explosion was only a ripple which did not really disturb the stream of discoveries and Pasteur resumed his scientific work immediately after his marriage on May 29. Many tragedies deeply affected his private life in subsequent years; the loss of his beloved father, the early deaths of two of his daughters and of his sister, the paralysis which struck him in 1868. But the ideal atmosphere of his conjugal life helped him to withstand these trials and to pursue uninterrupted the course of his productive life.

Marie Laurent was twenty-two at the time of her marriage – a gentle, graceful blue-eyed girl with a pleasant singing voice, whose joy of living mounted in silver tones as she went through her household duties. This gaiety of spirit she retained throughout the strenuous years ahead. When he proposed to her, Pasteur had nothing to offer but a life of study within modest material circumstances. At the most he could, in a moment of confidence, promise to gain immortality for their name. Madame Pasteur played her part in assuring this immortality by consecrating herself to her husband and to his dreams, and molding her behavior to fit the goal which he had formulated for their life. She accepted many limitations: a professor's small salary; his turning over to the purchase of scientific equipment the additional income derived from prizes; his odd mannerisms carried from the laboratory into the atmosphere of the home; and always the knowledge that work came first, even before the normal pleasures of a simple home life. She could write to her children in 1884: "Your father is absorbed in his thoughts, talks little, sleeps little, rises at dawn, and in one word continues the life I began with him this day thirty-five years ago."

All this, she understood and tolerated. The great part which she played in the achievements of the master has been described by Roux, who was Pasteur's associate for twenty years:

"From the first days of their common life, Madame Pasteur understood what kind of man she had married; she did everything to protect him from the difficulties of life, taking onto herself the worries of the home, that he might retain the full freedom of his mind for his investigations. Madame Pasteur loved her husband to the extent of understanding his studies. During the evenings, she wrote under his dictation, calling for explanations, for she took a genuine interest in crystalline structure or attenuated viruses. She had become aware that ideas become the clearer for being explained to others, and that nothing is more conducive to devising new experiments than describing the ones which have just been completed. Madame Pasteur was more than an incomparable companion for her husband, she was his best collaborator."

When she died in 1910 she was laid to rest near the companion with whom she had so completely identified her life. Because she was in truth the faithful partner of his human and divine mission, it is fitting that the Roman words *Socia rei humanae atque divinae* should have been engraved on her tomb.

The Strasbourg years reveal, in a forceful and often picturesque manner, the qualities which were to make of Pasteur one of the most adventurous and at the same time one of the most effective experimenters of his time. It is difficult, indeed, to visualize how the young and inexperienced professor could produce, against what would be for others the handicap of domestic happiness, such a varied harvest of new facts, scientific theories, and philosophical dreams. He had come to realize that the optical activity of organic substances could be used as a tool for the study of molecular structure; deep in his heart was also growing the hope that the study of molecular asymmetry would throw light on the genesis of life. To Chappuis he wrote in 1851: ". . . I have already told you that I am on the verge of mysteries, and that the veil which covers them is getting thinner and thinner. The nights seem to me too long, yet I do not complain, for I prepare my lectures easily, and often have five whole days a week that I can devote to the laboratory. I am often scolded by Madame Pasteur, whom I console by telling her that I shall lead her to posterity."

His scientific efforts increased with the broadening of his hopes and illusions. He undertook a strenuous trip through Central Europe in order to discover the natural origin of the paratartaric acid to which he had owed his first scientific triumph.

Back in Strasbourg, he used the money received from the Société de Pharmacie (as a prize for his synthesis of paratartaric acid) to secure additional laboratory equipment and the help of an assistant. His name was now widely known in chemical circles. Academic distinctions, the Legion of Honor, and even a proposal on the part of some of his admirers to introduce his name for membership in the Academy of Sciences, were all indices of the wide recognition gained by his chemical studies. Pasteur, however, had even larger dreams. Impressed by the fact that only living agents can produce optically active asymmetric compounds, he formulated romantic hypotheses on the relation of molecular asymmetry to living processes, and he undertook bold experiments aimed at creating life anew, or modifying it by introducing asymmetric forces in the course of chemical reactions. Thus, after ten years of disciplined work in the classical tradition, he had finally found a scientific outlet for his romantic mood. Madame Pasteur was referring to this phase of his work when she wrote to his father, obviously in reflection of her husband's most cherished hopes: "Louis . . . is always preoccupied with his experiments. You know that the ones he is undertaking this year will give us, should they succeed, a Newton or a

Galileo." Dumas, Biot and his other admirers tried in vain to discourage him from this search — worthy of an alchemist — but only the realization of his failure stopped him after a while, and he never forgot his exciting dreams. Even during the later crowded years, when he was involved in entirely different problems and in passionate controversies and struggles, he was to accept invitations to lecture on molecular asymmetry and never failed to reiterate the relation which he envisioned between this chemical property and the processes of life.

A combination of circumstances soon gave him the chance to direct his interest in the chemistry of life toward more attainable goals. In 1854, he observed that, in a solution of paratartrates infected with a mold, the "right" form of tartaric acid disappeared, whereas the opposite form persisted in the mother liquor. This revealed for the first time the close dependence of a physiological process — in this case the destruction of tartaric acid by a microorganism — upon the asymmetry of the chemical molecule. As Pasteur was pondering upon this extraordinary finding, a decree of the Minister of Public Education appointed him professor of chemistry and dean of science in the newly created Faculty of Lille — with the recommendation that he center his teaching, and his scientific activities, on the local industrial interests. The fermentation of beet sugar for the production of alcohol was one of the most important industries of the region of Lille, and Pasteur soon started to work on the problem of alcoholic fermentation. Madame Pasteur wrote to her father-in-law: "Louis . . . is now up to his neck in beet juice. He spends all his days in the distillery. He has probably told you that he teaches only one lecture a week; this leaves him much free time which, I assure you, he uses and abuses." Out of this episode came the celebrated studies on fermentations which brought Pasteur into intimate contact with the chemical phenomena of living processes, and eventually led him into the problems of disease. In 1857 he introduced the germ theory of fermentation before the Société des Sciences of Lille, stating his belief that all transformations of organic matter in nature would be found to be caused by various species of microorganisms, each adapted to the performance of a specific chemical reaction.

Pasteur's conception of his responsibilities went beyond his own scientific interests. With the most exacting conscience and driving energy, he adapted his teaching to the possible applications of chemistry to all the industries of Lille. He organized special laboratory demonstrations and exercises for the benefit of the young men who would soon pass from the university bench to the factory, and he arranged visits to centers of industrial activity in France and in Belgium. Within two years after his arrival at Lille, the scientific philosopher had been converted into a servant of society; and from that time on, most of his efforts were to be oriented, directly or indirectly, by the desire to solve the practical problems of his environment.

Late in 1857 Pasteur was appointed assistant director in charge of scientific studies and of general administration in his old Alma Mater, the Ecole Normale Supérieure in Paris. His duties included the supervision of housing, boarding, medical care and general discipline of the students, as well as the relations between the school and parents and other educational establishments. He did not take these new responsibilities lightly, as shown by his reports — in which he discussed, with thoroughness and vigor, the problems of household management, enforcement of disciplinary measures, and reorganization of advanced studies.

His new post did not provide him with either laboratory or research funds, for Balard had been replaced at the Ecole Normale by Sainte-Claire Deville, who had taken possession of the laboratories and of the credits allocated to the chair of chemistry. Undaunted by these difficulties, Pasteur found in the attic of the school two very small rooms abandoned to the rats, and he converted them into a laboratory which he equipped with funds from the family budget. The studies on alcoholic fermentation begun in Lille were completed in these miserable quarters. Their results were presented, in December 1857, before the Paris Academy of Sciences, with the conclusion that the conversion of sugar into alcohol and carbon dioxide was due to the activity of yeast — a microscopic plant. In the most emphatic terms, Pasteur stated that fermentation was always correlative with the life of yeast.

Eventually, he obtained from the authorities the appointment of an assistant whose time was to be given entirely to investigative work — an arrangement hitherto unheard-of. He was, furthermore, allowed to move his laboratory into a primitive "pavilion," consisting of five small rooms on two floors, which had been built for the school architect and his clerks. Crowded for space, and deficient in funds, he improvised under the stairway an incubator which could be reached only by crawling on hands and knees. Yet it was in this uncomfortable room that Pasteur daily observed, for long hours, the countless flasks with which he convinced the world that "spontaneous generation" was a chimera. After a few years, the small laboratory was enlarged by additional construction, and from these few rooms — so modest by modern standards — came the results of studies which made Pasteur's name famous in many fields of learning, and a household word wherever civilization prevails — a symbol of the benevolent power of science. To anyone familiar with the huge and palatial research institutes of today, there is a nostalgic charm in reading on a wall of the Rue d'Ulm, near a medallion portraying the master: *Ici fut le laboratoire de Pasteur.*

In 1860, the Academy of Sciences awarded Pasteur its prize for experimental physiology, in recognition of his studies on fermentation. Nothing could have given him keener pleasure — for, as he wrote to Chappuis and to his father, it was now his ambition to deal with "the mysteries of life and death." By these dramatic words, he implied the problem of "spontaneous generation" and the role of microorganisms in the transformation of organic matter and in the causation of disease.

Vainly had Biot and Dumas attempted to restrain Pasteur from entering the controversy on spontaneous generation — a problem which they considered too complex for experimental approach.

Pasteur persisted in his resolve, because he was convinced that the germ theory could not be firmly established as long as the belief in spontaneous generation persisted, and because he saw in the controversy a question of vast philosophical consequences. He had by then acquired such absolute confidence in his experimental skill, and was so well aware of his success as a scientific lecturer, that he no longer doubted his ability to deal with any problem, and to overcome any opposition. This complete faith in himself, often appearing as haughty conceit, is reflected in the supreme assurance with which he asserted that his results had final validity and were unassailable; in the scornful attacks which he directed at the claims of his opponents; in the way he challenged them to scientific debates and demonstrations before academic committees. The studies on spontaneous generation brought forth the first of the famous public controversies which are so peculiar an aspect of his scientific life. Henceforth, each one of the problems that he dealt with was the occasion of oratorical and literary debates in which he always triumphed over his opponents not only by the solidity of his facts, but also by his passionate vigor, and the eloquence and skill of his arguments. He became a crusader with an absolute belief in his creed, and also with an equal certainty that it was his mission to make it triumph.

This fighting spirit was not merely a manifestation of showmanship, but indeed an essential part of his scientific career. In many cases, Pasteur devised experiments to convince the scientific public of a truth which he had reached by intuitive perception; his most original demonstrations were often designed as blows to confound his adversaries. It appears best, therefore, to postpone until later a detailed account of these celebrated debates, as they contributed so much to the unfolding of the germ theory and to its introduction into the scientific consciousness of the nineteenth century.

In addition to the studies on spontaneous generation and on the distribution of microorganisms in the atmosphere, many spectacular findings crowded Pasteur's notebooks between the years

1860 and 1865: the discovery of butyric acid fermentation and of life without air, the role of yeasts and bacteria in the production of wine and vinegar and in the causation of their diseases; the demonstration that organic matter is decomposed through the agency of countless species of microorganisms; the teaching that "without the infinitely small, life would soon become impossible because death would be incomplete." Others before him had seen and described protozoa, fungi and bacteria; but it was Pasteur who had most clearly the prophetic vision of their importance in the economy of nature, and who revealed to the world "the infinitely great power of the infinitely small."

Long days in the laboratory, and heated debates in scientific academies, were not enough to satiate Pasteur's energy. He often pursued his studies in the field, wherever the demands of his problem led him. Experiments on the distribution of germs in the air were carried out in the quiet air of the cellars of the Paris Observatoire and on high peaks in the Alps. Many of the investigations on wines and their diseases took place at Arbois, in vineyard country, where a laboratory – not even supplied with gas – had first been improvised in a barroom, to the great surprise and confusion of the inhabitants and passers-by.

He lectured to chemical societies on the subject of molecular structure; to the vinegar manufacturers of Orléans on the scientific basis of their trade; to the lay public on the implications of the germ theory and of spontaneous generation. In grave, slow, low-pitched voice, he conveyed to his listeners the lucidity of his vision, the intensity of his convictions; like his fighting spirit, his eloquence was part of his scientific fiber. He was as eager to enlighten and convince the world as he was to discover the truth.

Despite all his triumphs, opposition did not abate and he was defeated twice for election in the Academy of Sciences. Finally, in December 1862, he was elected as a member in the mineralogy section, but with only thirty-six votes out of sixty. It is said that when the gates of the Montparnasse Cemetery opened next day, a woman walked towards Biot's grave with her hands full of flowers. It was Madame Pasteur, who was bringing them to the great teacher who had lain there since February 5, 1862, and who had loved Pasteur with so deep an affection. Biot had given to the young Pasteur the sanction of his learning and intellectual integrity. The disciple had now become an acknowledged master, and was to continue enlarging the patrimony of science for thirty more years, achieving a fame far beyond Biot's most loving dreams.

Nothing illustrates better the faith that some of his most distinguished contemporaries had in Pasteur's scientific prowess than the odd request made to him by Dumas in 1865. A catastrophic disease of silkworms was then ruining the production of silk in the south of France. Although Pasteur knew nothing of the disease, and had never seen a silkworm or a mulberry tree, Dumas now asked him to investigate the cause of the epidemic. Equally astounding is the fact that Pasteur dared to accept the challenge, and had the stamina to work on the problem for four consecutive years under the most strenuous conditions. The practical control of the silkworm epidemic demanded more than scientific perspicacity. To make his work of value to the silkworm breeders, Pasteur had to display the qualities of a successful industrialist concerned with economic necessities as well as with the technical problems; he had to be always ready to meet objections, always willing to adapt his language and procedures to the limited intellectual or scientific equipment of his public.

The silkworm campaign was a magnificent initiation into the problems of animal diseases, and it firmly convinced Pasteur that epidemics could be and therefore should be conquered. However, several years elapsed before he entered the field of human and animal pathology. This delay was due in part to his hesitation in dealing with the technical aspects of a problem for which he had no training, and which was the jealously restricted domain of physicians and veterinarians. Furthermore, unexpected circumstances forced him to limit his activity and to change for a time the direction of his interests.

In 1868 Pasteur was struck by a cerebral hemorrhage which endangered his life and caused a permanent paralysis of the left arm and leg. He was just beginning to recover his health when the Franco-Prussian War broke out, followed by the "Commune" uprising in Paris. While away from his laboratory, and even though distraught by national disasters and by worry over his son in the army, he still turned in thought back to his early scientific interests. As he had done twenty years previously, he planned experiments to introduce asymmetric forces in the course of chemical reactions and of plant growth. There was still present in his mind the hope that, to him, would be given the exciting adventure of modifying the course of living processes.

As the war ended, the immediate needs of his environment again took precedence over theoretical interests, and he considered it his duty to place his knowledge at the service of French economy. By a somewhat pathetic choice, he resolved to improve the quality of French beer, in order to show that French science could contribute to national recovery even in a domain where the superiority of Germany was obvious.

These studies on beer lasted from 1871 to 1876. With the financial help of industry, the laboratory of the Rue d'Ulm was transformed into a small experimental brewery; personal contacts with French and English brewers were established, and, within a short time, great progress was made toward the practical goal. This progress did not deal specially with the improvement of the *taste* of beer, but rather with the demonstration that — as had been found in the case of diseases of wine and vinegar — the spoiling of beer was caused by various foreign microorganisms. With this understanding, it became possible to minimize contaminations during the manufacture of beer, and to increase the keeping qualities of the finished product by the technique of "pasteurization."

These practical findings, it appears, took little time. But Pasteur seized the opportunity of this new contact with the problem of fermentation to probe more deeply into the chemical and physiological activities of yeast; to compare them with that of other living cells; and to reach, thereby, profound generalizations concerning the fundamental biochemical unity of living processes.

Once more, the natural philosopher claimed his right over the experimental technologist. The urge to understand nature had remained as pressing as the desire to answer the practical demands of society.

A few physicians had become aware of the potential significance of the germ theory for the interpretation of contagious diseases and epidemics. Most prominent among them was the Scotch surgeon Joseph Lister, who had been inspired to introduce the antiseptic method in surgical operations by Pasteur's demonstration of the widespread occurrence of microorganisms in the air. In 1873, Pasteur was elected associate member of the Paris Academy of Medicine, and immediately began an active participation in its debates, never tiring of pointing out to his colleagues the analogy between fermentation, putrefaction, and disease.

Medical science was then slowly approaching by a tortuous road a clear concept of infection, and was becoming aware of the part that microorganisms play in disease. In 1876 Robert Koch in Germany and Pasteur in France began independently the epochmaking investigations on anthrax from which historians date the germ theory of disease; their decisive experiments finally elucidated the riddle of contagion. By a prodigious effort through a period of ten years Pasteur established the fact that bacteria and filterable viruses can be the primary and sole cause of disease. He threw a flood of light on the mechanisms by which pathogenic agents spread through both animal and human communities and bring about, in susceptible hosts, those profound disturbances of normal physiology which may eventually culminate in death. Even more astounding, he recognized that prior contact with a microscopic agent of disease can render an otherwise susceptible host resistant to this agent; he worked out techniques by which the state of resistance — specific immunity — can safely be induced by first rendering the infective agent innocuous. The theory and the practices of immunization were applied by Pasteur himself to fowl cholera, anthrax, swine erysipelas and rabies, and they found widespread application to other diseases within his own time. When, in 1888, ill-health compelled him to abandon his tools, medical bacteriology and the sister sciences of immunology, public health and epidemiology had reached maturity, largely through his genius and devotion.

He had not solved the problem of life, but he had helped to push back the frontiers of death, and to render easier the sojourn of man on earth.

The mere recital of Pasteur's scientific achievements gives only a feeble idea of the intensity and fullness of his life. There were the ignorant to enlighten, the skeptics to convince, stubborn and prejudiced opposition to overcome. He never shirked from a fight, never accepted defeat, either in the laboratory, in the academies or in the field. He went to meet the physicians and surgeons in their hospitals, the veterinarians in the stables.

To convince farmers that protection of their cattle by vaccination was a practical possibility, he accepted, within a few months after the discovery of immunization, a challenge to submit his method to the severe test of field trial; this was at Pouilly le Fort, near Melun, and there in 1881 the survival of twenty-five vaccinated sheep made the world conscious that medicine had entered a new era. In July and October 1885, two peasant boys — Joseph Meister and Jean Baptiste Jupille — dangerously bitten by mad dogs, were brought to him in the hope that he could save them from rabies; he accepted the mental anguish of submitting the two boys to his method of treatment, which — without precedent in the annals of medicine, unorthodox in principle and unproven in practice — might have caused the death of those who had come to him as to a savior. Meister and Jupille survived, and the world went wild.

Antirabies treatment, as we shall see, may not be as effective or as practically important as was then believed; but by his courage, Pasteur had strengthened the faith of society in scientific medicine. Soon private and public funds were to flow into medical research.

It was not only to the promotion of his own work that Pasteur devoted his energies. Having recognized, with a sense of despair, that France was slowly losing her intellectual leadership through the neglect of her institutions of higher learning, he appealed to governmental authorities, and to the public, for the support of investigators and laboratories. Asked by Dumas to help in the preparation of a complete edition of Lavoisier's works, he undertook a thorough study of the great French chemist before writing an appreciation of his influence on the history of science. On the occasion of his election to the French Academy of Letters, he spent many days studying the philosophical faith and the life of his academic predecessor Littré, and took the occasion of the traditional eulogy to contrast with the exaggerated hopes of positivist philosophy his own conviction that philosophical and religious problems could not be analyzed by the methods of science. When the physiologist Claude Bernard was compelled by illness to abandon his studies for a year, Pasteur attempted to ease the forced retreat of his colleague by writing an enthusiastic account of Bernard's physiological and philosophical studies. Appointed professor of physics and chemistry at the School of Fine Arts, he refused to deal lightly with the subject. Instead, he prepared for his students critical analyses of the relation of architectural design to human comfort and health, scholarly accounts of the bearing of chemical knowledge on the practices of oil painting, and simple experiments to illustrate the properties of different oil pigments.

The exacting thoroughness which governed his behavior in the laboratory also characterized his participation in the affairs of the community. Unlike Faraday, who withdrew from the world to devote all his genius and energy to experimental science, and unlike most scientists, who abandon experimentation as soon as other responsibilities become too pressing, Pasteur managed to remain faithful to the laboratory while serving society. He was both a fervent scientist and an effective citizen.

National and personal tragedies brought into relief the full-blooded quality of his temperament. He had the narrowness and the exaltation of the patriot. The bombardment of Paris, and in particular of the Museum of Natural History, by the Prussians in 1871 inspired him to return, with words of anger and contempt, the honorary degree that he had received from the University of Bonn. Proud as he was of the glorious traditions of his country, he knew well that the France of the 1870's was no longer the leader of European thought which she had been in the eighteenth century. And yet, while he looked with envy and marvel at the vigor of civilization beyond the French borders, he retained unaltered his romantic attachment to his country: he always spoke of France with the same tender words that he used when speaking of his family.

Just as he had suffered in his patriotic affection, so he felt deeply the losses which bereaved his family. In 1865, while he was working on silkworm diseases in Alais, he received a telegram announcing that his father was very ill. He immediately started for Arbois, but arrived too late to see for one last time the man who had been his inspiration, his confidant, his guiding star – the symbol of family and country, of affection and duty. That night Pasteur, then forty-three years old, wrote to his wife from the old home where his character had been formed:

> DEAR MARIE, DEAR CHILDREN:
>
> Grandfather is no more; we have taken him this morning to his last resting place, close to little Jeanne's. In the midst of my grief I have felt thankful that our little girl had been buried there. . . . Until the last moment, I hoped I should see him again, embrace him for a last time . . . but when I arrived at the station, I saw some of our cousins all in black, coming from Salins; it was then that I understood that I could but accompany him to the grave.
>
> He died on the day of your first communion, dear Cécile; those two memories will remain in your heart, my poor child. I had a presentiment of it when, that very morning,

at the hour when he was struck down, I was asking you to pray for the grandfather at Arbois. Your prayers will have been acceptable unto God, and perhaps the dear grandfather himself knew of them and rejoiced with dear little Jeanne over Cécile's piety.

I have been thinking all day of the marks of affection I have had from my father. For thirty years I have been his constant care; I owe everything to him. When I was young, he kept me from bad company and instilled into me the habit of working and the example of the most loyal and best-filled life. He was far above his position both in mind and in character. . . . You did not know him, dear Marie, at the time when he and my mother were working so hard for the children they loved, for me especially, whose books and schooling cost so much. . . . And the touching part of his affection for me is that it never was mixed with ambition. You remember that he would have been pleased to see me the headmaster of Arbois College? He foresaw that advancement would mean hard work, perhaps detrimental to my health. And yet I am sure that some of the success in my scientific career must have filled him with joy and pride; his son! his name! the child he had guided and cherished! My dear father, how thankful I am that I could give you some satisfaction!

Farewell, dear Marie, dear children. We shall often talk of grandfather. How glad I am that he saw you all again a short time ago, and that he lived to know little Camille. I long to see you all, but must go back to Alais, for my studies would be retarded by a year if I could not spend a few days there now.

In this letter appears Pasteur's profound sentimentality, in which familial love, religious belief and sense of duty are so inextricably associated. But in the last sentence is also revealed another dominant aspect of his personality: the will to work and the urge to create, that no sorrow and no handicap could overcome. He had ignored the most extreme material difficulties in his garret at the Ecole Normale; he also ignored physical infirmity when he became partially paralyzed in 1868.

As soon as he began to regain his faculties, a week after the attack of paralysis, he dictated a scientific communication to his student Gernez, who was watching over him during the night. Within a few weeks, he started again for Alais to resume his studies on silkworm diseases, despite the difficulties of the trip, the lack of comfort of his southern quarters, and contrary to the advice of his physicians. For Pasteur was before all a man of indomitable will. It was not only his opponents that he wanted to overpower; it was also nature – it was himself. He was an adventurer and a conqueror, but one whose goal was to serve the inner God – the "enthusiasm" from which originate all great human actions. He had hoped that the mysteries of life and death would be revealed to him at the end of his journey. But, failing this romantic goal, there were still worth-while lands to discover and to conquer. In 1888, as he opened the new research institute to be called after him, he dedicated it with the following words:

". . . Two contrary laws seem to be wrestling with each other nowadays: the one, a law of blood and of death, ever imagining new means of destruction and forcing nations to be constantly ready for the battlefield – the other, a law of peace, work and health, ever evolving new means for delivering man from the scourges which beset him."

It was to serve peace, work and health that he had labored, fought and suffered with so much passion.

Pasteur's first published work dates from 1847, the last one from 1892; thus for almost half a century, the dauntless warrior had been before the scientific world, tirelessly working at the solution of theoretical and practical problems.

By 1885, he was an immensely famous man, honored by academies, entertained by princely and democratic rulers, acclaimed by specialists. But it was the antirabies treatment which assured his place in the heart of all civilized men, which made of him a hero in the golden legend of science. Within a short time after the treatment of Meister and Jupille an international subscription was opened to accumulate funds for the creation in Paris of an institute devoted to the treatment of rabies, and to the prosecution of microbiological and biochemical research. Of this building,

Pasteur could say that "every stone of it is the material sign of a generous thought. All virtues have co-operated to raise this dwelling of labor."

On December 27, 1892, his seventieth anniversary was the occasion of a solemn jubilee in the great amphitheater of the Sorbonne, attended by the President of the French Republic and by delegations of French and foreign institutions of learning. As emphasized by one of the official orators, it was not merely a great scientist who was the hero of the day, but a man who had devoted all his strength, his heart and his genius to the service of mankind.

". . . Who can now say how much man owes to you and how much more he will owe to you in the future? The day will come when another Lucretius will sing, in a new poem on Nature, the immortal Master whose genius engendered such benefits.

"He will not describe him as a solitary, unfeeling man, like the hero of the Latin poet; but he will show him mingling with the life of his time, with the joys and trials of his country, dividing his life between the stern enjoyment of scientific research and the sweet communion of family intercourse . . ."

Unable to speak for emotion, and compelled to extend his thanks through the voice of his son, Pasteur then expressed for a last time in public his conviction that science would some day bring happiness to man.

". . . Delegates from foreign nations, who have come from so far to give France a proof of sympathy: you bring me the deepest joy that can be felt by a man whose invincible belief is that Science and Peace will triumph over Ignorance and War, that nations will unite, not to destroy, but to build, and that the future will belong to those who will have done most for suffering humanity."

Addressing the students, he recalled the rich satisfactions which he had derived from his years of toil and expressed his undying confidence in the power of the experimental method to improve the lot of man on earth.

Ceci est mon testament
Je laisse à ma femme
tout ce que la loi
me permet de
lui laisser

Puissent mes
enfants ne jamais
s'écarter de la voie
du devoir et garder
toujours pour leur
mère la tendresse
qu'elle mérite

L. Pasteur

Paris ce 29 mars 1877

Arbois ce 25 août 1880

This is my testament:

I leave to my wife everything that the law allows me to leave her. May my children never depart from the line of duty and always retain for their mother the love that she deserves.

Paris, March 29, 1877
Arbois, August 25, 1880

"Young men, have faith in those powerful and safe methods, of which we do not yet know all the secrets. And, whatever your career may be, do not let yourselves be discouraged by the sadness of certain hours which pass over nations. Live in the serene peace of laboratories and libraries. . . ."

Now, his strength was gone. He entered his new Institute an ill and exhausted man, broken by time and by endless toil. He had still a few years to live. To anyone else, these might have brought the happiness of well-deserved rest and recognition, reward of a rich and productive life. The microbiological sciences which he had done so much to create were growing before his eyes; the great Institute which bore his name was a humming hive of research and an international center of learning; honors came to him from everywhere and a happy family surrounded his leisurely days. But how empty was his life now that scientific creation was no longer permitted him! How tragic the vision of the passionate adventurer and conqueror, now armed with the material means that he had lacked in the past, his mind still clear, his dreams still living, but his body too weak to start again on the endless trail!

On November 1, 1894, he was seized with a violent attack of uremia from which he only partially recovered. On the following New Year's Day, he could enjoy in the laboratories of the Pasteur Institute a display, especially prepared for him, of the flasks, cultures and other specimens, companions of his celebrated studies. The bacilli of diphtheria and of bubonic plague – recently isolated – were also on exhibition as symbols of the magnification of his own work. His interest in science was still alive, also his patriotic fervor. When asked if he would accept from the German Emperor the badge of the Order of Merit, he refused. He had not forgotten 1871. The old fiery heart was still burning.

On June 13 he left Paris for a period of recovery at a branch of the Pasteur Institute at Villeneuve l'Etang, in the park of Saint-Cloud. For a few weeks, he could continue his dreams under the noble trees of the park, surrounded by his family and disciples.

Rapidly, however, his paralysis and weakness increased; his speech became more and more difficult. It is reported that on September 27, as he was offered a cup of milk, he refused it with the words: "I cannot." These were his last words. For the first time, he yielded; not to obstacles, not to opposition, not to men — only to a power greater than man, to the Death which he had fought with all his genius and all his heart. The next day, September 28, 1895, in the late afternoon, he died, his body almost entirely paralyzed, one of his hands in that of his wife, the other holding a crucifix.

The monastic simplicity of the room in which he passed away is an expression of the austerity of his life; and the gorgeously adorned chapel, in which his tomb was set, a symbol of the place that he occupies in the memory of men.

CHAPTER III

Pasteur in Action

> The painter or draughtsman should be solitary, so that physical comfort may not injure the thriving of the mind, especially when he is occupied with the observations and considerations which ever offer themselves to his eye and provide material to be treasured up by the memory. If you are alone, you belong wholly to yourself; and if you are accompanied even by one companion, you belong only half to yourself; and if you are with several of them, you will be ever more subject to such inconveniences.
>
> — LEONARDO DA VINCI

SPEAKING at the unveiling of Jean Baptiste Dumas's statue in 1889, Pasteur contrasted the rich life of his revered master with that of the scientists who keep aloof from the social implications of their activities:

"Among superior men, there are those who, isolating themselves in their studies, have for the public turmoil of ideas only disdain, pity and indulgent condescension. Unconcerned with the general public opinion, they aim at exerting a direct influence only on narrow, selected circles. Should this elite fail them, they still find in the spectacle of their own intelligence an acute and lasting pleasure. . . .

"There are a few men who are equally at ease in silent labor and in the debates of the large assemblies. Above and beyond their personal investigations, which assure them a special place in posterity, they keep their minds attentive to all general ideas, and their hearts open to generous sentiments. These men are the guides and protectors of nations. . . .

"Still others, finally, carried away by the eagerness to see their ideals triumph, throw themselves into the struggles of public life."

Pasteur probably thought of himself in pronouncing these words. He had labored in the silence and solitude of libraries and laboratories, but he had also shared the practical problems of his time in the shops and in the fields, fighting whenever necessary at the tribunes of academies, in the technical press and before the lay public. Like Dumas he had been a great scientist, a great teacher, but also an effective man of action and organizer. Unlike him, however, and unlike most scientists who abandon the laboratory bench when the call of administrative responsibilities becomes urgent, he did not pass successively through these different phases; he lived all of them at the same time. Single-handed, he could carry a given problem from the level of abstract concept, through the exacting discipline of the experimental method, to the hustle and bustle of practical life. The whole of science was his province — its dreams, discipline, controversies, struggles, triumphs, and practical realizations. Home and the market place, as well as laboratories and academies, saw him function in all the expressions of the scientific way of life. This was perhaps the most characteristic aspect of his genius.

For many years, he worked alone. When, later, young men came to join him, they participated in the execution of his work, but rarely contributed to the elaboration of his thoughts. He often left his assistants completely ignorant of the strategy of his investigations, revealing to them only the part essential to the task of the day. "He kept us remote from his thoughts," said Duclaux, his most intimate student and associate, who also spoke of "the Olympian silence with which he loved to surround himself until the day when his work appeared to him ripe for publicity. He said not a word about it, even in the laboratory, where his assistants saw only the exterior and the skeleton of his experiments, without any of the life which animated them.

"Very briefly, without unnecessary explanation, Pasteur would indicate to each his task and send him away to attend to his observations."

Loir, who was Pasteur's technical assistant from 1884 to 1888, has recently confirmed Duclaux's account. "He wanted to be alone in his laboratory and never spoke of the goal he had in mind. Pondering over his notebooks, he would write on small cards the experiments that he wanted to have done and then, without explaining anything, would ask his assistants . . . to do them."

Even during the periods of greatest activity, there were but few assistants; each had his room, or his corner in the main laboratory, where he worked in silence, without disturbing the master except when called upon for discussion. When Bertin became director of sciences at the Ecole Normale, he urged Pasteur to study the effect of physical agents on the activity of microorganisms. Pasteur appeared interested and Bertin appointed the young physicist Joubert as assistant. Soon, however, Joubert had to handle a Pravaz syringe to share in the work of the laboratory. As this was outside his own line of interest, he left, to be replaced by other young physicists also introduced by Bertin. Several came in and soon left, discouraged by the little interest that Pasteur took in their presence. For Pasteur, the choice of a collaborator was a matter of little concern, in fact a very secondary thing. There had to be one, since the position was open; but who it was mattered little, provided he did faithfully and in silence the work which was asked of him.

Pasteur's meditations could proceed only in silence, and the presence of any visitor foreign to his occupations was sufficient to disturb him; only persons working on his problems were welcome in the laboratory. Once when he had gone to visit Wurtz at the School of Medicine, he found the chemist at work amidst his pupils, in a room full of activity, like a humming beehive. "How," exclaimed Pasteur, "can you work in the midst of such agitation?" "It excites my ideas," answered Wurtz. "It would put mine to flight," retorted Pasteur.

The laboratory was opened to very few and one could penetrate it only by ringing the bell at the main door. Pasteur was not cordial even to his friends when he was at work; to interrupt him was to make him unhappy. "I can still see him," wrote Roux, "turning toward the intruder, waving his hand as if to dismiss him, and saying in a despairing voice, 'No, not now, I am too busy.' And yet, he was the most simple and most hospitable of men; but he could not understand how anyone could dare to disturb a scientist at work on his notes. When Chamberland and I were in the course of an interesting experiment, he would watch around us, and seeing from far through the window our friends coming to fetch us, he would meet them himself at the door to send them away."

He was not eager to accept physicians in his laboratory, even during that period when he was engaged in medical research. He felt that the demands upon them were too varied to allow them to focus attention on specific problems, and to achieve the concentration essential to investigative work. He was irritated also by the trend in medical circles at that time to discuss any subject in florid and eloquent language, instead of by the factual statements of experimental science. Moreover, "Physicians are inclined to engage in hasty generalizations. Possessing a natural or acquired distinction, endowed with a quick intelligence, an elegant and facile conversation . . . the more eminent they are . . . the less leisure they have for investigative work. . . . Eager for knowledge . . . they are apt to accept too readily attractive but inadequately proven theories."

Physicians may interpret this statement as a manifestation of inferiority complex on the part of the chemist toward the art of medicine. In reality, it merely expresses the traditional feud between investigator and practitioner. Claude Bernard, although trained in the Paris School of Medicine during its period of greatest clinical glory, shared Pasteur's irritation at his medical colleagues. Have you noticed, he would say, how physicians, when walking into a room, always carry about themselves an air that seems to imply "Look at me, I have just saved another life"? This was in the middle of the nineteenth century. Today Pasteur and

Bernard might find material for their scorn in the scientist who, unmindful of the long history of the world prior to his efforts, entertains the illusion that his last experiment will open a new era in thought.

Pasteur did not want the hustle and bustle of medical life to disturb the peace of his laboratory. He had arranged for the clinical aspects of the work on rabies, administered by Dr. Grancher, to be carried out in an annex a few blocks away from the Ecole Normale, and he saw to it that the casual interest of the medical visitor should not introduce confusion in the disciplined and meditative atmosphere of his sanctuary. He was extremely shocked at learning from Loir that Grancher's laboratory had two fine armchairs, one of them a rocking chair. "Pasteur could not understand that one could feel the need of physical comforts in a laboratory. It confirmed his conviction that he should keep his own quarters closed to people with such ideas." Even smoking was an unwelcome dissipation, that could be indulged in by Pasteur's assistants only while he was away.

Silence was especially imperative while the master was formulating the next phase of his experiments — his "working hypotheses." For days he would then absorb himself in the study of his notebooks, remaining isolated from everybody and everything, ignoring the presence of his collaborators, and not even raising his head for hours in succession. Thus he could recapture from his past experience the immense wealth of observations and imaginative thoughts recorded in the tidy pages crowded with his small writing. From them would emerge those fragments which — attracted and held as in a field magnetized by the intensity of his thoughts — organized themselves into new and unexpected patterns.

When he had completed the study of his notes, the new phase would begin. Now he would construct the tentative plot of his next scientific story — the "working hypothesis," from which would be born the project of the next experiment At that time, he would pace the floor for hours without speaking; he would even continue his silent monologues at home, walking back and forth from one room to the other. These solitary meditations lasted for days. During that period he was so absorbed in his thoughts that he was unaware of the presence of persons around him. Duclaux often had to wait long hours before being asked the object of his visit. Anyone who had to bring up an urgent matter found it necessary to insist in order to force his attention. Then he reacted as if waking out of a dream, but never with impatience; he slowly turned toward the interrupter, passing his hand over his face several times in a familiar gesture.

After his ideas had taken shape, he re-established contact with his collaborators, telling them only enough of his dreams, goals and plans to formulate the technical details of the experiments. Exploratory tests were few in number, but designed with extreme care to determine whether the hypothesis had a factual foundation. If the results were negative, the tentative ideas were immediately rejected from his mind, and it became useless to bring them back to his attention; he would not even remember them. If, on the contrary, positive results suggested that the hypothesis might be valid, experiments were tirelessly multiplied to explore and develop its possibilities.

Experiments were usually carried out as soon as they had been sufficiently discussed and prepared. This rapid passage from conception to execution accounts, in part, for Pasteur's phenomenal scientific productivity. He was never discouraged by obstacles, a quality that he regarded as one of his greatest assets – "Let me tell you the secret which has led me to the goal. My only strength resides in my tenacity" – a judgment which has been confirmed by Roux: "How many times, in the presence of unforeseen difficulties, when we could not imagine how to get out of them, have I heard Pasteur tell us, 'Let us do the same experiment over again; the essential is not to leave the subject.' "

The unmatched reproducibility of Pasteur's findings – either of the period when he worked alone or after paralysis forced him to delegate the execution of his experiments to others – is sufficient evidence of the precision of his work. He was not, however, much interested in laboratory techniques as such, but demanded only that they be well adapted to answer his questions, to settle the truth of his hypotheses, and to permit the formulation of effective and dependable procedures. Above all, the answer must be unequivocal, for he wanted to be able "everywhere and always to give the justification of principles and the proof of discoveries."

Most of his experiments were simple in design and execution, but all details were carried out, observed and recorded with the most exacting attention. The loving care with which he prepared and handled crystals of organic substances for the studies on molecular structure becomes the more impressive when it is realized that all measurement of the angles of the crystalline facets, and of the deflection of optical activity, had to be carried out with homemade instruments. The isolation, transfer, and cultivation of microbial cultures had to be done without the benefit of bacteriological equipment; autoclaves did not exist; the use of gelatin and agar plates had not yet been introduced; an incubator had to be improvised in a corner of the stairway. Although many of Pasteur's conclusions rested on microscopic studies, he used only the most simple techniques and equipment to familiarize himself with the microbial world. All cultures were examined directly in the living state until 1884, when stains and the oil immersion lens were introduced into his laboratory from Germany. Nevertheless, with his primitive means of observation, and without any prior training in biological microscopy, Pasteur recognized new species of microorganisms, differentiated their physiological states, and could diagnose diseases of vinegar, wine and beer, as well as of man and animals, with an accuracy unexcelled at the time.

When his attack of paralysis in 1868 deprived him of the use of his left hand he had to depend upon assistants for the performance of most of his experiments. Himself a masterful laboratory worker, he was very exacting of others. On hearing his assistants point out that an experiment he wanted to have done presented special difficulties, he would say: "It is your responsibility; arrange it anyway you like, provided it is done, and well done." The description left by Loir shows with what care he supervised the technical phases of his work.

"At the time of transferring cultures, Pasteur and I went to the incubator with a small tray for transporting the flasks. They were brought to a special, small room which was never opened on any other occasion and they were kept there for exactly two hours in order to allow them to reach room temperature undisturbed. After that time had elapsed, we returned to the small room, still without speaking and with a minimum of motion. I sat at the table and Pasteur sat on a chair behind me, a little to the side and two feet behind, in order to be able to see everything I did. On the table was placed a wire basket containing long, sealed sterile pipettes. I took one, broke the tip, flamed it . . . before using it for inoculation. The platinum wire [1] did not come in use in the laboratory until 1886."

Like the performance of the experiment, the observation of results was a ritual of which no detail could be slighted. "It is necessary to have seen Pasteur at his microscope," Roux said, "to form an idea of the patience with which he could examine a preparation. In fact, he looked at everything with the same minute care. Nothing escaped his nearsighted eye; and jokingly, we used to say that he could see the microbes grow in his bouillons."

Long hours of silent observation were also devoted to the infected animals, their surroundings, and their behavior. He would stand in a corner of the basement (where the animals were kept) with a card in his hand, watching for hours the motions and attitudes of an infected chicken. If perchance anyone should go down without knowing that the master was already there, he would signal to remain silent, and continue his observations.

Then back at his small desk he would stand, writing everything he had observed. He demanded of his collaborators an exact account of their own phase of the work, asking for the most minute

[1] Loops made of platinum wire are now often used instead of glass pipettes for the transfer of bacterial cultures.

details; he insisted on writing down, himself, all available information, as if to make more completely his own — part of his very flesh and mind — everything pertaining to the work. "He did not leave to anyone the responsibility of keeping the laboratory notebook up-to-date. He himself took down the information that we gave him, in all its details. How many pages he thus covered, with small irregular and crowded handwriting, with drawings in the margins, side and footnotes, the whole entangled and difficult to read for anyone not used to it, and yet kept with extraordinary care. Nothing was written down that had not been duly observed, but once it was written, it became for Pasteur an incontestable truth. When, during a discussion, he would bring the argument 'It is in the notebook,' no one would dare to discuss the problem further.

"Once the notes were taken, we would discuss the experiments to be undertaken, Pasteur standing at his desk, ready to write down what would be agreed upon, Chamberland and I facing him, leaning against a cabinet. This was the important time of the day; each would give his opinions, and often an idea at first confused would be clarified by the discussion and lead to one of those experiments which dissipate all doubts. At times we disagreed and the voices would be raised; but although Pasteur was regarded as opinionated, one could express one's mind to him; I have never seen him resist a reasonable opinion.

"A little before noon, someone came to call Pasteur for lunch; at half-past twelve he returned to the laboratory and, usually, we found him motionless near a cage, observing a guinea pig or a rabbit. Around 2 P.M. Madame Pasteur sent for him, as he would otherwise forget the meetings of academies and committees of which he was a member. . . . He returned around 5 P.M., wanted to be informed immediately of what had been done, took down notes on it and verified the labels of the experiments. Then, he would report on the most interesting papers heard at the Academy and would discuss the work in progress."

It is this extreme devotion to all the details of the work, this complete knowledge and mastery of all the facts pertaining to his experiments, which gave Pasteur such absolute confidence in his own results and assured their reproducibility in all cases. Because of this confidence, he never hesitated to challenge his opponents before academic commissions, as he knew that he could always duplicate his results; because of it also, he accepted the incredibly drastic terms of the public test on the vaccination of sheep against anthrax – "What succeeded with fourteen sheep in the laboratory will succeed just as well with fifty at Melun." And it did – not only in Melun, but wherever his detailed instructions were followed to the letter, not only in the case of anthrax vaccination, but in all cases where investigators had enough energy, patience and loving care to respect in all their details the instructions issued from the infallible notebooks.

The investigation into fermentations brought Pasteur into contact with the practical world and he soon developed an acute awareness of the power of the scientific method in increasing the effectiveness of technological operations. He did not share the common belief that pure science and applied science correspond to two independent forms of intellectual activity, demanding different gifts from those engaging in them. He felt that sound training in the theoretical disciplines was an adequate preparation for the task of giving scientific findings practical application, and expressed these views on many occasions – for example, in a letter written in 1863 to discuss the organization of professional teaching:

> There are no applied sciences . . . There are only . . . the applications of science, and this is a very different matter. . . .
>
> . . . We must place professional teaching in the hands of professors as well trained as possible in the theory, principles and methods of pure science, but of whom we shall ask, in addition, that they show an interest in the applications of science. Is it possible rapidly to procure professors with these qualifications without resorting to expensive innovations?
>
> . . . Yes, certainly, because the study of the applications

of science is easy to anyone who is master of the theory of it. . . .

The studies on beer, started in a brewery near Clermont-Ferrand after the Franco-Prussian War, were continued in Paris. There a pilot plant was established in the laboratory of the Rue d'Ulm. Chemical studies went on in the large room of the ground floor, while boilers and fermentation vats crowded the basement. The austere atmosphere of the laboratory was mellowed for a time by the aroma of fermenting barley and hops, while on the days of degustation the clinking of glasses and the laughter of Bertin (who acted as expert beer taster) dispelled the atmosphere of silence of the sanctuary.

Much of the work on vinegar and wine was carried out in Orléans and at Arbois in direct contact with producers. The technical development of pasteurization required many consultations with them to make sure that the degree of heating was not such as to spoil the taste of the product, and with engineers in order to work out the practical aspects of the method. Pasteur's publications on the subject present detailed specifications with drawings of equipment for carrying out pasteurization on an industrial scale. He did not neglect to consider the cost of operation and to discuss other economic aspects of the preservation of foods and beverages by heating.

In order to study the diseases of silkworms Pasteur trained himself, his assistants and even his family, in the practical operations involved in the production of silk. He was not satisfied with establishing only the scientific validity of the egg-selection method; he wanted also to prove that it was practically feasible and economically profitable. For several months every year between 1866 and 1870 he behaved as if he were the director of a commercial enterprise and sent his assistants all over the south of France to teach his method to the silkworm breeders.

The vaccination against anthrax and swine erysipelas brought forth responsibilities similar to those of the control of silkworm diseases. To meet the cost of experimentation with farm animals, it was necessary to enlist the interest and support of governmental bodies and agricultural societies. The laboratory was moved to farm or pasture whenever the problem called for studies in the field. For the first time, bacteriological research was being carried out on a large scale as a part of national economy. The commercial production and distribution of the vaccine was turned over to Chamberland, who established an annex of Pasteur's laboratory at Rue Vauquelin close to the Ecole Normale. It was not easy to convince skeptics that vaccination was a profitable operation because it entailed the risk of a few animals dying following the injection of the attenuated vaccine. Pasteur proposed the organization of an insurance company to protect farmers against unavoidable losses, and thus help in overcoming their resistance.

After 1875, the experimental brewery in the cellar of the laboratory was dismantled to be replaced by a small animal house and hospital, for the study of contagious diseases. With the initiation of the work on rabies larger animals' quarters became necessary – but they were not easy to secure, as everywhere the public was terrified at the thought of rabid dogs being housed in the vicinity of residential dwellings. Finally, however, large kennels were established at Garches, in a former state domain close to the park of Saint-Cloud.

The successful outcome of Meister's and Jupille's treatment brought about a sudden demand for immunization of persons bitten by rabid animals. The preparation of the rabies vaccine, the treatment of patients, the collection and analysis of statistical evidence, all received Pasteur's personal supervision. Makeshift arrangements had to be improvised for the maintenance of large numbers of rabbits and for the desiccation of the infected spinal cords of rabbits used in the preparation of the vaccine. The medical administration of the new treatment for which there was no precedent required difficult decisions. The housing of patients arriving from all parts of the world, often without adequate resources, presented unexpected problems which were solved by emergency measures.

"Many," wrote Duclaux, "have described the strange spectacle offered by the laboratory and the courtyard near to it, where assembled a picturesque and polyglot crowd of bitten individuals come to beg of science the end of their apprehensions and the certainty of tomorrow. But what has not been mentioned enough is the contagious confidence which spread through all the newcomers – and made of them believers whose faith contributed to their recovery.

"Laboratory and consultation room soon became too small; we had to leave the hospitable Rue d'Ulm to establish ourselves on larger grounds borrowed from the former Collège Rollin. It was while we were camped there that the international subscription was opened which resulted in the creation of the Pasteur Institute."

The Institute was organized to provide more extensive facilities for the treatment of rabies and also for the prosecution of microbiological, biochemical and physiological sciences. Pasteur's dream of large research laboratories with adequate resources for investigation had finally come true, but as he entered the magnificent institute, his strength failed him; he was "a man vanquished by time." Yet he continued to haunt the laboratories, following with eagerness the work of his disciples. He was a symbol of the great creators who, despite poverty and at the cost of sacrifices and suffering, establish the foundations of science that less gifted men may continue to add slowly to the great structure arising from the struggles of genius.

Theoretical studies in the laboratory and practical tests in the field were not sufficient to satisfy Pasteur's eagerness to prove the validity of his convictions. He acknowledged three steps in the establishment of evidence: first to try "to convince oneself . . . then to convince others . . . the third, probably less useful, but very enjoyable, which consists in convincing one's adversaries." As a vigorous fighter, he derived great satisfaction from overcoming his opponents. Because he believed in the importance of his work, he was eager to see it known and accepted everywhere, and was often unwilling to wait for the judgment of time.

His discoveries and observations were quickly reported through brief communications to learned societies and in letters to his masters and colleagues. In addition, he carried his message to the world in the form of polished and formal lectures to scientific academies, as well as to technical and lay audiences. In limpid, forceful and at times eloquent language, he summarized on these occasions his experimental studies, and also their philosophical and practical implications.

He wrote long and detailed letters to clarify public opinion on matters which he considered of importance, to defend his work and his viewpoints, also to educate his followers as well as his opponents in theoretical principles and technical procedures. These letters were to individuals and to the press, and he often elected this latter channel to publicize the application of his discoveries.

For example, he reported in a trade journal the manner in which he convinced the Mayor of Volnay, M. Boillot, of the effectiveness of controlled heating as a means of preserving wines:

I beg the Society to allow me to publish my interview . . . in dialogue form. The teachings to be derived from this conversation will thus reach more effectively those who could profit by them. . . .

PASTEUR: Do you heat your wines, M. Mayor?

M. BOILLOT: No, sir . . . I have been told that heating may affect unfavorably the taste of our great wines.

PASTEUR: Yes, I know. In fact, it has been said that to heat these wines is equivalent to an amputation. Will you be good enough, M. Mayor, to follow me in my experimental cellar . . . ? Here are rows of bottles of your great vintages which have been heated, and there, bottles of the same vintages not heated. The comparative experiment dates from 1866, more than seven years ago . . .

[For two pages, Pasteur describes in great detail how the mayor, after tasting heated and unheated wines, had to acknowledge the superior keeping quality of the former, even in the case of the products from his own vineyards.]

> M. Boillot: I am overwhelmed. I have the same impression as if I were seeing you pouring gold into our country.
>
> Pasteur: There you are, my dear countrymen, busy with politics, elections, superficial reading of newspapers, but neglecting the serious books which deal with matters of importance to the welfare of the country, indeed to your own interests. I suppose you consider it might demand too much effort to understand and follow the wise advice of those who labor on your behalf, often at the sacrifice of their own health.
>
> M. Boillot: Do not be mistaken, sir. I had read in the Proceedings of the Academy that your process preserves and improves our wines, but I have read also on the following page the statement by some of your colleagues that heating can spoil the flavor. How do you expect us, poor vintners, to decide?
>
> Pasteur: . . . You are revealing there one of the bad traits of our national character. . . . Our first inclination is to doubt the success of others. And yet, M. Mayor, had you read with attention, you could have recognized that everything I wrote was based on precise facts, official reports, degustation by the most competent experts, whereas my opponents had nothing to offer but assertions without proof.
>
> M. Boillot: . . . Do not worry, sir. From now on, I shall no longer believe those who contradict you and I shall attend to the matter of heating the wines as soon as I return to Volnay . . .

Even more peculiar to Pasteur were the passionate and celebrated controversies which—because of his vitality, his conviction and his genius—gave such a picturesque and often dramatic flavor to the heroic age of microbiology. In subsequent chapters, we shall consider in detail the controversies on problems of theoretical interest—with Liebig on the germ theory of fermentation; with Pouchet and Bastian on spontaneous generation; with Claude Bernard and Berthelot on the intimate mechanisms of alcoholic fermentation; with Colin on anthrax of chickens; with Koch on the efficacy of anthrax vaccination; with Peter on the treatment of rabies. There were also conflicts involving priority rights, or those arising simply from the clash of incompatible personalities. Whatever the cause of the argument, scientific or personal, Pasteur handled with the same passion those whom he believed to misrepresent truth, or to be prejudiced against him.

When young, he communicated his feelings to the members of his family. Thus at the occasion of his first defeat for election to the Paris Academy of Sciences he wrote to his wife of his contempt for the "mediocrities who control the election," looking forward to the day when he could "read before them a fine memoir while thinking: *Fools that you are, try to do as well.* . . . I am now speaking of that fool of — and of — and of so many other nonentities who have arrived where they are only because there was no one else, or by sheer luck."

In 1862, as his name was presented to the Academy for the third time, attempts were made in certain groups to minimize the significance of his studies on tartaric acids. The situation was critical as the vote was expected to be close. Duclaux's account of Pasteur's immediate reaction will serve to introduce this "spirit of combativity which constitutes one of the facets, and not the least curious, of his scientific temperament."

"That evening there was to be a meeting of the Société Philomathique where it was likely that many important scientists would be present. . . . I was dispatched to a cabinetmaker and came back with screws, files and a long block of pine lumber. It was ten years since M. Pasteur had touched the problem of the tartrates, but he still had their crystalline forms at his finger tips. A few strokes of the saw, guided by him with a marvelous surety, sufficed to transform the lumber into a series of crystalline forms with their faces and facets . . . which were rendered more readily distinguishable being covered with different colored papers.

"His exposition began as a lesson . . . But in ending, M. Pasteur challenged his contradictors to confess their ignorance, or their bad faith . . . telling them, in essence, 'If you understood the question, where is your conscience? And if you did not understand it, how did you dare speak of it?' M. Pasteur has since won many oratorical victories, but I know of none better deserved than the one gained by this penetrating improvisation. He was still in ebullition as we walked toward the Rue d'Ulm, and I remember making him laugh by asking him why . . . he had not thrown his wooden crystals at the heads of his opponents."

Pasteur became more and more inflammable as time went on. Not satisfied with challenging his opponents to disprove his claims, he heaped scorn upon their ignorance, their lack of experimental skill, their obtuseness or even their insincerity. From his desk at the Academy of Medicine, he pointed out to clinicians the emptiness of their debates, the uncertainties of their premises and conclusions, and contrasted the vagueness of the clinical art with the assured power of the new physiological and microbiological sciences. In sentences which often betrayed an irritating haughtiness under their outward pretense of humility, he lectured to physicians on the germ theory of fermentation, its application to putrefaction, gangrene and contagious disease, and the "fruitful fields of the future" which it opened to medicine. He told the verbose and facile clinician Poggiale, his colleague at the Academy of Medicine: "I refuse the right to verify, question or interpret my findings . . . to anyone who is satisfied with reading my studies in a superficial manner, his feet at the fireplace." To those who, like Colin, misunderstood or misapplied the principles of the experimental method, he scornfully pointed out that one single positive finding of his was worth more than a hundred negative experiments. "M. Colin," he told the Academy, "looks into 98 obscure closets and concludes from this that light is not shining outside." Or again, "There is only one road which leads to truth, and one hundred that lead to error; M. Colin always takes one of the latter."

When he despaired of convincing his colleagues of the Academy, he would address, over their heads, the young physicians and students who attended the meetings:

"Young men who . . . are perhaps the medical future of our country, do not come here to be entertained by the excitement of polemics, but to learn of methods. Know then, that I denounce as an example of the most detestable of methods the reasoning which leads M. Colin to conclude from a negative observation that there exists in the inoculation material of anthrax a virulence factor other than the bacteria . . .

"I denounce as a reasoning worthy of a Molière comedy, of Molière ridiculing the medical spirit of his time, the following paragraph from one of M. Colin's replies. 'I do not know exactly what the anthrax bacteria are. It is not absolutely sure that they are living beings. . . . Is it impossible that they are of the same origin as the anatomical structures . . . ?' "

Irritated by the tendency of his colleagues to trust in argument and eloquence rather than in the accurate statement of facts, he even dared lecture the Academy of Medicine as a body on the proper manner of conducting scientific debates.

"You were asking yourself how the Academy could introduce . . . the true scientific spirit in its works and discussions. Let me give you a method which would not be a panacea but which certainly would be useful. We should resolve never to call this desk a tribune, or a communication presented from it a discourse, or the one who is speaking an orator. Let us leave these expressions to political assemblies, deliberating on topics which do not lend themselves to factual demonstration. These three words — tribune, discourse, orator — appear to me incompatible with scientific rigor and simplicity."

Roux has described the intensity with which Pasteur reacted to the famous discussions on the germ theory of disease, which took place in the Paris Academy of Medicine.

"He would leave the meetings in a great state of emotion. M. Vallery-Radot, Chamberland and I often waited for him as he came out. 'Have you heard them! To experiments they reply only by speeches,' he would say. His irritation slowly subsided as we walked home; and he imagined further experiments, to bring more light, for contradictions excited him to new investigations . . .

"Pasteur's passion for science sometimes carried him to conclusions of an amusing naïvety. For him, a man guilty of a bad experiment or of unsound reasoning could not be trusted in any way. Once when he was reading to us in the laboratory a piece of work which he considered particularly bad, he exclaimed with irritation: 'I should not be surprised if a man capable of writing such nonsense should beat his wife.' [2] As if conjugal cruelty were the utmost in scientific misbehavior."

He was aware of his lack of serenity, and sometimes spoke of this "lively and caustic manner . . . which I recognize to be peculiar to me in the defense of truth." "Moderation! This is a word which is rarely applied to me. Yet, I am the most hesitating of men, the most fearful of committing myself when I lack evidence. But on the contrary, no consideration can keep me from defending what I hold as true when I can rely on solid scientific proof." Once more at the occasion of his jubilee in 1892, he assured his colleagues: "If I have at times disturbed the calm of your academies by discussions of too great an intensity, it is only because I wanted to defend the cause of truth." Pasteur has been criticized for the violence of his attacks on the enemies of the germ theory. It must be remembered, however, that he was fighting almost singlehanded against the official doctrines of the day. Darwin was fortunate in having for his disciple such a master of exposition as Huxley. The latter constituted himself "Darwin's bulldog," and just as Darwin's motto was peace at any price, so Huxley's was war, whatever the cost. In contrast, Pasteur was for a long time almost the only articulate champion of the germ theory, and he had to act both Darwin's and Huxley's roles.

* * *

[2] Another account reads "should be untrue to his wife."

The defense of his own discoveries was not the only cause which excited his fighting spirit; science, and its contribution to the welfare of mankind and to the power of the nation, inspired some of his most passionate tirades. Shocked at the neglect of scientific research in France, he attempted to capture the interest of governmental bodies, by letters and by affixing to his studies dedications to sovereigns, by lectures and demonstrations designed to entertain as well as instruct the court and society. Because he respected traditional and legal authority, it was with expressions of reverence, and almost humility, that he first pleaded the cause of science. His words and person were shown polite interest, but when no action was forthcoming in answer to his request for funds – while millions of francs were spent on the new Paris Opera House, and not even a few thousand could be found for laboratories – he lost patience. Much as he had appealed to the young physicians from his desk at the Academy, he decided to appeal directly to public opinion by sending to the official newspaper *Le Moniteur* an impassioned plea for the support of scientific research. The article was rejected by the editorial committee as subversive, but was finally published in the *Revue des Cours Scientifiques* in 1868 under the title *Le budget de la science.* In the meantime, it had reached the Emperor, who was sufficiently moved by it to take immediate and personal action for the reorganization of French science; but the Franco-Prussian War soon interrupted the execution of the new plans.

Pasteur saw in the defeat of France a tragic vindication of his attitude, and in 1871 reissued his warning in an enlarged form, under the title *Quelques réflexions sur la science en France.* In it, he lamented the material circumstances which prevented young French scholars from devoting their energies to investigation; he contrasted the miserable state of laboratories in France with the magnificent support they were receiving abroad and particularly in Germany; he recalled the prominent part played by French science in allowing the country to overcome the onslaught of Europe during the Revolution and the Napoleonic wars.

"The public bodies, in France, have long ignored the correlation between theoretical science and the life of nations. Victim of her political instability, France has done little to maintain, spread, and develop, the progress of sciences. . . . She has lived on her past, believing herself great through the discoveries of science, because of the material prosperity which she owed to them, but failing to realize that she was allowing the sources of wealth to go dry.

"While Germany multiplied her universities, created a healthy competition between them, surrounded her teachers and doctors with honor and consideration, organized vast laboratories with the best equipment, France, enervated by revolutions, always preoccupied with the sterile search for the best form of government, paid only distant attention to her institutions of higher learning.

"In the present state of modern civilization, the cultivation of the highest forms of science is perhaps necessary even more to the moral state of a nation than to its material prosperity. . . .

"Our disasters of 1870 are present in the mind of everyone. . . . It is too obvious, unfortunately, that we lacked men adequately prepared to organize and utilize the immense resources of the nation . . .

"If, at the moment of the supreme peril, France could not find the men to take advantage of her power and of the courage of her children, it is, I am certain, because she has neglected the great labors of thought for half a century, particularly in the exact sciences."

He described with enthusiasm how, by virtue of her leadership in scientific research during the fifty years before the first Revolution, the France of 1792 had multiplied her forces through the genius of invention and had found, wherever needed, men capable of organizing victory. And in words of overwhelming conviction he exclaimed, "Oh my country! You who so long held the scepter of thought, why did you neglect your noblest creations? They are the divine torch which illuminates the world, the live source of the highest sentiments, which keep us from sacrificing everything to material satisfactions. . . .

"Take interest, I beseech you, in those sacred institutions which we designate under the expressive name of laboratories. Demand that they be multiplied and adorned; they are the temples of wealth and of the future. There it is that humanity grows, becomes stronger and better. There it learns to read in the works of nature, symbols of progress and of universal harmony, whereas the works of mankind are too often those of fanaticism and destruction . . ."

Pasteur carried home his scientific preoccupations, and his family was witness to his silent meditations, audience to his dreams. As we have seen, this intimate association between the home and the life of study had begun during childhood with his parents. In the Pasteur family, learning was not a passing phase, to be disposed of as soon as possible in order to enjoy an idle summer vacation or devote the leisure of adulthood to trivial chat and the reading of the daily newspaper; learning was a never-ending component of one's life, changing not in intensity with the seasons and the years, but only as to its nature according to one's social responsibilities and place in the world.

In order to be closer to his work Pasteur and, whenever possible, his assistants had their living quarters near the laboratory. The working day began at 8 A.M., lasted until 6 P.M., and holidays were rare. "I would consider it a bad deed," said he, "to let one day go without working." Evenings were devoted to reading, correspondence, and the preparation of scientific papers. Madame Pasteur copied everything "with her beautiful handwriting, so easy to read." Never did a manuscript go to the printer except in the neatest form, with all amendments carefully pasted in.

Pasteur carried into social and home life mannerisms which grew out of his scientific problems. As much as possible he avoided shaking hands, for fear of infection. At the dinner table he would wipe glassware and dinnerware in the hope of removing contaminating dirt. Loir has described the odd behavior that arose from his habit of intense and detailed observation.

"He minutely inspected the bread that was served to him and placed on the tablecloth everything he found in it: small fragments of wool, of roaches, of flour worms. Often I tried to find in my own piece of bread from the same loaf the objects found by Pasteur, but could not discover anything. All the others ate the same bread without finding anything in it. This search took place at almost every meal and is perhaps the most extraordinary memory that I have kept of Pasteur."

During periods of great preoccupation, he remained completely silent even with members of his family. Nothing could erase the tenseness of his expression until he had solved his problem. Once the solution had been found, however, he became exuberant, and wanted everyone around him to share his hopes and joy; wife and children had to participate in both his anguish and his triumph. He would pursue at mealtime his controversies of the afternoon at the Academy, or with some distant opponent. To him, it was inconceivable that a subject worthy of intense discussion at a scientific meeting should not remain the center of attention at a social gathering.

Pasteur's only masters were Work and Science. In truth, he could tell the musicians and artists assembled to honor his Jubilee in 1892 that he was seeing them all for the first time; and he was sincere when he wrote, "Let us work, this is the only thing which is entertaining." What were the motives which powered this incessant activity, this dedication to a life of toil, this sacrifice of the small pleasures of existence?

One hears now and then that Pasteur was avid for money, that a desire for fortune motivated his enormous expenditure of energy. As a small French bourgeois, issuing from an environment of struggle, he certainly longed for financial security; but so do most men. There is, however, no evidence that the urge for money played a significant role in directing his activities. After he had developed techniques for the preservation of vinegar, wine and beer by the use of heat, he took patents to protect the rights to his discovery. That there were discussions within his family concerning the possible financial exploitation of these patents is revealed in one of his letters: "My wife . . . who worries concerning the future of our children, gives me good reasons for overcoming my scruples." Nevertheless, Pasteur decided to release his patents to the public, and did not derive financial profit even from the development or sale of industrial equipment devised for pasteurization.

The discovery of immunization against anthrax and swine erysipelas again presented the opportunity of large monetary rewards. Following the Pouilly le Fort experiment, Pasteur had turned over to his laboratory the profits coming from the sale of the anthrax vaccine in France, reserving for himself and his collaborators only the income from the sale to foreign countries. In 1882 a Dutch financier offered him one hundred thousand francs, a large sum at the time, for the exclusive right to use the technique in South Africa. Pasteur contemplated accepting the offer, and began to dream with his family and close associates of the use he would make of this fortune. However, a word of caution from Dumas, pointing out the disgraceful manner in which Liebig had allowed his name to be used as an advertisement for meat extract, sufficed to stop him, and he went no further with his plan.

The practice was common in France for one individual to hold several major teaching appointments in order to increase his income. Pasteur strongly objected to this, as interfering with investigative work; true to his own preaching, he gave up in 1868 the post of professor of physics and chemistry at the School of Fine Arts, which he had held for three years, and he refused to teach chemistry at the Ecole Normale while he held the chair at the Sorbonne.

According to Loir, there had been many discussions concerning the advisability of using the manufacture and sale of vaccines as a source of income for the laboratory. This policy was finally adopted under Duclaux's influence and against Pasteur's judgment in order to facilitate the financing of further scientific work. In fact, Pasteur willingly left to others the management of affairs whenever money was involved, and his own salary was paid directly to his wife. His daily needs were small; the laboratory and afternoon meetings at the Academy bounded his life outside of his home. When on a scientific trip, he took just enough money to meet the immediate necessities and depended on his wife to supply him with further funds when the need arose.

It is, indeed, incredible that the mere urge for money could have been the incentive for the expenditure of so much energy and talent, and only small men with empty hearts and without imagination could explain Pasteur's ambitions in such simple terms. At the most, financial independence was the expression of security and even more a symbol of the place which he considered his due in society. Official decorum and the proper type of carriage for important occasions were, for him – as was neat and conservative clothing – more a matter of decency than of comfort or enjoyment. Like most creative workers, he respected the traditions and conventions which lay outside the field of his own endeavor, probably because he lacked the time to reassess their meaning and value, and because he was too conscientious to formulate lightly a personal code of ethics. He certainly had social ambitions, but not so much to participate in the brilliant and entertaining life of his time as to be recognized as a leader of his community, in fact as the most elevated expression of its genius. He once said, "Cities should be aware that they are remembered in the course of ages only through the genius or the valor of a few of their children." He wanted to be one of those for whom cities are remembered.

He was jealous of his right to his discoveries. Once, very early in his career, when he had hurried publication for fear of losing priority, he wrote with candor:

"How disturbing to lose by hasty publication the charm of following a fruitful idea with calm and prolonged meditation! And yet I would be even more disturbed if M. Marbach . . . should arrive first at the general idea which I follow.

"I therefore incline toward the immediate publication of all the positive facts I know."

In later years, he combatted with endless evidence and argument the accusation that he had borrowed ideas or facts from others.

Concern for priority is undoubtedly one of the most sensitive points among scientific workers. Thus Humphry Davy, who had refused to patent his discovery of the miners' safety lamp, or to receive money for it, became very angry at the assertion that Stephenson deserved priority for the invention. And yet, most scientists are unwilling to acknowledge this jealousy, and pretend that their only interest is in the advancement of science irrespective of recognition. Pasteur has spared us from these false claims because he did not know how to conceal the great pride he took in his discoveries, as well as in the honors poured upon him by his scientific peers, by his country, by the world. For him, social recognition was the symbol that he had fulfilled his calling. The longing to transcend the primary needs and satisfactions of one's vegetative life, to become one of the greatest actors in the unfolding of the future of one's community or of humanity, is an urge probably widespread among men. In this sense, it is true that many great achievements are motivated by generous impulses and that the acknowledgment of it by society is often a source of gratification to aspiring men. Pasteur differed from most of them only by displaying in public, often with a childish naïveté, the pleasure which he derived from fame.

He always attributed his eagerness for achievement and for recognition to a desire to brighten the glory of France. His patriotism was strengthened by the memory of the glorious past of his country, as we have seen, and also by his awareness that France had lost the leading position she had occupied in Europe during the preceding century. This identification between personal urges and national glory had been instilled into him by his father, who had never forgotten the intoxicating days when the flags of Revolutionary and Imperial

France waved over the capitals of Europe. Speaking of him, Pasteur once evoked the influence of this atmosphere on his own life. "I still can see you reading by the lamp after the day of labor some account of battle from one of those books of contemporary history which recalled the glorious era of which you had been witness. While teaching me to read, you were careful to make me aware of the greatness of France." And he was certainly sincere when he asserted: "Science has been the dominating passion of my life. I have lived only for it and the thought of France supported my courage during the difficult hours which are an inevitable part of prolonged efforts. I associated her greatness with the greatness of science." When, following the French disasters of 1870–1871, he was offered by the Italian Government a chair of chemistry at the University of Pisa, with a high salary and very great personal advantages, after much hesitation he refused. "I should feel like a deserter if I sought, away from my country in distress, a material situation better than that which it can offer me."

Granted that love of country and urge for social recognition can serve as stimuli to human endeavor, they are of little importance in deciding the direction of one's efforts. And, in fact, the one problem in which the desire to serve his country was the most direct motivation of Pasteur's choice, namely the improvement of French beer, was also the most trivial. The book which crowned this phase of his work, the *Studies on Beer,* reveals Pasteur's attitude. It contains little concerning brewing practice except information related to the microbial alterations of beer; but it discusses at great length problems which had been very close to his heart for many years — the distribution of microorganisms in nature and the mechanism of fermentation.

Indeed, Pasteur knew well that real science is of equal relevance to all men, whatever their nationality. It is true that the scientist, as a citizen, can add to the fame of his community by the distinction of his achievements. Few, however, are the cases where national pride alone appears as an adequate inducement to scientific pursuit.

"I am imbued with two deep impressions; the first, that science knows no country; the second, which seems to contradict the first, although it is in reality a direct consequence of it, that science is the highest personification of the nation. Science knows no country because knowledge belongs to humanity, and is the torch which illuminates the world. Science is the highest personification of the nation because that nation will remain the first which carries the furthest the works of thought and intelligence.

"The conviction of having attained truth is one of the greatest joys permitted to man, and the thought of having contributed to the honor of one's country renders this joy even deeper. If science knows no country, the scientist has one, and it is to his country that he must dedicate the influence that his works may exert in the world."

Although Pasteur spoke now and then of the disinterested search for truth, he had only little hope that science would ever reveal the ultimate nature of things. He regarded science as an instrument of conquest which permitted man to gain mastery over the physical world, rather than as a technique for understanding the universe. Repeatedly, he expressed gratification at seeing that his labor would help to better the lot of man on earth. "To him who devotes his life to science, nothing can give more happiness than increasing the number of discoveries, but his cup of joy is full when the results of his studies immediately find practical applications."

On many occasions, also, he stated in terms of unquestionable warmth and sincerity his desire to alleviate the sufferings of his fellow men. "One does not ask of one who suffers: What is your country and what is your religion? One merely says: You suffer, this is enough for me: you belong to me and I shall help you."

And at the occasion of his jubilee, he summarized his creed by the oft-quoted words:

"I am utterly convinced that Science and Peace will triumph over Ignorance and War, that nations will eventually unite not to destroy but to edify, and that the future will belong to those who have done the most for the sake of suffering humanity."

He believed that great was the "part played by the heart in the progress of sciences" – and indeed, there is no question that the desire to be useful to his fellow men, his awareness of the problems of his community, determined to a large extent the fields of endeavor in which he gained his greatest popular triumphs. By natural endowment and training, he was qualified to pursue far and profitably the theoretical problems which had occupied his younger years and which haunted the rest of his days. Instead, he chose to devote much of his energy to the practical affairs of man. By so doing, he did not consider that he was sacrificing intellectual distinction. For, according to him, the search for knowledge which has direct bearing on the practical problems of human life could not be readily differentiated from the search for abstract truth.

Among theoretical scientists, there are those who pretend that the desire to contribute to human welfare plays no part in the motivation of their efforts. They regard pure curiosity, or at the most some concern with the dignity of the human mind, as sufficient to explain and justify the pursuit of science for its own sake. By so doing, they place their activities above the level of the ordinary preoccupations of their fellow men, an attitude which gives them the illusion of occupying an exalted place in the social structure. It is probable that conceit or blindness, rather than intellectual superiority, is often the true inspiration of this philosophy, for social pressure exerts a greater influence on the orientation of their activities than most scientists are willing to acknowledge.

Nevertheless, there certainly exists a curiosity to understand the universe and the significance of life, which is independent of the immediate needs of society, and which accounts for much intellectual exertion. It is this longing which the English biologist Joseph Arthur Thomson expressed in the following words. "The scientific worker has elected primarily to know, not do. He does not *directly* seek, like the practical man, to realize the ideal of exploiting nature and controlling life; he seeks rather to idealize

— to conceptualize — the real, or at least those aspects of reality that are available in his experience . . ." In fact, as we have seen, Pasteur's passionate engrossment with research began long before he could visualize his scientific efforts as having a direct effect on the welfare of mankind. None of his subsequent achievements gave him greater emotional pleasure than did the discovery of a correlation between optical activity and the morphology of the tartaric acid crystals.

The act of discovery, independent of its consequences, remained for Pasteur a never-ending enchantment: enchantment arising from the emotion associated with treading over land heretofore unknown to man; enchantment from the new vistas suddenly opened to the discoverer, and the promise of more adventure.

"It is characteristic of science and of progress that they continuously open new fields to our vision. When moving forward toward the discovery of the unknown, the scientist is like a traveler who reaches higher and higher summits from which he sees in the distance new countries to explore."

Pasteur often returned to his earlier publications. Turning the pages of his writings, he would marvel at the lands that he had revealed by dispelling the fogs of ignorance and by overcoming stubbornness. He would live again his exciting voyages, as he told Loir in a dreamy voice: "How beautiful, how beautiful! And to think that I did it all. I had forgotten it."

These had been the great adventures of his life. If he spoke of laboratories in endearing terms, if he demanded that they be adorned, it was not only because he saw in them the temples of the future, but also because he had known happiness and enchantment within their walls.

Whatever the initial motivation of his efforts — idle curiosity or social compulsion; whatever the target at which he aims — discovery of a natural law or solution of a practical problem — the scientist of genius perceives the far-reaching implications of the isolated facts which come within his field of vision, and recognizes between them relations of broad generality. Pasteur exhibited from the beginning of his scientific career, and retained throughout his life, this ability to recognize large theoretical issues, to the extent of translating practical achievements into the terms of general laws of nature. From simple observations on the optical activity of tartrates he derived an interpretation which encompassed the problem of molecular structure. The study of wine and beer led him to see in the life of yeast a microcosm illustrating the biochemical unity of life; while studying fowl cholera he discovered the principles of immunization and the broader fact that any animal coming into contact with a foreign substance becomes indelibly altered by this experience.

The eagerness to control nature for practical ends, and the longing to conceptualize, were always simultaneously present in his mind. The fact that both tendencies remained equally powerful throughout his life constitutes perhaps the most characteristic trait of his scientific career, and accounts for its somewhat erratic course. Moreover, there grew very early, deep in his heart, the secret desire to accomplish some prodigious feat. The ever-recurring evocation of his early studies on the optical activity of organic molecules, with its possible bearing on the genesis of life, supports the view that practical problems never completely monopolized his mind. Speaking of the late afternoon conversations in the laboratory, Roux has stated that Pasteur's imagination reached its highest peak whenever these early studies were mentioned. "He would speak in poetical terms of molecular asymmetry and of its relation to the asymmetric forces of nature. On these days, Pasteur forgot the dinner hour; Madame Pasteur had to have him called several times, or had to come herself to fetch him."

By reason of his very success as an experimenter, he became the prisoner, almost the slave, of limited and practical tasks. But beyond the daily problems, his gaze was fixed on the romantic hope that he would some day penetrate the secret of life. The alchemist never entirely ceased to live and function within the academician.

By upbringing, schooling and self-discipline, Pasteur was made to behave as a bourgeois and to accept the rigid code of experimental science; but he was by temperament an adventurer. With so many worlds to conquer, others yet to be discovered and even more to imagine, why invoke money or social distinction to account for his labors? Rewards and honors of all sorts came to him, and he enjoyed them. But how pale they must have been, compared with the glowing visions – of insight, divination, adventure and power – which he experienced at the Ecole Normale and under the trees of the Luxembourg Gardens, in the company of ghosts of so many other dreamers!

CHAPTER IV

From Crystals to Life

> The men of experiment are like the ant; they only collect and use: the reasoners resemble spiders, who make cobwebs out of their own substance. But the bee takes a middle course, it gathers its material from the flowers of the garden and of the field, but transforms and digests it by a power of its own.
>
> — FRANCIS BACON

PASTEUR spent the first ten years of his scientific life, from 1847 to 1857, investigating the ability of organic substances to rotate the plane of polarized light, and studying the relation of this property to crystal structure and molecular configuration. These studies provided the basis on which the new science of stereochemistry was built during his own lifetime. From them also arose his intuitive belief that fermentations are the manifestations of living processes, a belief that eventually led him to the germ theory of fermentation and of disease. There is no indication that Pasteur entered the field of crystallography in the hope of solving great theoretical or practical problems of physics, chemistry or biology. As a serious student he was eager to participate in the investigations of some of his respected teachers and to follow the line of their interests. The problem that was to be of such momentous consequence — for his career, for the future of chemical and biological sciences, and for the welfare of society — was the outcome of the personal associations which he enjoyed while a student at the Ecole Normale.

At the beginning of the century Jean Baptiste Biot, who was

soon to play such an important role in Pasteur's life, had recognized that crystals of quartz rotate the plane of polarized light, traversing them in the direction of their long axis. He had also noticed that certain quartz crystals rotate light to the right, whereas others of the same thickness rotate light to the same extent to the left. At about the same period, Haüy and his pupil Delafosse had observed on quartz some crystal faces — called hemihedral facets — which were inclined sometimes in one direction, sometimes in the other, with reference to the edges of the crystal that bore them. It was the English astronomer John Herschel who had the idea of combining the purely crystallographic observations of Haüy and Delafosse on the existence of right- and left-handed facets in quartz crystals with the physical observations of Biot on right- and left-handed rotation of light by the same crystals. He was able to establish that the ability of quartz to rotate the plane of polarized light is an expression of the configuration of the crystal.

It was well known that quartz exhibits its characteristic effect on light only when in the crystalline state. In 1815, Biot had discovered that certain natural organic substances such as sugar, camphor, tartaric acid, oil of turpentine, protein, and the like could also rotate the plane of light but that, in contrast with quartz, they exhibited their optical activity in the liquid state and in solution.

Pasteur was well acquainted with all these facts as his professor of mineralogy at the Ecole Normale was the same Delafosse who had made a special study of the facets present on the quartz crystals. Another accidental association further increased his familiarity with the problem of the relation of optical activity to crystalline structure. At the end of 1846, there came to work in Balard's laboratory, where Pasteur was now an assistant, a young and intense chemist — Auguste Laurent — whose direct influence on the eager student can be traced in a manuscript note left by Pasteur.

". . . One day it happened that M. Laurent — studying, if I mistake not, some tungstate of soda, perfectly crystallized —

showed me, through the microscope, that this salt, apparently very pure, was evidently a mixture of three distinct kinds of crystals, easily recognizable with a little experience of crystalline forms. The lessons of our modest and excellent professor of mineralogy, M. Delafosse, had long since made me love crystallography; so, in order to acquire skill in using the goniometer, I began to study carefully the formations of a very fine series of compounds, all very easily crystallized; tartaric acid and the tartrates. . . . Another motive urged me to prefer the study of those particular forms. M. de la Provostaye had just published an almost complete work concerning them; this allowed me to compare, as I went along, my own observations with those, always so precise, of that clever scientist."

Thus, it appeared as if fate had brought together many influences to prepare Pasteur for his first scientific adventure. He had become attracted by the "subtle and delicate techniques involved in the study of these charming crystalline forms"; influenced by Herschel's observations and by his association with Delafosse and Laurent, he had in mind constantly the relation between optical activity and the orientation of facets in quartz crystals; finally he had become very familiar with the optical activity of tartrates and with their crystalline characteristics. Although he knew that the optical activity of organic substances was an expression of the molecule in solution, and not of the crystalline structure as in the case of quartz, he assumed, probably under the influence of Delafosse, that there might be something external—for example, facets in the tartrate crystals—which would indicate the arrangement of the atoms within the tartrate molecule.

As he was pondering over these facts, Biot communicated to the Academy of Sciences in 1844 a note in which the German chemist, Mitscherlich, described some very curious findings which startled Pasteur and launched him on that voyage of discovery which was thereafter coextensive with his life.

Two different forms of tartaric acid were then recognized.

One, the true tartaric acid, had long been known to occur as a constant component of tartar in the wine fermentation vats. The other, first seen in 1820 by Kestner, an industrialist of Thann, occurred among the large crystals of true tartaric acid in the form of needlelike tufts which resembled oxalic acid crystals. It had been called "paratartaric acid," also "racemic acid," to recall its origin from the grape (*racemus*). Mitscherlich had discovered that the two forms of tartaric acids and their respective salts, the tartrates and paratartrates, "have the same chemical composition, the same crystal shape with the same angles, the same specific gravity, the same double refraction, and therefore the same angles between their optical axes. Their aqueous solutions have the same refraction. But the solution of the tartrate rotates the plane of polarization, while the paratartrate is inactive."

Pasteur saw immediately an incompatibility between the finding that the two tartrates behaved differently toward polarized light, and Mitscherlich's claim that they were identical in every other particular. He was convinced that there *had* to be some chemical difference between the two substances, and he hoped that this difference would express itself in the shape of the crystals. This incompatibility provided him with the first specific question, the first well-defined problem, on which to test his skill as an experimenter. In so doing, he demonstrated one of the most fundamental characteristics of the gifted experimenter: the ability to recognize an important problem, and to formulate it in terms amenable to experimentation.

Imbued with the idea that optical activity must be associated with irregularities in the crystalline shape, Pasteur began a systematic observation of the crystals of tartaric acid and of the various tartrates which he had laboriously prepared. He saw at once on the tartrate crystals small facets, similar to those present on quartz crystals, which had escaped the attention of his predecessors – a telling example of the part played by a working hypothesis in the process of discovery. Acute observers as they were, Mitscherlich and de la Provostaye had failed to see the small facets on the tartaric acid crystals because they were not interested in seeing them. Pasteur, on the contrary, looked for them because he had postulated their existence, and he detected them at once. So do preconceived ideas influence our perception of natural phenomena – as they do our judgment of economic, social and moral problems.

The nineteen tartrates prepared and studied by Pasteur were found to exhibit the typical facets on one side of their crystals and all rotated polarized light in the same direction. He naturally inferred that crystal shape and optical activity were linked in the case of tartrates as they were already known to be in the case of quartz, notwithstanding the fundamental difference that quartz possesses optical activity only in the crystalline state, whereas the tartrates retain this property in solution. To clinch the correlation, it was now necessary to determine whether the crystals of paratartaric acid, found by Mitscherlich to be optically inactive, differed from the optically active tartrates either in not possessing any facets at all or in having symmetrical pairs of facets. To this end, Pasteur set about examining the crystals of paratartaric acid and its salts and found, in accordance with his anticipations, that they did not possess the facets characteristic of the true tartrates.

Mitscherlich had made his allegation with respect to one particular substance, namely the sodium-ammonium paratartrate, and Pasteur therefore examined it with especial care. To his intense surprise and disappointment, and quite contrary to anticipation in the light of his hypothesis, he found facets similar to those present on the crystals of optically active tartrates. Still intent on finding a difference, he noticed that while the facets were all turned towards the right in the tartrate, in the case of paratartrate crystals some were turned to the right, and some to the left. Obeying the promptings of his hypothesis, he sedulously picked out the right-handed crystals and placed them in one heap, and the left-handed crystals in another, dissolved each group in water, and then examined the two solutions separately in the polarimeter – with results which made of this simple operation one of the classical experiments of chemical science. The solution of the right-handed crystals turned the plane of polarization to the right, the solution of the left-handed crystals to the left. When the two solutions were mixed in equal amounts, the mixture proved optically inert.

It is easy to recapture the dramatic quality of the situation and the intense excitement which it must have caused in the young investigator. Pasteur was so overcome with emotion by his finding that he rushed from the laboratory, and, meeting one of the chemistry assistants in the hall, embraced him, exclaiming, "I have just made a great discovery. . . . I am so happy that I am shaking all over and am unable to set my eyes again to the polarimeter!"

Pasteur retained throughout his life a vivid memory of this first scientific triumph and never tired of referring to it in conversations and in lectures. As late as 1883, almost forty years after the event, he described it again in the course of a lecture delivered before the Société Chimique de Paris.

"I was a student at the Ecole Normale Supérieure, from 1843 to 1846. Chance made me read in the school library a note of the learned crystallographer, Mitscherlich, related to two salts: the tartrate and the paratartrate of sodium and ammonium. I meditated for a long time upon this note; it disturbed my schoolboy thoughts. I could not understand that two substances could be as similar as claimed by Mitscherlich, without being completely identical. To know how to wonder and question is the first step of the mind toward discovery.

"Hardly graduated from the Ecole Normale, I planned to prepare a long series of crystals, with the purpose of studying their shapes. I selected tartaric acid and its salts, as well as paratartaric acid, for the following reasons. The crystals of all these substances are as beautiful as they are easy to prepare. On the other hand, I could constantly control the accuracy of my determinations by referring to the memoir of an able and very precise physicist, M. de la Provostaye, who had published an extensive crystallographic study of tartaric and paratartaric acid and of their salts.

"I soon recognized that . . . tartaric acid and all its combinations exhibit asymmetric forms. Individually, each of these forms of tartaric acid gave a mirror image which was not superposable upon the substance itself. On the contrary, I could not find anything of the sort in paratartaric acid or its salts.

"Suddenly, I was seized by a great emotion. I had always kept in mind the profound surprise caused in me by Mitscherlich's note on the tartrate and paratartrate of sodium and ammonium. Despite the extreme thoroughness of their study, I thought, Mitscherlich, as well as M. de la Provostaye, will have failed to notice that the tartrate is asymmetric, as it must be; nor will they have seen that the paratartrate is not asymmetric, which is also very likely. Immediately, and with a feverish ardor, I prepared the double tartrate of sodium and ammonium, as well as the corresponding paratartrate, and proceeded to compare their crystalline forms, with the preconceived notion that I would find asymmetry in the tartrate and not in the paratartrate. Thus, I thought, everything will become clear; the mystery of Mitscherlich's note will be solved, the asymmetry in the form of the tartrate crystal will correspond to its optical asymmetry, and the absence of asymmetry in the form of the paratartrate will correspond to the inability of this salt to deviate the plane of polarized light. . . . And indeed, I saw that the crystals of the tartrates of sodium and ammonium exhibited the small facets revealing asymmetry; but when I turned to examine the shape of the crystals of paratartrate, for an instant my heart stopped beating: all the crystals exhibited the facets of asymmetry!

"The fortunate idea came to me to orient my crystals with reference to a plane perpendicular to the observer, and then I noticed that the confused mass of crystals of paratartrate could be divided into two groups according to the orientation of their facets of asymmetry. In one group, the facet of asymmetry nearer my body was inclined to my right with reference to the plane of orientation which I just mentioned, whereas the facet of asymmetry was inclined to my left in the other. The paratartrate appeared as a mixture of two kinds of crystals, some asymmetric to the right, some asymmetric to the left.

"A new and obvious idea soon occurred to me. These crystals asymmetric to the right, which I could separate manually from the others, exhibited an absolute identity of shape with those of the classical right tartrate. Pursuing my preconceived idea, in the logic of its deductions, I separated these right crystals from the crystallized paratartrate; I made the lead salt and isolated the acid; this acid appeared absolutely identical with the tartaric acid of grape, identical also in its action on polarized light. My happiness was even greater the day when, separating now from the paratartrate the crystals with asymmetry at their left, and making their acid, I obtained a tartaric acid absolutely similar to the tartaric acid of grape, but with an opposite asymmetry, and also with an opposite action on light. Its shape was identical to that of the mirror image of the right tartaric acid and, other things being equal, it rotated light to the left as much in absolute amount as the other acid did it to the right.

"Finally, when I mixed solutions containing equal weights of these two acids, the mixture gave rise to a crystalline mass of paratartaric acid identical with the known paratartaric acid."

The enthusiastic Balard quickly broadcast into scientific circles the news of these unexpected findings, and Pasteur was thus brought into contact with Biot, who, throughout his long and laborious career, had contributed so much knowledge to the problems of crystallography and optical activity. It was Biot who had presented to the Academy, three years before, the note by Mitscherlich which had so much perplexed Pasteur. It was he again who was asked to present the new discovery. But before doing so, the skeptical veteran submitted the young man's almost suspiciously plausible results to a stringent verification.

Pasteur has left the following account of his first dealings with Biot:

"He (M. Biot) sent for me to repeat before his eyes the several experiments and gave me a sample of racemic acid which he had himself previously examined and found to be quite inactive toward polarized light. I prepared from it, in his presence, the sodium ammonium double salt, for which he also desired himself to provide the soda and ammonia. The liquid was set aside for slow evaporation in one of the rooms of his own laboratory, and when thirty to forty grams of crystals had separated, he again summoned me to the Collège de France, so that I might collect the dextro and levorotatory crystals before his eyes, and separate them according to their crystallographic character — asking me to repeat the statement that the crystals which I should place on his right hand would cause deviation to the right, and the others to the left. This done, he said that he himself would do the rest. He prepared the carefully weighed solutions, and at the moment when he was about to examine them in the polarimeter, he again called me into his laboratory. He first put into the apparatus the more interesting solution, the one which was to cause rotation to the left. Without making a reading, but already at the first sight of the color tints presented by the two halves of the field in the Soleil polarimeter, he recognized that there was a strong levorotation. Then the illustrious old man, who was visibly moved, seized me by the hand, and said, 'My dear son, I have loved science so deeply that this stirs my heart.' "

Thus, the first phase of Pasteur's experimental investigations had established the existence of three tartaric acids, differentiated by the orientation of facets on their crystals and by the corresponding optical activity, but otherwise identical in chemical properties. Two lectures which Pasteur presented in 1860 before the Société Chimique de Paris give us his interpretation of the new phenomena and particularly the mental picture which he had formed of the molecular configuration responsible for optical activity. The intellectual achievement involved in the formulation of this picture appears the more striking when it is remembered that the science of structural organic chemistry was not yet born and that the concept of the asymmetric carbon atom was still several decades away.

"In isomeric bodies, the elements and the proportions in which they are combined are the same, only the arrangement of the atoms is different. The great interest attaching to isomerism lies in the principle that bodies can be and really are distinct, through possessing different arrangements of their atoms within their molecules. . . . We know, on the one hand, that the molecular arrangements of the two tartaric acids are asymmetric, and, on the other hand, that these arrangements are absolutely identical, excepting that they exhibit asymmetry in opposite directions. Are the atoms of the dextro acid grouped in the form of a right-handed spiral, or are they placed at the apex of an irregular tetrahedron, or are they disposed according to this or that asymmetric arrangement? We do not know. But there can be no doubt that we are dealing with an asymmetric arrangement of the atoms, giving a non-superposable image. It is equally certain that the atoms of the levo-acid are disposed in an exactly opposite manner. Finally, we know that racemic acid is formed by the union of these two groups of oppositely arranged asymmetric atoms.

"Quartz . . . you will say at once . . . possesses the two characteristics of asymmetry – hemihedry in form, observed by Haüy, and the optical activity discovered by Arago! Nevertheless, molecular asymmetry is entirely absent in quartz. To understand this, let us take a further step in the knowledge of the phenomena with which we are dealing.

"Permit me to illustrate roughly, although with essential accuracy, the structure of quartz and of the natural organic products. Imagine a spiral stairway whose steps are cubes, or any other objects with superposable images. Destroy the structure of the stairway and the asymmetry will have vanished. The asymmetry of the stairway was simply the result of the mode of arrangement of the component steps. Such is quartz. The crystal of quartz is the stair complete. It is hemihedral. It acts on polarized light by virtue of this. But let the crystal be dissolved, fused, or have its physical structure destroyed in any way whatever; its asymmetry is suppressed and with it all action on polarized light, as it would be, for example, with a solution of alum, a liquid formed of molecules of cubic structure distributed without order.

"Imagine, on the other hand, the same spiral stairway to be constructed with irregular tetrahedra for steps. Destroy the stairway, and the asymmetry will still exist, since we are dealing with a collection of tetrahedra. They may occupy any positions whatsoever, yet each of them will, nevertheless, have an asymmetry of its own. Such are the organic substances in which all the molecules have an asymmetry of their own, revealing itself in the crystalline form. When the crystal is destroyed by solution, there results a liquid of molecules, without arrangement, it is true, but each having an asymmetry in the same sense, if not of the same intensity in all directions."

Pasteur then summarized his views in the following conclusions:

"When the atoms of organic compounds are asymmetrically arranged, the molecular asymmetry is betrayed by crystalline form exhibiting non-superposable hemihedrism. The presence of molecular asymmetry reveals itself by optical activity. When this non-superposable molecular asymmetry appears in two opposed forms, as in the case of 'dextro' and 'levo' tartaric acids and all their derivatives, then the chemical properties of the identical but optically opposite substances are exactly the same, from which it follows that this type of contrast and analogy does not interfere with the ordinary play of the chemical affinities."

Although the investigation of tartrates had been suggested and guided by the preconceived idea that molecular asymmetry must find expression both in optical activity and in asymmetry of crystalline shape, Pasteur himself was soon to find out that the relationship does not always hold true. He had been fortunate in beginning his investigation with the tartrates — for, among substances endowed with optical activity, they present in the simplest form the relation between chemical structure, crystalline morphology and deviation of polarized light. One might be tempted, therefore, to attribute his success to luck. However, so often was Pasteur helped by apparent "luck" in the subsequent course of his scientific career that the reason for his success must be found elsewhere. Throughout his life, he displayed an uncanny gift in selecting the type of experimental material best adapted to the solution of the problem under investigation. This gift, which is common to all great experimenters, certainly consists in part of an intuitive wisdom based upon a large background of knowledge. Good fortune is offered to many, but few are they who can recognize it when it is offered in a not too obvious manner.

Pasteur could have been thinking of many vital experiences of his own when he reiterated, time and time again, "In the field of experimentation, chance favors only the prepared mind."

So it was that, by the age of twenty-seven, Pasteur had already given evidence of the qualities which were to make him a great investigator. He had had the independence and the audacity to question the validity of statements made by a scientist of acknowledged authority. He had formulated a bold working hypothesis in terms amenable to study by available experimental methods. With industry and thoroughness, he had prepared himself to use these methods, not for the sake of their mastery, but in order to obtain an answer to the questions which he had in mind. The dominance of problems over technical procedures is one of the most striking aspects of his experimental genius. Once he had recognized and formulated a problem, he was able to bring to bear on its solution any available technique, be it physical, chemical or biological; he was always willing to devote himself to the mastery of the experimental methods best suited to give an answer to his questions.

Neither the pressure of his teaching duties, nor his marriage in 1849, could deter Pasteur from exploiting the scientific vein which he had uncovered. In later years, he referred to youth as "the time when the spirit of invention flourishes." In his case, the flowering was a burst of experiments and of ideas, a few of them faulty, some visionary, but all equally interesting. Never did the spirit of invention germinate in the form of more unexpected and at times fantastic flowers. So much happened over such a short period of time, that a chronological recital would give only an impression of confusion and would indeed be untrue to the intellectual processes which powered Pasteur's activities. As we know, the initial discoveries on tartaric acids opened a number of independent channels of investigation, and Pasteur attempted to split his energy so as to follow all of them simultaneously. We shall be less ambitious, and at the cost of losing much of the excitement of the chase, follow, one after the other, a few of the trails which he cleared between 1850 and 1857.

As mentioned earlier, Pasteur had made a fortunate choice in using the tartaric acid series to master the art of crystallography and to investigate the relation of crystalline structure to optical activity. Chance had also favored him in making paratartaric acid become available shortly before he began his scientific career. For unknown reasons, however, this peculiar acid had now become extremely scarce and failed to appear again, even in the Thann factory where it was formerly so abundant. Moreover, no chemical technique was known to produce it in the laboratory – and for this reason, the Société de Pharmacie of Paris instituted a prize of 1500 francs for the first chemical synthesis of the mysterious substance. Pasteur was naturally much perplexed by this situation and wondered, in particular, what could be the origin of the acid to which he owed his first scientific laurels.

In 1852, Mitscherlich having asked for an opportunity to meet the young Pasteur, Biot arranged at his home a first meeting which was followed by a dinner at Baron Thénard's. Pasteur's account of this event in a letter to his father glows with his naïve pride at being in such distinguished company. "You will like to see the names of the guests: Messrs. Mitscherlich, Rose, Dumas, Chevreul, Regnault, Pelouze, Péligot, Prévost and Bussy. As you will notice, I was the only outsider; they are all members of the Academy." Of especial interest for him was his discovering at that dinner that a Saxony manufacturer was again producing paratartaric acid. Mitscherlich believed that the tartars from the wine fermentation vats employed by this manufacturer came from Trieste; in 1820, the Thann manufacturer Kestner had received his crude tartars from Naples, Sicily or Oporto. Whence came paratartaric acid? and why had it disappeared? The problem, chemical in nature, demanded the techniques of the explorer and of the detective for its solution. Without hesitation and despite financial difficulties, Pasteur started on a hunt over Central Europe to trace the origins of the mysterious acid.

In detailed letters to his wife, the traveling neophyte marveled that the world outside of France should be so civilized and polished. He visited professors and manufacturers in Leipzig, Freiburg, Vienna, Prague, proud to see his name already well known to fellow scientists. He spent his spare time studying in hotel rooms the game collected in his scientific hunt — and finally, in Prague, obtained convincing evidence that paratartaric acid was always present in the mother liquor remaining after purification of the crude tartars, but was being progressively eliminated from the true tartaric acid in the course of purification. Apparently it was only an accident of manufacture which had allowed it to accumulate in relative abundance in the Thann factory a few decades earlier. Homesick, tired and divested of funds, Pasteur returned to Strasbourg, satisfied with having achieved his ends. An account of his journey in the newspaper *La Vérité* contained a sentence which amused everyone, Pasteur included. "Never was treasure sought, never adored beauty pursued over hill and dale with greater ardor."

It is somewhat difficult today to justify so much enthusiasm. One wonders whether Pasteur did not judge of the importance of paratartaric acid in the realm of chemistry from what it had meant for his own life, a common and forgivable sin. If this trip has not remained one of the great journeys of science, as its hero considered it to be, at least it illustrates the determination and energy with which Pasteur approached any problem. It gave him, moreover, the opportunity of investigating the claims of German workers who thought that they had succeeded in producing paratartaric acid by chemical synthesis. He had no difficulty in disproving these claims and returned to Strasbourg determined to be the first to perform this chemical feat. In fact, he succeeded, in June 1853 — synthesizing paratartaric acid by maintaining tartrate of cinchonine at a high temperature for several hours. That he was intoxicated by what he considered a world-shaking discovery is revealed in a letter to his father. "My dear father, I have just sent the following telegram: 'Monsieur Biot, Collège de France, Paris. I transform tartaric acid into racemic acid; please inform MM. Dumas and Sénarmont.' Here is, at last, that racemic acid (which I traveled to Vienna to find), artificially obtained from tartaric acid. I long believed this transformation to be impossible. This discovery will have incalculable consequences."

The consequences were indeed by no means unimportant, although they were hardly the ones that he may have had in mind. The same operation which had yielded synthetic paratartaric acid also gave a fourth form of tartaric acid — mesotartaric acid — which, although optically inactive, was quite distinct from paratartaric acid in not being susceptible of resolution into left and right tartaric acids.

The work on tartrates had made it of obvious interest to establish whether the relation between optical activity and crystalline structure was a property peculiar to the tartaric acid series, or whether other organic substances exhibiting optical activity would also present evidence of morphological asymmetry in the crystalline state. Unfortunately, few organic substances gave crystals adequate for study by the methods then available. Among those which formed beautiful crystals were asparagine and its derivatives: aspartic acid and malic acid. Pasteur therefore made haste to study them.

As asparagine was then a rare substance, he prepared large amounts of it from vetch — which he grew in the gardens and cellars of the University at Strasbourg — and recognized with satisfaction that asparagine crystals exhibited facets and were endowed with optical activity. Aspartic acid and malic acid, derived from asparagine, also deviated the plane of polarized light, but in these cases the relation of optical activity to crystalline structure was either lacking or less clear than in the case of the tartrates. The different active and inactive aspartates, although very similar chemically, were entirely different from the point of view of crystallography, even presenting apparently incompatible shapes. The active and inactive malates, also very similar chemically, failed to give evidence of correlation between the inclination of the facets on the crystals and the direction of the rotation of light. Thus, the beautiful relation shown by the tartrates between optical activity and crystalline structure was not always obvious in the case of other related substances. Nevertheless, so convinced was Pasteur of the significance of his early findings that he was not too much concerned with the apparent inconsistencies which he was now encountering. Instead, he assumed with a tranquil assurance that crystalline shape was of only secondary importance, and while not abandoning his belief that there existed some subtle correlation between crystal structure and optical activity, he concluded that the latter property was the more constant and fundamental expression of asymmetry within the molecule itself. Furthermore, he reasoned that if there is a common atomic grouping between the right-handed tartaric acid of the grape and natural malic acid, the atomic grouping of the left-handed tartaric acid must also have its counterpart in a malic acid still unknown which, when discovered, would be a levo compound. The growth of the science of organic chemistry was soon to prove him right. Thus even before stereochemistry had developed its doctrines and its body of factual knowledge, Pasteur was bold enough to anticipate its unborn concepts and forecast the existence of "levo" malic acid.

Pasteur had first separated the oppositely active components of the inactive paratartrate by picking them manually, taking advantage of the orientation of their facets. This laborious method was slightly modified later by his student Gernez, who achieved the separation by introducing into the supersaturated solution a crystal of just one of the active components, a procedure sometimes resulting in the selective crystallization of the added component.

Pasteur soon arrived at a second fundamentally different method, which he introduced with the following reasoning:

"The properties of the two tartaric acids and their derivatives appear identical as long as they are brought into contact with . . . potash, soda, ammonia, lime, baryta, aniline, alcohol, the ethers, in short with substances devoid of asymmetry. . . .

"On the contrary, if they are subjected to the action of . . . asparagine, quinine, strychnine, brucine, albumin, sugar, *et cetera,* asymmetric like themselves, then the two tartrates exhibit different behavior. The solubility of their salts is different. If combination takes place, the products differ from each other in crystalline form, in specific gravity, in the amount of water of crystallization, in stability on heating, in fact just in the same way as the most distantly related isomers can differ from each other.

"The knowledge that salts of dextro and levo tartaric acid can acquire such different properties only through the optical activity of the base with which they combine, justified the hope that this difference . . . would afford a means of splitting racemic acid into its components. Many are the abortive attempts which I have made in this direction, but I have at length achieved success with the help of two new asymmetric bases, quinidine and cinchonidine, which I can very easily obtain from quinine and cinchonine respectively. . . .

"I prepare the cinchonidine racemate by first neutralizing the base and then adding as much acid again. The first crystals to separate are perfectly pure cinchonidine levo tartrate. The whole of the dextro tartrate remains in the mother liquor, as it is more soluble; gradually this also crystallizes out, but in forms which are totally distinct from those of the levo tartrate."

It was in 1857, at the end of the period we are now considering, that Pasteur discovered a third method for the fractionation of optically inactive compounds into their "right" and "left" components. This method, which is in some respects the most remarkable of the three, was the result of one of those chance occurrences which are observed and seized upon only by the prepared mind.

It had long been known that impure solutions of calcium tartrate occasionally became turbid and were fermented by a mold during warm weather, and Pasteur noticed one day that a tartrate solution of his had become thus affected. Under the circumstances, most chemists would have poured the liquid down the sink, considering the experiment as entirely spoiled. But an interesting problem at once suggested itself to his active mind; how would the two forms composing the paratartaric acid be affected under similar conditions? To his intense interest, he found that whereas the dextro form of tartaric acid was readily destroyed by the fermentation process, the levo form remained unaltered. Pasteur's own account of the changes which occurred in a solution of paratartrate infected with fermenting fluid reveals his power to proceed from trivial observations to broad theoretical concepts. As we shall see, the selective destruction of dextro tartaric acid was the first link in the chain of arguments which led him into the study of fermentations and of contagious diseases.

"The solution of paratartrate, at first optically inactive, soon becomes perceptibly levorotatory, and the rotation gradually increases and attains a maximum. The fermentation then stops. There is now no trace of the dextro tartaric acid left in the liquid, which on being evaporated and treated with an equal volume of alcohol yields a fine crop of crystals of ammonium levo tartrate.

"Two different aspects of this phenomenon require emphasis. As in every true fermentation, there is a substance undergoing chemical alteration, and corresponding with this there is the development of a moldlike organism. On the other hand, and it is precisely to this point that I should like to draw your attention, the levo salt is left untouched by the mold which causes the destruction of the dextro salt, and this despite the identity of the physical and chemical properties of the two salts that prevails as long as they are not submitted to asymmetric influences.

"Here then we see molecular asymmetry, a property peculiar to organic matter, influencing a physiological process, and influencing it, moreover, by modifying the chemical affinities. . . .

"Thus, the concept of the influence of molecular asymmetry of natural organic products is introduced into physiological studies, through this important criterion (optical activity), which forms perhaps the only sharply defined boundary which can be drawn at the present time between the chemistry of dead and that of living matter."

Even before 1857 Pasteur had become convinced that one of the fundamental characteristics of living matter was its asymmetric nature. This view became one of the cardinal tenets of his biochemical thinking, one to which he returned throughout his life; thus in 1886, thirty years after he had abandoned experimentation on molecular asymmetry, he discussed a paper on the two asparagines of opposite optical activity presented by one of his colleagues in the Academy of Sciences, and he called attention to the fact that one of the asparagines is sweet to the taste while its optical antipode is insipid, suggesting that this difference might be due to a different action of the two asymmetric antipodes on the asymmetric constituents of the gustatory nerve.

Biot, it will be recalled, had discovered that certain organic materials – sugar, albumin, turpentine, and the like – rotate the plane of polarized light. On the basis of his own experience, Pasteur was soon in a position to remark that, in contrast to the behavior of the majority of these naturally occurring substances, the artificial products of the laboratory are without optical activity; and he became possessed by the idea that molecular asymmetry can be produced only through a vital agency. His preconceived views appeared shattered when in 1850 Dessaignes, a French chemist from Vendôme, announced that he had obtained aspartic acid identical with the natural product by heating the ammonium salts of fumaric and maleic acids, two substances known to be optically inactive. Dessaignes thus claimed to have accomplished what Pasteur firmly believed to be impossible, namely the preparation by chemical means of an optically active molecule (aspartic acid) from inactive ones (fumaric or maleic acids). Pasteur hastened at once from Strasbourg to Vendôme and obtained from Dessaignes a specimen of his artificial aspartic acid; he soon had the satisfaction of showing that the latter substance was optically inactive and therefore not really identical with the natural aspartic acid. Furthermore he converted this inactive aspartic acid into malic acid, and in accordance with his anticipations, found that this malic acid differed from the natural one in being also inactive toward polarized light.

Pasteur's triumph was apparently complete, but it was now necessary to account for the existence of these new forms of synthetic aspartic and malic acids which were different from the natural products. The work involved in their study led to a comedy of errors in which Pasteur made a spectacular discovery on the basis of false theoretical concepts. Because of his preconceived idea that optically active substances could not be synthesized by chemical methods, he did not even consider that the synthetic malic acid which he had produced might be a mixture in equal amounts of the right and left malic acids. How could he have imagined the simultaneous synthesis of two optically active compounds when he considered the synthesis of one as unrealizable? In order to account for the production of optically inactive malic acid by synthesis, he therefore postulated the existence of a new class of substances in which the asymmetry was abolished by some internal rearrangement of the atoms, instead of by the mixture of two molecules of opposite character as in the case of paratartaric acid. This interpretation was proved wrong when it was shown that, contrary to his prejudiced notion, the inactive malic acid which he had synthesized was nothing but a mixture of optically active antipodes.

Nevertheless, Pasteur's hypothesis of a new type of molecule was soon confirmed—under unexpected circumstances. While working on the synthesis of paratartaric acid from tartaric acid, he obtained in addition to the former a new form of optically inactive tartaric acid—mesotartaric acid—that could not be resolved into left and right components; this new compound corresponded to the molecular structure which he had postulated. Thus, in the very act of putting the wrong interpretation on the nature of inactive malic acid, he had made a most important addition to the theory of molecular structure. He had postulated and demonstrated the existence of a new type of optically inactive molecules, namely one in which the compound is neither dextro nor levorotatory, nor formed by the union of these. It is now well known that this type of molecule does not exist for every asymmetric compound, as was erroneously assumed by Pasteur, but only for those which, like tartaric acid and the tetra- and hexahydric alcohols and the hydroxy dibasic acids, can have their molecules divided into two symmetrical halves, the optical effect of the one half neutralizing the optical effect of the other within the molecule.

In 1860, Perkin and Duppa prepared from succinic acid by synthetic reaction a substance which Pasteur himself identified as paratartaric acid, and they achieved thereby the synthesis not only of one, but actually of two asymmetric molecules. Had Pasteur slightly modified the terms of his dogma of the impossibility of producing asymmetric molecules by synthesis, it might still stand unchallenged. For even today, the synthesis from symmetric materials of a single asymmetric molecule, without at the same time building up an equal number of asymmetric molecules of configuration opposite to the first, can be accomplished only by introducing an asymmetric element into the reaction. The relationship between optical activity and life still appears to be as close as suspected by Pasteur. In a certain measure, optically active compounds may be compared with living organisms, for just as they can convert inanimate material into organic, similarly it is possible to synthesize from optically active compounds other optically active substances which bear a certain qualitative relationship to the original active compound.

Pasteur affirmed to the last that, in final analysis, asymmetric bodies are always the products of living processes. In order to account for the selective accumulation of either the right or the left form of a given substance during the course of living processes, one might conceive that the two forms are produced simultaneously but one of them is utilized or destroyed as fast as produced. Pasteur adopted another view which, in fact, is not exclusive of the first. He boldly connected the asymmetry of natural products with the presence of asymmetry in the forces acting upon them at the time of their formation. Although the earth is round, he pointed out, it would be symmetrical and superposable on itself only if motionless. As soon as it turns on its axis, its image in a mirror no longer resembles it, as the image turns in a different direction. If there is an electric current flowing along the equator and presiding over the distribution of magnetism, this current also turns in opposite directions in the earth and in its image. In short, the earth is an asymmetrical whole from the point of view of the forces which make it live and function, and of the things which it produces. It is for this reason, Pasteur felt, that the substances produced by living creatures are asymmetrical and endowed with optical activity. If some asymmetric forces were in operation at the time of the genesis of a plant, it might initiate an asymmetric process which would then render asymmetric all subsequent biological operations. Protoplasm, in other words, would thus be made asymmetrical when first produced.

These imaginings fired Pasteur's mind and he spoke of them with uncontrolled enthusiasm on several occasions.

"The universe is an asymmetrical whole. I am inclined to think that life, as manifested to us, is a function of the asymmetry of the universe and of the consequences it produces. The universe is asymmetrical; for, if the whole of the bodies which compose the solar system moving with their individual movements were placed before a glass, the image in the glass could not be superposed upon the reality. Even the movement of solar light is asymmetrical. A luminous ray never strikes in a straight line upon the leaf where plant life creates organic matter. Terrestrial magnetism, the opposition which exists between the north and south poles in a magnet and between positive and negative electricity, are but resultants of asymmetrical actions and movements. . . .

"Life is dominated by asymmetrical actions. I can even imagine that all living species are primordially, in their structure, in their external forms, functions of cosmic asymmetry."

Nor did Pasteur simply propound these questions. Instead, he was bold enough to attempt experimentation in this highly speculative domain, hoping to duplicate in the laboratory the asymmetrical effects which he assumed to preside over the synthesis of organic materials in nature. For a while he considered the possibility that the barrier which separates the synthetic products of the laboratory from those formed under the influence of life might not be an impassable one. If it be true that nature elaborates the living substances by means of asymmetrical forces, why should not the chemist attempt to imitate nature? Why should he not bring asymmetrical forces to bear upon the production of chemical phenomena instead of limiting himself to methods founded upon the exclusive use of symmetrical forces? While in Strasbourg, Pasteur had powerful magnets constructed with a view to introducing asymmetrical influences during the formation of crystals. At Lille, in 1854, he had a clockwork arrangement made with which he intended, by means of a heliostat and reflector, to reverse the natural movement of the solar rays striking a plant, from its birth to its death, so as to see whether, in such an artificial world – in which the sun rose in the west and set in the east – the optically active substances would not appear in forms opposite to those occurring in the normal order of nature. His faithful advisers warned him that he risked sacrificing much energy, time and resources to no avail on these fantastic experiments. Worldly-wise enough to realize that he had a better chance of success in working toward more limited objectives, Pasteur eventually abandoned his ambitious projects without having achieved any results. But he never forgot his romantic ideas, his alchemist dream of unraveling the chemical riddle of life. Time and time again, he referred to them in conversations, lectures and unpublished notes. During and after the war of 1870, while he was forced to remain away from his laboratory, he returned once more to the thoughts of his early years, and discussed them in a letter to his assistant Raulin.

"I have begun here some experiments on crystallization which will open great prospects, should they lead to positive results. As you know, I believe that there exists in the universe an asymmetrical influence which presides constantly and naturally over the molecular organization of principles immediately essential to life. In consequence, the species of the three kingdoms bear a definite relation to the movements of the universe by their structure, by their form, by the disposition of their tissues. For many of these species, if not for all, the sun is the *primum movens* of nutrition; but I believe in another influence which would affect the whole organization, for it would be the cause of the molecular asymmetry proper to the chemical components of life. I would like to grasp by experiment a few indications of the nature of this great cosmic asymmetrical influence. It may be electricity, magnetism. . . . And, as one should always proceed from the simple to the complex, I am now trying to crystallize double racemate of soda and ammonia under the influence of a spiral solenoid."

Intoxicated by his imagination, Pasteur thus attempted to alter the course of chemical synthesis, and even to create or modify life by means of asymmetrical forces. What would a world be, he wondered, in which sugar, cellulose, albumins, and other organic materials would consist of molecules oriented differently from the ones which we know? Although he never gave up these dreams, he became increasingly aware in later years of the difficulties presented by their experimental realization, especially after his studies on spontaneous generation had convinced him of the overwhelming directional effect of the "germ" on the development of living things. It seems that only one of his collaborators and disciples, Duclaux, has spoken with warmth and sympathy of these projects, to the extent of himself elaborating on a statement made by the master in 1874. "In order to introduce into a cell proximate principles different from those which exist there, it would be necessary to act upon it at the time of its greatest plasticity, that is, to take the germ cell and try to modify it. But this cell has received from its parents a heredity in the form of one or several active substances, the presence of which is sufficient to render it rebellious to certain actions and . . . to impart to its evolution a definite direction. This cell contains in the beginning not only its *being* but also its *becoming*, and it constitutes therefore an initial force that . . . gives its own direction to new forces which appear every day in the little world it governs. . . .

"Ah, if spontaneous generation were possible! If one . . . could cause a living cell to evolve from inactive mineral matter! How much easier it would be to give it a direction, to make these asymmetries . . . enter into its substance and thence into its vital manifestations. I am adding something to what Pasteur has written on these captivating questions, but I do not believe that I have gone beyond what was in his thought in my effort to show how . . . he arrived at two of the problems which it was fated that he should solve: the question of fermentations and that of spontaneous generations."

Pasteur carried with him into the grave the dream of his scientific youth — the fantastic vision of developing techniques for the creation or the modification of life by introducing asymmetrical forces into chemical reactions. Duclaux was probably right in regarding the subsequent studies on spontaneous generation and fermentations as consequences of the early visions of the master. Indeed, it is a striking fact, perhaps worthy of the attention of psychoanalysts, that Pasteur devoted much of his later life to demonstrating that nature operates as if it were impossible to achieve what he — Pasteur — had failed to do. For he proved that all claims of the creation of life out of lifeless material were based on faulty observation or unskilled experiments, that spontaneous generation has never been observed; that, as far as one knows, life always comes from a "germ," from life. He demonstrated also that this "germ" imparts upon the new life which it creates a directional force so intense that each living being, however simple, possesses a specificity of property and functions peculiar to it. Each microbe, Pasteur would show, is the specific agent of a particular fermentation, of a particular disease. Just as he had failed in his attempts to create or modify life, so he proved that others, who had claimed to be successful where he had failed, had been merely the victims of illusion.

It was one of Pasteur's characteristics that, while often dreaming romantic concepts, he possessed to an extreme degree the ability of observing small concrete facts which soon brought his activity back to the level of the possible. In the midst of his concern with the philosophical consequences of the optical activity of organic matter, he once observed that optically active amyl alcohol was a constant product of alcoholic fermentation. This was enough to suggest that amyl alcohol was another of the products of life to be added to the list of optically active organic materials first studied by the venerated Biot. Alcoholic fermentation, then, was a manifestation of life. It was this conviction that launched Pasteur on a new sea of experiments, on the tempestuous voyage from which he was to bring back the germ theory of fermentation and of disease.

CHAPTER V

The Domestication of Microbial Life

There is a devil in every berry of the grape.
— THE KORAN

SHORTLY after Pasteur's arrival in Lille, the father of one of his students — Bigo by name — came to consult him concerning difficulties that he was experiencing with the alcoholic fermentation of beet sugar in his distillery. Pasteur agreed to investigate the matter, spent some time almost daily at M. Bigo's factory and, as shown by his laboratory notebooks, began to study alcoholic fermentation in November of the same year. This experience awoke in him an interest in the broader aspects of the fermentation problem and, by the spring of 1857, he was investigating the production not only of alcohol, but also of lactic, butyric and tartaric acids by fermentation reactions.

The word "fermentation" was then loosely applied to the spontaneous changes which often occur in organic solutions and which result in the production of spirituous or acidic substances. The production of alcohol during the making of wine, beer, or cider was called "alcoholic fermentation"; the conversion of wine or cider into vinegar, the "acetic acid fermentation"; the souring of milk, during which milk sugar is converted into lactic acid, the "lactic acid fermentation." It was also well known that many natural materials, such as meat, eggs, bouillons, could spontaneously undergo other types of changes which were designated by the word "putrefaction." Putrefaction was generally assumed to be closely related to fermentation, but to differ from it in the products formed, as evident from the ill-smelling emanations which accompany the former process.

The man of science as well as the layman of 1850 regarded fermentations and putrefactions as caused by chemical agents – the "ferments" – complex and obscure, true enough, but no more obscure than those involved in the chemical reactions classified by the Swedish chemist Berzelius under the name of "catalytic processes." According to Berzelius, the catalyst, or the ferment, acted by its mere presence to start the reaction without becoming a part of it, much as lightning or a hot cannon ball can start a fire, without supplying the fuel which keeps it going. Berzelius regarded catalysts (or ferments) as "bodies that were capable, by their mere presence . . . of arousing affinities ordinarily quiescent at the temperature of the experiment, so that the elements of a compound body arrange themselves in some different way, by which a greater degree of electrochemical neutralization is attained." Thus there were alcoholic, lactic acid, butyric acid, acetic acid, putrefaction ferments – all catalysts capable in some mysterious manner of bringing about the formation of the substance for which they were named. The ferment responsible for the production of alcohol was also known under the name of yeast, as was the leaven active in the rising of the dough during the making of bread.

The processes of fermentation constitute some of the first successes of technology. Even before the dawn of history, man learned to use yeast to transform the difficult-to-digest starch paste into light and savory bread, an achievement which contributed much to the pleasure of life and perhaps to the development of our civilization. With yeast, also, he learned to produce, from sweet and spiritless solutions, the volatile and stimulating liquids which have received in all languages names suggesting spirits and the power of life because of their multiple and strange virtues. All ancient folklores have associated the activity of yeast with the phenomena of life; indeed, bread and wine became the symbol of Life Eternal in the Mediterranean religions. In addition to the magic of their results, the nature of the processes initiated by yeast caught the fancy of philosophers and chemists. The bubbling that takes place spontaneously in the mass of vintage, or in the flour paste, appeared to them as the manifestation of some living spirit; fermentation became one of the favorite subjects for meditation and for experiments by the alchemists, and they derived from its study much of their language and ideology. The subtle changes in property that occur in the mass of fermenting material seemed to them the symbol of those mysterious forces which, instrumented by the philosopher's stone, could convert the baser metals into gold.

With the advent of rational science, it became the ambition of natural philosophers — of scientists — to explain fermentation in more understandable terms. The chemists of the eighteenth and nineteenth centuries attempted to formulate alcoholic fermentation by means of the chemical reactions and symbols which were proving so successful in describing other phenomena of nature. Whatever their philosophical or religious faith, the natural philosophers believed that nothing could better demonstrate the ability of the human mind to unravel the riddle of life than to succeed in explaining these mysterious fermentations. And in fact, they were essentially right — if not in their surmises, then at least in their general view of the future course of science; for much of our understanding of the biochemical reactions of living processes has evolved from the study of yeast and of alcoholic fermentation. It is the enviable privilege of yeast and of the products of its activities that they have, directly or indirectly, fed the dreams and the follies of man, inspired poets, and challenged philosophers and scientists to meditation and creative thinking.

Lavoisier, Gay-Lussac, Thénard, Dumas — the high priests of Pasteur's scientific cult — had studied the transformation of cane sugar into alcohol by the methods of quantitative chemistry. They had reached a formulation which appeared so exact as to give the illusion that the fundamental nature of the phenomenon had been finally discovered.

According to Lavoisier, the sugar was split into two parts, one of which was oxidized at the expense of the other to form carbonic acid, while the second part, losing its oxygen to the first, became the combustible substance, alcohol "so that if it were possible to recombine these two substances, alcohol and carbonic acid, sugar would result." What could be clearer than this simple relationship? Lavoisier's formula was satisfactory except for the fact that there was in it no place for yeast. And yet, all chemists accepted as a fact that yeast always accompanied and probably initiated fermentation. Although it was the prime mover of the reaction, yeast did not appear to take part in it. Berzelius explained away the puzzling question by the word "catalysis."

The interpretation of alcoholic fermentation held by the nineteenth-century chemists is, in certain respects, a forerunner of the modern physicochemical interpretation of living processes. One by one, all the activities of living things are being described in terms which, as they come to be better defined and their interrelationships better understood, give an ever simpler and more satisfactory account of the forces and processes which together constitute life. But insofar as we know, the integration of all these physicochemical processes depends upon prior life. The twentieth-century biochemist does not know how to introduce life into his equilibrium reactions any better than the nineteenth-century chemist knew where to place yeast in the formula of alcoholic fermentation. However, so great were the triumphs of physicochemical science during the "wonderful century" that many scientists had enough confidence, or perhaps merely enough intellectual conceit, to ignore the difficulty, and to refuse to recognize the existence of an unsolved mystery in fermentation and putrefaction. Pasteur was willing to reintroduce mystery in the problem by stating that yeast was a living being and fermentation an attribute of its life. He did not have to devise any radically new experimental approach to demonstrate this concept. In fact, the true relation of yeast to fermentation had been clearly stated by at least four experienced investigators before him, and probably recognized many more times by obscure and timid men who, awed by the authority of established science, had not dared to project their observations beyond the walls of their studies.

In 1835, Cagniard de la Tour had observed that the yeast produced during fermentation consisted of living cellular organisms which multiplied by budding, and he suggested that the life of these cells was intimately associated with the process of fermentation. Independently of Cagniard de la Tour, and almost simultaneously, Schwann in Germany had published experiments which substantiated the former's suggestions. Following Gay-Lussac's work around 1835, it had been believed that the introduction of oxygen into a fermentable fluid was sufficient to initiate fermentation or putrefaction. Schwann, on the contrary, showed that the production of alcohol and of yeast cells did not take place when grape juice or other sugar solutions which had been boiled were brought into contact with air previously heated; some organic matter, preferably yeast, had to be added to the system to initiate the fermentation process. Schwann concluded from these findings that living microorganisms played an important part in fermentation. To further support his belief, he tried to impede the production of alcohol by adding toxic substances to the fluid. Finding that nux vomica, so toxic to animals, did not retard fermentation, whereas arsenic interrupted it, he concluded that the living agents responsible for alcohol production were more plant than animal-like. He also confirmed Cagniard de la Tour's observations that the deposit produced during fermentation consisted of budding yeast cells, and he finally showed that fermentation commenced with the appearance of these yeast cells, that it progressed with their multiplication, and ceased as soon as their growth stopped. Another memoir on the same topic was published in 1837 by Kützing who, like Cagniard de la Tour, founded his opinions on microscopical observations; he recognized yeast as a vegetable organism and accurately described its appearance. According to him, alcoholic fermentation depended on the formation of yeast, which increased in amount whenever the necessary elements and the proper conditions were present for its propagation. "It is obvious," said Kützing, "that chemists must now strike yeast off the roll of chemical compounds, since it is not a compound but an organized body, an organism."

These three papers were received with incredulity and Berzelius, at that time the arbiter of the chemical world, reviewed them all with impartial scorn in his *Jahresbericht* for 1839. He refused to see any value in the microscopical evidence and affirmed that yeast was no more to be regarded as an organism than was a precipitate of alumina. He criticized Schwann's experiments on the ground that the results were irregular and therefore did not prove anything concerning either the nature of yeast or the effect of heating on the presence or absence of fermentation. This criticism was justified to some extent by Schwann's own honest confession that his results were not always predictable.

To the scorn of Berzelius was soon added the sarcasm of Wöhler and Liebig. At the request of the Académie des Sciences, Turpin of Paris had repeated in 1839 Cagniard de la Tour's observations and confirmed their accuracy. Stimulated by this publication, Wöhler prepared an elaborate skit, which he sent to Liebig, who added to it some touches of his own and published it in the *Annalen der Chemie,* following immediately upon a translation of Turpin's paper. Yeast was here described, with a considerable degree of anatomical realism, as consisting of eggs which developed into minute animals shaped like distilling apparatus. These creatures took in sugar as food and digested it into carbonic acid and alcohol, which were separately excreted — the whole process being easily followed under the microscope!

The facts and theories presented by Cagniard de la Tour, Schwann, Kützing and Turpin to support the vitalistic theory of fermentation and putrefaction appear so simple and so compatible with the knowledge then available that it is difficult to understand why they did not immediately gain wide acceptance, but were instead neglected and even forgotten. It was not the difficulty of imagining the existence of microscopic living agents that proved an obstacle to the adoption of the vitalistic theory of fermentation. Ever since 1675, when the Dutch lens grinder Leeuwenhoek had first shown their presence in water and in body fluids, microscopic organisms had often been seen by naturalists. Before Pasteur's time, Christian Ehrenberg, Ferdinand Cohn and other botanists had described and classified the bacteria – microorganisms smaller than yeast, much harder to see and to study.

Schwann's results were ignored because it was not always possible to duplicate them. In particular, when he worked with infusions of animal tissues instead of simple sugar solutions, putrefaction often ensued when heated air was brought into contact with the heated infusion, a fact which seemed to prove that putrefaction could occur even in the absence of living agents. In 1843, Hermann Helmholtz, at that time a young medical student, and one who was to become immensely famous among the physiologists and physicists of the century, made his scientific debut with a paper in which he concluded that putrefaction of nitrogenous substances was independent of germ life and that, even in alcoholic fermentation, germs probably occupied only a secondary and subordinate place. He was willing to grant only that putrid and fermentable materials possibly provided an attractive food substance to certain germs, and that when present, these germs might be capable of modifying to some extent the course of the fermentation and putrefaction processes, without being responsible for their initiation.[1]

There were other more profound reasons which made physiologists wary of accepting the vitalistic theory of fermentation and putrefaction. The belief that living things were the cause of these processes was in conflict with the *Zeitgeist*, the scientific and philosophical temper of the time. Mathematics, physics and chemistry had achieved so many triumphs, explained so many natural phenomena, some of them pertaining to the mystery of life itself, that most scientists did not want to acknowledge the

[1] This was merely a restatement of a view formerly expressed by Liebig: "As to the opinion which explains the putrefaction of animal substances by the presence of microscopic animalcula, it may be compared to that of a child who would explain the rapidity of the Rhine current by attributing it to the violent movement of the many millwheels at Mainz."

need of a vital force to account for these commonly occurring processes. As fermentation could be described by a simple chemical reaction, it appeared pointless to explain it in terms of a living agent, instead of by the simple play of physical and chemical forces. Little by little, science had expelled living forces from the domain of physiology and everyone believed that it was capable of pursuing this course even further; to appeal to a living agent as the cause of a chemical reaction appeared to be a backward step. In the kingdom of science, there were probably many who did not share the official optimism, and who did not believe that the time had come when everything could be accounted for in terms of known physicochemical forces. But the priests of the new faith — Berzelius, Liebig, Wöhler, Helmholtz, Berthelot and others — were the supreme rulers of scientific thinking. The power of their doctrine and of their convictions, and the vigor of their personalities, smothered any voice that ventured to express an opinion in conflict with their own philosophy. As a reaction against the romantic and confused airings of the German *Natur philosophie,* the new prophets had pronounced anathema on anyone who preached the doctrine of vitalism.

Justus von Liebig, the dean of biochemical sciences, did not ignore the facts disclosed by Cagniard de la Tour, Schwann, Kützing and Turpin. He was even willing to admit that yeast might be a small plant and that it played a secondary role in fermentation. But he pointed out that in other decompositions of sugar, the lactic and butyric fermentations for example, nothing resembling yeast was to be found; nor had germs been seen participating in the putrefaction of meat. If yeast then contributed to alcoholic fermentation, it was not as a living thing but only because, on dying, it released in solution albuminoid material which imparted a vibration to the sugar molecule, a movement which caused it to break down into alcohol and carbonic acid. "The yeast of beer, and in general all animal and plant substances undergoing putrefaction, impart to other substances the state of decomposition in which they find themselves. The movement which is imparted to their own elements, as the result of the disturbance of the equilibrium, is communicated equally to the elements of the substances which come in contact with them."

Liebig had no observed fact or theoretical basis to substantiate this description of the fermentation process. His hypothesis was a mere jumble of words, and Goethe could have well said of him:

> At the point where concepts fail,
> At the right time a word is thrust in there.
> With words we fitly can our foes assail,
> With words a system we prepare,
> Words we quite fitly can believe,
> Nor from a word a mere iota thieve.

However, Louis Thénard had made an observation which appeared compatible with Liebig's argument. He had seen that, by adding twenty grams of yeast to one hundred grams of cane sugar in solution in water, a rapid and regular fermentation was obtained, after which the remaining yeast, collected on a filter, weighed only thirteen grams. Added to a new quantity of pure sugar solution, this residual yeast produced fermentation more slowly than the original yeast, after which it was reduced to ten grams and became even less capable of causing the fermentation of sugar. This appeared as proof that yeast destroyed itself in the course of its own fermenting activity.

It does not appear profitable to pursue the complex structure of all the arguments which brought Liebig's views to the status of official dogma. Vague as it was, his theory agreed with the spirit of the age and served to incorporate the phenomena of fermentation and putrefaction into the fold of the physicochemical doctrine. To overcome the theory, it was not sufficient to oppose facts to facts, and interpretation to interpretation; it was necessary to bring to battle enough energy, talent and conviction to challenge and override the entrenched official position.

As a fighter, Pasteur proved more than a match for Liebig.

There is no indication that Pasteur had given any systematic thought to the problem of fermentation before he arrived in Lille. The first entries in his laboratory notebooks of experiments dealing with this subject date from September 1855, and yet, by August 1857, he was ready to present before the Société des Sciences de Lille, in his *Mémoire sur la fermentation appelée lactique*, a complete statement of the germ theory with a proposed methodology of experimentation. Unfortunately, few facts are available to account for this magnificent intellectual performance, to reveal in particular how he overcame, in his own mind, the pressure of authoritative scientific opinion and came to adopt a biological interpretation for a phenomenon which was described in all textbooks as a chemical reaction.

Even though trained as a chemist, Pasteur always had an eye for the biological implications of his work. As early as 1854, for example, after observing that the part of a crystal which has been damaged grows faster than the other parts, as if with the purpose of restoring the integrity of crystalline structure, he stated that this phenomenon was analogous to "those exhibited by living beings which have received a wound. The part which has been damaged slowly recovers its original form, and the process of tissue growth is, at this point, much more active than under normal conditions." We have also emphasized in the preceding chapter how possessed he was by the thought that molecular asymmetry is related to living processes. The expression "The great problem of life . . ." often appeared, in one sentence or another, in letters to his friend Chappuis or to his father, or in his lectures and notes. This unusual concern with the ultimate nature of living phenomena accounted in part for his receptiveness to any fact susceptible of a vitalistic interpretation.

The production of optically active molecules, in particular of optically active amyl alcohol, seems to have been the specific fact that led him to regard fermentation as brought about by living agents. Amyl alcohol, a well-known product of distillation, had been the first exception which he had encountered to the correlation between asymmetry in crystal structure and optical activity, and he had studied it with especial care. The manner in which this study launched him into the analysis of yeast fermentation with a preconceived idea opposite to Liebig's doctrines is clearly set forth in his own writings. Liebig assumed that the optical activity of amyl alcohol was a consequence of the asymmetry of the sugar, from which it was derived during fermentation; Pasteur on the contrary felt convinced that the molecule of amyl alcohol was too remotely related to that of sugar to have preserved the asymmetry of the latter. "Every time," he says, "that we try to follow the optical activity of a substance into its derivatives, we see it promptly disappear. The fundamental molecular group must be preserved intact, as it were, in the derivative, in order for the latter to be optically active. . . . The molecular group of amyl alcohol is far too different from that of the sugar, if derived from it, to retain therefrom an asymmetrical arrangement of its atoms." In consequence, Pasteur regarded the asymmetry of amyl alcohol as due to a new creation, and such creation of an asymmetric molecule was, according to him, possible only through the intervention of life. As a corollary, fermentation had to be a vital process, and not the purely chemical transformation which Liebig assumed it to be.

Although Pasteur worked intensively on alcoholic fermentation through 1855 and 1856, his first communication on the germ theory dealt primarily with the conversion of sugar into lactic acid, the reaction which is responsible for the souring of milk. The selection of this subject to introduce the new doctrine is the more surprising because lactic acid fermentation is less important than alcoholic fermentation from the industrial point of view, was then less well known scientifically, and had less historical glamour. It appears possible that Pasteur's choice was dictated in part by a shrewd sense of the strategy best adapted to the defeat of Liebig's chemical theory. Alcoholic fermentation had already lost its bloom, for Liebig and his partisans were willing to regard yeast as a living organism. Their great argument was always: What role can you attribute to yeast, when so many other related fermentations, the lactic fermentation for example, take place without the presence of anything which resembles it? . . . In a certain sense, therefore, lactic fermentation was the *champ clos* in which the battle had to be fought. Another reason may also have influenced Pasteur. Of all fermentations, none is simpler chemically than the conversion of sugar into lactic acid. Essentially it involves the breakdown of one molecule of sugar into two halves which are the lactic acid molecules. If it could be shown that this simple process was really carried out by a living agent, then, reasonably, it would become unnecessary to fight, one after the other, all the battles of the individual fermentations. Lactic acid fermentation could rightly serve as a general pattern for this class of phenomenon.

In reality, Pasteur's *Mémoire sur la fermentation appelée lactique* does not offer a rigorous demonstration of the germ theory. It states only that the gray material deposited during the conversion of sugar into lactic acid does, in its mode of formation and in many of its properties, present some analogy to yeast. If a bit of this gray deposit is added to a new sugar solution, it increases in amount as lactic acid is produced. Like yeast, this lactic ferment has also an organized shape, although it is different, smaller and more difficult to see. To make yeast grow, it is the practice to add an albuminoid material to the fermenting sugar; similarly more rapid production of lactic acid, accompanied by more rapid and abundant production of the gray deposit, is assured if one adds some albuminoid material to the sugar solution along with the lactic ferment. Whereas Liebig and his school regarded the albuminoid material as the ferment itself, Pasteur claimed it was nothing but food for the yeast, or for the lactic ferment, both of which needed it to grow and thereby to cause the fermentation.

Side by side with these affirmations concerning the vital nature of the fermentation process, Pasteur described in the same short memoir a methodology which even today forms the basis of bacteriological technique. He showed that one can grow the ferment in a clear nutrient bouillon, where it multiplies to give rise to a population of microscopic living beings, all the individuals of which resemble one another. Once grown in adequate amount, and in the pure state, it accomplishes with extraordinary rapidity the chemical transformation over which it presides, namely the production of lactic acid. The memoir presents, also, an exact statement of the influence of the acidity, neutrality, or alkalinity of the liquid on the course of fermentation. Whereas yeast prefers acid media, the lactic ferment grows best at neutrality; and it is for this reason that the lactic fermentation is favored by the addition of chalk to the sugar solution. There is also a hint, as it were an omen, of the effect of antiseptics. "The essential oil of onion juice completely inhibits the formation of the yeast of beer; it appears equally harmful to infusoria. It can arrest the development of these organisms without having any notable influence on the lactic ferment." Antiseptics, then, can be used for separating the different ferments one from the other.

The idea of a specific ferment associated with each fermentation, of disproportion between the weight of the ferment produced and the weight of matter transformed, of vital competition between two organisms which simultaneously invade the same medium, resulting in the dominance of the one best adapted to the culture conditions — all these ideas, which the future was to develop so thoroughly, are found clearly set forth in this paper. Its fundamental spirit can be summarized in Pasteur's own words: "The purity of a ferment, its homogeneity, its free unrestrained development by the aid of food substances well adapted to its individual nature, these are some of the conditions which are essential for good fermentation."

It is a remarkable fact that this preliminary memoir, which presented in such specific terms the credo and the ritual of the new doctrine, has stood the strain of all subsequent experimentation without showing any defect. And yet, its claims were presented without unequivocal evidence. Indeed, the criteria as to what constitutes evidence for the causal participation of a living agent in a chemical or pathological process are so ill-defined, even today, that the history of microbiology offers countless examples of claims of etiological causation which subsequent experience has failed to verify. Pasteur himself was well aware of this difficulty, and stated at the end of his memoir, "If anyone should say that my conclusions go beyond the established facts,

I would agree, in the sense that I have taken my stand unreservedly in an order of ideas which, strictly speaking, cannot be irrefutably demonstrated."

Pasteur's claims did not deal with the ultimate nature of the fermentation process. They only expressed the view that, under the conditions used by Lavoisier, Gay-Lussac, Thénard and Liebig, the formation of lactic acid, of alcohol, of butyric acid – of all products of fermentation – was always dependent upon the life of yeast or of bacteria. Fermentation was, as he said, "correlative with life"! In consequence, the further understanding of the fermentation process demanded a knowledge of the conditions which affect the life of the germs and their physiological activities. The demonstration that ferments were living germs was the first problem that had to be solved before a following generation could undertake the detailed analysis of the composition of these germs, of their enzymatic equipment, of the specific chemical reactions for which they are responsible. The specificity of enzymes and their mode of action is to our age what the specificity of germs and the conditions of their life was to the middle of the nineteenth century. Pasteur's role was to define the problem of fermentation in terms that were scientifically meaningful for his time. Such a limited scope may not be sufficient for the philosopher, but the experimenter has to be satisfied with it, for as is said in Ecclesiastes: "To every thing there is a season, and a time to every purpose . . ."

Although the *Mémoire sur la fermentation appelée lactique* was the Manifesto of the germ theory, it was only in the *Mémoire sur la fermentation alcoolique*, published in its preliminary form in 1857 and *in extenso* in 1860, that Pasteur summoned the experimental evidence which demonstrated the participation of living agents in the phenomena of fermentation.

It was by then generally accepted that yeast was a necessary accompaniment of alcoholic fermentation, but a few workers still doubted that it was a living, organized structure. Berzelius considered yeast as some amorphous organic material, precipitated during the fermentation of beer and mimicking the morphology of simple plant life; but mere form, Berzelius pointed out, does not constitute life. Even those who were willing to regard yeast as a microscopic plant believed, with Liebig, that it induced fermentation not as a living agent, but because its death liberated into the sugar solution albuminoid material which imparted a molecular vibration capable of breaking down the sugar molecule into alcohol and carbon dioxide. Pasteur finally demolished this thesis, by two independent lines of evidence. He showed that the products of alcoholic fermentation are more numerous and complex than indicated by the simple terms of Lavoisier's chemical formula; and he succeeded in causing fermentation to occur in a purely mineral medium, under conditions where fermentative activity and yeast multiplication went hand in hand.

Unimpressed by the authority of tradition, Pasteur first established that, in addition to the alcohol and carbon dioxide demanded by the classical formula of alcoholic fermentation, there are produced significant amounts of other substances – such as glycerine, succinic acid, amyl alcohol. It was, he pointed out, because the supporters of the chemical theory of fermentation were prejudiced in favor of a simple reaction that they had neglected to look for these substances, which are always present.

Although the multiplicity of these side products – several of them optically active – made it plausible that fermentation was a complex process due to the agency of life, this did not prove the vitalistic theory. In order to link the production of alcohol and the multiplication of yeast by a cause-effect relationship, Pasteur undertook an experiment which, for the time, was as bold and original in its concept as it appears today simple and obvious.

Liebig had supported his view – that yeast produces alcohol only after its death and when undergoing decomposition – by quoting the experiment in which Thénard had found that the weight of yeast decreased during fermentation. Liebig also emphasized that ammonia is often released by yeast as it decomposes while fermenting the sugar. These statements were based on correct observations, and described facts which were of common occurrence in those days when, unknown to the investigators, yeast fermentation often took place with the presence of contaminating bacterial growth. These observed facts made it therefore not illogical to conclude that fermentation and the death of yeast were causally related. Unfortunately, logic is an unreliable instrument for the discovery of truth, for its use implies knowledge of all the essential components of an argument – in most cases, an unjustified assumption. Nor is the experimental method the infallible revealer of pure and eternal fact that some, including Pasteur, would have us believe. The validity of a theory is usually proven more by its consequences than by conscientious effort and brilliant intellect. The observations quoted by Liebig were correct, but the very fact that yeast is a living plant had introduced so many complicating and unknown factors in the experiments that some of the most significant facts of the reaction had escaped him. It is because Pasteur had intuitively guessed the true nature of fermentation that he was able to find the flaw in Liebig's intellectual edifice and to arrange conditions for demonstrating that alcohol was a product of the life of yeast.

Armed with his conviction, Pasteur was bold enough to reverse Liebig's reasoning and to arrange his own experiments in such a way that, instead of ammonia being given off during fermentation, the yeast would be produced from ammonia added to the mixture. Technically, the problem was to grow the yeast in a nutrient liquid devoid of organic nitrogen – a liquid containing only sugar, ammonia to provide nitrogen, and some mineral salts to supply the yeast globules with their structural elements. Pasteur had the ingenious idea of adding also to his nutrient medium the ashes of incinerated yeast in addition to the salts of phosphoric acid, potassium, magnesium and iron, hoping to supply thereby the unknown mineral elements required by the small plant. He had to acknowledge that, even under these conditions, yeast grew less readily than in the juice of the grape or in beer brew, probably because it had to synthesize all its tissue constituents instead of finding many of the metabolic factors ready-made in the natural organic fluids. Nevertheless, he could report in 1860 that he had obtained fermentation in his synthetic medium inoculated with minute amounts of yeast, and that the amount of alcohol produced ran parallel with the multiplication of the yeast. Realizing that this was the crucial test of his thesis, he returned to it over and over again. By perfecting the medium and employing a more vigorous strain of yeast, he succeeded in obtaining more rapid fermentation by a technique which he described in the *Etudes sur la Bière* in 1876. This momentous experiment established once and for all that, contrary to Liebig's claims, organic material in decomposition is not necessary to start alcoholic fermentation. An imperceptible trace of yeast, introduced into a liquid containing only sugar and mineral salts, makes the sugar ferment while the yeast develops, buds and multiplies. All the carbon of the new yeast globules is derived from the sugar; all their nitrogen from the ammonia.

What, then, was the meaning of that odd experiment in which Thénard had found that the amount of yeast decreases when large amounts of it are added to a sugar solution to make it ferment? That meant, according to Pasteur, that many of the old yeast globules died in a sugar solution depleted of nitrogen and minerals. In breaking down, they released into solution some of their own cellular constituents which were then used by the new young globules to multiply and to ferment the sugar. There were not enough new yeast globules formed to balance the loss of weight in the old ones due to the dissolution. However, if the weight of the organic matter which had gone into solution was added to that of the formed yeast, then the total weight was found to *increase* during fermentation, because there is always a little sugar which is transformed into yeast. Thus, whatever the conditions employed, whether the ferment was introduced into synthetic mineral media, or into fluids rich in organic matter derived from grape, from barley, or from the decomposing yeast itself, the production of alcohol was always dependent upon the life of yeast. And Pasteur concluded his 1860 memoir with these incisive and uncompromising words: "Alcoholic fermentation is an act correlated with the life and with the organization of these globules, and not with their death or their putrefaction. It is not due to a contact action in which the transformation of the sugar is accomplished in the presence of the ferment without the latter giving or taking anything from it." These were the very words with which he had ended his preliminary announcement of the germ theory of lactic fermentation in 1857. Subsequent experimentation had served only to add evidence to the conclusion which he had intuitively reached at the beginning of his studies.

Pasteur now returned to the lactic acid fermentation and here again succeeded in causing it to proceed in a simple medium. Although the lactic ferment was smaller than that of yeast, its needs were not less, only different. Like yeast it had its own specific requirements, as all living things have.

Within a few years, the generality of the view that ferments are living beings was established. Pasteur himself showed that the germs responsible for the production of tartaric, butyric and acetic acids could be readily cultivated in synthetic media. Each one of these germs was a living microorganism, characterized by a definite morphology, definite nutritional requirements and susceptibilities to toxic influences, as well as by the ability to carry out a specific type of biochemical performance. Raulin, the first of Pasteur's assistants, added weight to the new doctrine by defining with unsurpassed completeness and precision the growth requirements of the mold *Aspergillus niger*, and by revealing the influence of nutritional factors, and particularly of rare mineral elements, in the life of microorganisms.

One cannot exaggerate the importance of these studies for the evolution of biochemical sciences. As early as 1860, Pasteur himself pointed out that the findings made in his laboratory would permit physiology to attack the fundamental chemical problems of life. The bodies of plants and animals consist of an immense number of cells, whereas in microorganisms, the living agent is reduced to the single-cell level. By studying microbial physiology, therefore, it became possible to analyze the chemical phenomena which determine the function of the individual cell — the fundamental unit of life — be it that of a plant, a microorganism, an animal, or even a man.

By 1859 Pasteur had sufficiently mastered the techniques of pure culture and of preparation of selective media to be able to bring about at will one or another type of fermentation, and to determine its causative agent and its chemical mechanism. For example, when he added to a solution containing a salt of lactic acid a drop of a liquid undergoing butyric acid fermentation, there soon ensued a transformation of the lactic acid into butyric acid, with evolution of a gas consisting of a mixture of carbon dioxide and hydrogen. Simultaneously with the new chemical process, a new microscopic population appeared in the fluid. It consisted of short rods which, curiously enough, moved rapidly to and fro with undulating movements. Because of the view then prevalent that motility was one of the differential characteristics between the animal and plant kingdoms, Pasteur was at first inclined to regard these motile beings as minute animals, and for this reason, he referred to them as "infusoria." For this reason also, he was for a while reluctant to believe that they could be the real cause of butyric acid fermentation — as this activity was universally considered more plant- than animal-like. "I was far from expecting such a result," said he; "so far, indeed, that for a long time I thought it my duty to try to prevent the appearance of these little animals, for fear they might feed on the microscopic plants which I supposed to be the true butyric ferment, and which I was trying to discover. . . . Finally, I was struck by the coincidence which my analyses revealed between the infusoria and the production of this [butyric] acid."

Pasteur's hesitation in accepting a motile microorganism as the cause of butyric acid fermentation, because motility was thought to be the prerogative of animal life, illustrates the range of difficulties which he had to overcome in defining the place of microbial life in natural processes. Within a few years, his own studies on anaerobic metabolism were to show that there exist many chemical reactions which are common to all types of life. But in 1860, biologists and biochemists had not yet realized that, at the microscopic level, animal and plant life merge and cannot be differentiated by any simple criteria. Moreover, Pasteur was not a naturalist, and he was working alone, without the support of a scientific tradition, without associates who could share his doubts and his surprises at the unexpected phenomena that he was discovering wherever he turned. The many and brief accounts of new observations which occur repeatedly throughout his writings leave the impression of a child running to and fro in a forest, overwhelmed with a sense of wonder at the signs of unknown life which he sees or only perceives, intoxicated at discovering the diversity of the Creation. Butyric acid fermentation, which so disturbed him by revealing intense motility in microbial life, also led him to discover new and unexpected forms of biochemical processes and a new haunt of life.

Life in the absence of air leaped, so to speak, into his field of vision while he was examining under the microscope a drop of fluid undergoing butyric fermentation. It was his practice to take a drop, place it on the glass slide, cover it quickly with a cover slip, and examine the preparation through the microscope. While examining, with the care that he applied to everything, one of these little flattened drops of liquid undergoing butyric fermentation, he was astonished to see that the bacteria became nonmotile on the margins of the drop although they continued to move with agility in the central portion. This was a spectacle quite the reverse of that which he had observed in the case of other infusions in which the animalcules often left the central portions of the drop to approach the margin, the only place where there was enough oxygen for all. In the presence of this observation, he asked himself whether the butyric microorganisms were trying to escape from the oxygen; and he soon found that it was indeed possible to retard or even arrest butyric fermentation by passing a current of air through the fermenting fluid. Thus was introduced into science the idea that there exists a form of life which can function in the absence of oxygen, although this gas had until then been believed to be an essential requirement of all living creatures. We shall see how Pasteur developed this idea later. For the moment, we must be content with saluting its dawn.

To facilitate discussion of the problem, Pasteur devised the words "aerobic" and "anaerobic" to designate respectively life in the presence and in the absence of oxygen. How could the anaerobic beings, which fear oxygen, live and multiply under natural conditions in the culture broths of the laboratory, which were all in contact with the oxygen of air? Without hesitation and without proof, Pasteur guessed that he had introduced into his cultures, along with the anaerobic butyric ferment, other microscopic germs which could use up the oxygen in solution and form on the surface of the fluid a film of growth below which the gaseous environment became compatible with anaerobic life. On the basis of limited observations, and without extensive experimental evidence, he also became convinced that similar phenomena occur during putrefaction, and that the evil-smelling decomposition of beef bouillon, egg albumin, or meat is the result of the anaerobic life of specialized germs that attack proteins under the protection of aerobic forms capable of removing the oxygen from the environment. He soon arrived at the view that gas production during butyric fermentation and putrefaction is the manifestation of life in the absence of oxygen, and he suspected an intimate relationship between fermentation processes and anaerobic life. Several years were to elapse, however, before he could define these problems in terms concrete enough to give them a clear meaning.

Pasteur never attempted to work out in detail the mechanisms by which nitrogenous organic matter is destroyed during putrefaction. He probably judged that this process was too complex and obscure from the chemical point of view to lend itself to elegant experimental analysis. He chose instead the production of vinegar to illustrate further the activities of microorganisms.

Vinegar production was then well known to result from the oxidation of alcohol to acetic acid. In the French process, as practiced in Orléans, alcohol was allowed to undergo slow oxidation in casks standing on end in piles, and about two-thirds full of a mixture of finished vinegar and new wine. In the German process, the vinegar was made from a weak alcohol solution, to which was added some acetic acid and some acid beer or sharp wine, or other organic matter in course of alteration; this mixture was poured over a hollow column several meters high, containing loosely piled beech shavings, and was allowed to trickle down slowly against an upward current of air.

The German vinegar process appeared to be readily explained in terms of chemical theories based on the catalytic oxidation of alcohol in the presence of platinum. Concentrated alcohol allowed to fall on finely divided platinum is spontaneously oxidized to aldehyde and acetic acid, with production of much heat. Although platinum is not altered in the course of the process it activates the reaction between oxygen and alcohol, thus behaving in Berzelius's terminology as a true catalyst. Liebig, therefore, appeared to be on firm ground when he assumed that other oxidation processes occurring in nature – the oxidation of ammonia to nitrate in soil, or the oxidation of drying oils such as linseed oil, for example – were the outcome of similar reactions, utilizing catalysts other than platinum. In the case of vinegar making, he considered that the beech shavings simply played more economically the role of platinum. Like platinum they seemed to act by their mere presence, being still intact and effective after ten to twenty years of use. The acid beer or sharp wine added to the alcohol mixture was there, according to Liebig, merely to set the process in motion. The reasoning by which Liebig believed he had demonstrated the mechanism of the formation of acetic acid in the German vinegar industry appeared very convincing. In his words: "Alcohol, when pure or diluted with water, does not change to acid in the presence of air. Wine, beer . . . which contain foreign organic matter in addition to alcohol, slowly become acid in contact with air . . . Dilute alcohol undergoes the same transformation when one adds to it certain organic matter, as germinated barley, wine . . . or even ready-made vinegar. . . .

"There cannot be any doubt concerning the role of nitrogenous substances in the acidification of alcohol. They render it capable of absorbing oxygen, which alone, by itself, cannot be absorbed. The acidification of alcohol is absolutely of the same order as the formation of sulfuric acid in the lead chambers; just as oxygen is fixed on sulfurous acid through the intermediary agency of nitrous oxide, similarly the nitrogenous substances, in the presence of acetic acid, absorb oxygen in such manner as to render it susceptible of being fixed by alcohol. . . . When wet, wood shavings absorb oxygen rapidly and rot. . . . This property of absorbing oxygen remains when the shavings are wetted with dilute alcohol, but in this case, the oxygen is carried over to the alcohol instead of to the wood, thus giving rise to acetic acid."

Despite its simple and logical appeal, Liebig's theory was wrong. It was based only on analogy and logic, but Nature, as if to humble man, demands that he return every time to firsthand experience if he wishes to discover the truth. Analogy and logic provide exhilarating intellectual entertainment, but they rarely constitute dependable guides for the exploration of reality.

In the Orléans vinegar process, there is produced on the surface of the liquid in the casks which behave properly a fragile pellicle known as "mother of vinegar," which the vinegar maker takes great pains not to disturb and not to submerge, because he considers it a precious ally. Experience having taught that the pellicle needs air for its development, windows are open at the top of the cask, above the surface of the liquid. Vinegar making goes well as long as the pellicle remains spread over the surface of the liquid, but stops if it is broken and falls to the bottom. It is then necessary to produce a new mother of vinegar to start the process again.

As early as 1822, Persoon had suspected the living nature of the "mother of vinegar" and had given to it the name "mycoderma" to suggest its plant nature. In 1837, Kützing had even seen the bacterial cells which constitute the mycoderma skin and had postulated a connection between the life of the bacteria and the production of vinegar. Similarly Thompson in 1852 had stated his conviction that the production of acetic acid was due to the "vinegar plant." But here, as in the case of the relation of yeast to alcoholic fermentation, the authority of Berzelius, Wöhler and Liebig had squelched the voice of those who were trying to bring life back into an area which chemistry believed it had conquered. And here again, it was Pasteur who dared to challenge Liebig's autocratic mandate and conjure the experimental evidence on which was established the vitalistic theory of acetic acid fermentation.

Pasteur convinced himself that, in the Orléans process as well as in laboratory reactions, the conversion of wine into vinegar depended upon the development of a thin layer consisting of microscopic bacteria named *Mycoderma aceti* which were capable of floating on the surface of the fluid because of their fatty nature. Moreover, he found that minute amounts of the Mycoderma, transferred to a synthetic solution containing dilute alcohol, ammonia, and mineral salts, increased in abundance, and simultaneously produced acetic acid. He also detected a barely visible film of the same *Mycoderma aceti* on the surface of the wood shavings used in the German process. New shavings, or active shavings which had been heated to destroy the Mycoderma present on their surface, were found by him to be unable to convert alcohol into acetic acid, however slow the flow of alcohol over them. Neither did alcohol fix oxygen when allowed to flow along a clean rope, but as soon as the shavings or the rope were wetted with a fluid containing the Mycoderma, alcohol was converted into acetic acid as in the towers of beech shavings.

These observations were of great practical importance in giving a rational basis to vinegar manufacture, and Pasteur was thus led to advocate modifications of the time-honored industrial procedures. In 1864, he was asked to outline his new method in a speech before the Chamber of Commerce of Orléans. There, in simple, precise, and clear words, he presented the theoretical basis of his concept and its practical applications.

Mycoderma aceti, he pointed out, will grow best at fairly high temperatures, and for this reason the transformation of wine into vinegar occurs most rapidly in rooms heated at 15°–20° C. Mycoderma needs nitrogenous materials, phosphates of magnesium and potassium, and other nutrients. It is most active at acidic reaction, a property that explains the practice of adding preformed vinegar to wine before starting the manufacture. These exacting requirements account for the fact that *Mycoderma aceti* is not capable of converting pure alcohol diluted with water into acetic acid, as this fluid lacks the nutritional elements required for growth. For this reason a little sharp wine, or acid beer, or other organic matter, is added to the dilute alcohol solution used in the German process before allowing it to trickle down the hollow column of beech shavings. This organic matter is not added to act as a ferment and set the phenomenon in motion as thought by Liebig, but only to serve as food for the Mycoderma.

The oxidation to acetic acid of ten liters of alcohol requires more than six kilograms of oxygen, which has to be supplied by more than fifteen cubic meters of air. It is the role of *Mycoderma aceti* to transfer this oxygen from the air to the alcohol. Any agent or condition interfering with this oxygen transfer, which takes place through the surface pellicle of the mother of vinegar, also interferes with the production of acetic acid, and Pasteur made clear to the vinegar manufacturers that many of their failures could be explained in these terms.

When the Mycoderma pellicle falls to the bottom of the vat and is submerged, it becomes unable to convert alcohol into acetic acid. On the other hand if the supply of oxygen is too abundant and the concentration of alcohol too low, the Mycoderma destroys the formed acetic acid by oxidizing it further to carbon dioxide and water.

The precise understanding of the mechanism by which wine is converted into vinegar had immediate practical consequences. Knowing the living nature of the "mother of vinegar" and its physiological requirements, the vinegar makers no longer had to submit blindly to its heretofore unpredictable vagaries. Instead of depending upon the spontaneous but slow and erratic appearance of the Mycoderma pellicle on their vats, they became its master and could sow it on the surface of the fluid. The traditional method made it necessary to keep the fermentation going by constantly feeding more wine to the casks, lest the Mycoderma veil should sink to the bottom, become starved for oxygen, and lose its activity. By hastening or retarding growth at will, the producers could now make the process more dependable, easier to control and could adjust the production of vinegar to the demands of the market. Biological science had found its place in industrial technology.

In 1866, Pasteur published a book entitled *Etudes sur le vin, ses maladies, causes qui les provoquent. Procédés nouveaux pour le conserver et pour le vieillir.*[2] He came from a wine-producing district, and had much to say concerning the factors affecting the taste, appearance, and nutritive qualities of the beverage.

As everyone knows, aging alters the properties of wine, increasing its mellowness, decreasing its opacity, changing its color. Whereas all young wines are raw and thick, a properly aged wine acquires refinement and sometimes distinction. On the other hand, wine can lose completely its strength and character if the aging process is carried too far. Pasteur therefore asked himself the question, what goes on in a wine that becomes old normally, in the absence of disease microorganisms, and how can the aging process be controlled?

Practical wine makers as well as chemists had always been convinced that uncontrolled access of air was detrimental during the making of wine. Chemical analysis revealed that oxygen disappeared rapidly from the atmosphere with which wine was in contact in casks or bottles. As admission of more air was often associated with loss of quality, concomitantly with absorption of

[2] *Studies on Wine. Its Diseases; Causes that Provoke Them. New Procedures to Preserve It and to Age It.*

oxygen, it appeared to be a justifiable conclusion that spoiling of the flavor was a result of oxidation. In consequence, care was taken throughout the operations of wine making to expose the wine to the air only so long as absolutely necessary for the decanting. Pasteur showed that the role of air in the aging of wine is in reality a very complex one and consists of at least two independent effects. On the one hand, oxygen exerts an adverse effect on quality by encouraging the growth of certain contaminating microorganisms. On the other hand, when acting alone, free from any microbial action, oxygen may have a beneficial effect. It takes away the acidic and rough taste of new wine, rendering it more fit to drink; it precipitates slowly some of its dark coloring matter, giving it finally the onionskin tint, so praised when the right degree has been achieved. Oxygen brings about the quality of old wine, but if its aging action is allowed to proceed too far, it ends by spoiling the very wine which it first improved.

Pasteur devised simple experiments to illustrate the relation of oxygen to the aging of wine.

Suppose that wine saturated with carbon dioxide is introduced into a glass bottle so as to fill it completely, under such precautions that the liquid never comes into contact with air; suppose also that the bottle is sealed hermetically with wax. Such a wine will preserve its original color and savor; it will not age and will remain raw new wine, because glass and wax protect it completely from the access of air. If, however, the bottle is left half empty and merely stoppered with an ordinary cork, a significant degree of gas exchange takes place through the cork, and oxygen slowly gains access to the wine. An amorphous deposit consisting of the red coloring matter appears slowly, and the flavor simultaneously changes. The wine may even fade completely in color – and in flavor – if the amount of air in the bottle is too high.

Perusal of all the pages that Pasteur devoted to the aging process leaves little doubt that he greatly enjoyed studying the factors affecting the quality of wines. The *Etudes sur le vin* describe extensive use of chemical techniques – in determining the concentration of alcohol, glycerine, tartaric acid, succinic acid, gums, and sugar present in the wine, and in following the changes in the composition of air exposed to it. New and old methods of wine making are discussed in great detail, either because they bear a relation to the problem at hand, or merely for the sake of interest. Thus, we learn that in certain regions it used to be the practice to stir the crushed grapes vigorously, before fermentation, in order to achieve in advance a form of aging of the wine. Pasteur also mentions his habit of discussing with peasants their empirical procedures in an attempt to find a rational basis for their time-honored practices.

"In all countries and all epochs, as appears from the writings of the Latin agronomists, wine makers have recognized a relation between the life of wine and that of the grapevine. They pretend that when the grape flowers, around June 15 in the Jura, wine is in travail, and again in August, when the grape begins to ripen. They are inclined to believe that there exists some mysterious correlation between these circumstances. In reality, these are the periods when variations occur in the temperature of the cellars, and the changes in fermentative activity probably find their explanation in these changes of temperature. But what does it matter if the peasants credit a myth? It is only the fact itself that we need consider, because it serves as a guide in certain practices of vinification.

"The most ancient writings recommend that the first draining-off be done in the month of March, when the north wind blows and not the south wind, which is the wind of rain, at least in the Jura. Do not dismiss the practice as prejudice. . . . In my opinion, it has a rational basis. Wine, especially young wine, is supersaturated with carbon dioxide. If the barometric pressure is very low for several days, wine will let the gas escape. There will arise from the bottom of the casks small bubbles which carry up with them some of the fine deposit. The wine will then be less clear than if we draw it off on a day when the barometric pressure tends to increase the solubility of gases in liquids. . . ."

As a wine maker of Arbois had assured him that the north wind affected both wine and the water of the river La Cuisance, Pasteur made haste to look for an explanation.

"The river Cuisance which goes through Arbois has its source a few miles away in a chalky country. Its water is loaded with calcium carbonate dissolved by virtue of the carbon dioxide that it contains. On rainy days, the river water becomes less limpid. . . . Sometimes the moss in its bottom is seen to raise. On the contrary, let the north wind blow, and one can see a needle at a depth of several feet. Does not this confirm the explanation which I just gave you concerning the advisability of drawing the wine by the north wind?"

Despite Pasteur's obvious interest in the varied technological aspects of wine making, the most important part of this book deals with the study of microorganisms which interfere with the normal course of fermentation.

Under natural, that is, uncontrolled conditions, the different fermentation processes often give rise to undesirable products. For example, acidification may occur where it is not wanted and may spoil alcoholic beverages, or the conversion of alcohol to vinegar may be accompanied by the production of volatile substances with a suffocating odor. There is no clearer evidence of the revolution introduced by Pasteur in the biochemical sciences than to compare the approach to the study and control of these "diseases of fermentations" before and after his work. Before him, the appearance of undesirable products was assumed to be the result of faulty chemical reactions. Liebig regarded the diseases of wine as due to the changes that wine was constantly undergoing. Under optimum conditions, he taught, the wine reached the end of fermentation in such a state that its sugar and the organic matter serving as ferment were equally exhausted. If there had been too little ferment in the beginning, a portion of the sugar remained unchanged, and the wine was sweet, that is to say, incomplete. If there had been too little sugar, on the contrary, some ferment remained which continued to work and to produce vitiations of the flavor. This explanation was universally accepted, and paraphrased in all textbooks.

Pasteur, being convinced that each type of fermentation is

caused by a specific microorganism, contributed to the problem two independent concepts which found immediate application in practical technology. He recognized that bad fermentations are commonly due to contaminating microorganisms which generate undesirable products. He also emphasized that the activities of a given microorganism are conditioned by the physicochemical conditions of its environment and that in consequence undesirable products may be formed even by the right organism if the conditions of fermentation are not adequately controlled. Thus, in vinegar making, the oxidation of alcohol by *Mycoderma aceti* may either fall short or go beyond the ideal point. If the oxygen supply is not adequate, incomplete oxidation gives rise to aldehydes, intermediate between alcohol and acetic acid, that impart to the product a suffocating odor. The Mycoderma lives with difficulty under these conditions and may even die. Under other conditions, it may carry the oxidation too far and ruin the vinegar by converting the acetic acid into water and carbonic acid; this is likely to happen when all the alcohol in the nutrient fluid has been exhausted. These examples show that the chemical performance of microorganisms is conditioned by the nutritional and respiratory conditions under which they live, and Pasteur did not miss the occasion to suggest that his observations on the biochemical behavior of Mycoderma might have a bearing on the disturbances of oxidations that occur in animal tissues.

The possibility that unfavorable environmental conditions may cause physiological disturbances in other microbiological processes was always present in Pasteur's mind, and recurs in casual remarks scattered throughout his writings. Had not circumstances led him to become involved in those diseases — either of fermentations or of animals and man — which are brought about by foreign microorganisms, he could certainly have traveled with success the road to the modern concepts of physiological and metabolic diseases.

A statement left by M. Bigo's son clearly shows that Pasteur soon learned to correlate many of the difficulties in the fermentation process with the presence of abnormal formed elements mixed with the yeast globules in the fermenting fluids. "He had noticed by microscopic examinations that the globules were round when fermentation was healthy, that they lengthened when alteration began, and were quite long when fermentation became lactic. This very simple method allowed us to watch the process and to avoid the fermentation failures which were then so common." While on vacation at Arbois, in September 1858, Pasteur had occasion to submit some spoiled Jura wines to microscopic examination and saw in them a microorganism presenting morphological similarity with the lactic acid organism which he had just discovered. This observation, and his experience in the Lille distillery, probably helped him to conclude that the diseases of fermentations were caused by foreign organisms which competed with yeast in the fermenting fluid. His studies on the production of acetic acid provided further evidence for this view. Souring is one of the most common types of deterioration affecting wine; and there was no difficulty in tracing it to an oxidation of alcohol to acetic acid similar to that carried out by *Mycoderma aceti* during the making of wine vinegar. In addition to souring, there are many other types of alterations that affect unfavorably the quality of wines; the Bordeaux wines "turn," the Burgundy wines become "bitter," the Champagnes become "ropy." Fortunately, Pasteur was well placed to test by experimentation his general thesis that these diseases were also due to contamination by foreign organisms, for some of his childhood friends owned well-stocked cellars at Arbois. There, in an improvised laboratory, he submitted to systematic microscopic examination all the healthy and diseased wines that were submitted to him. From the very beginning success rewarded his efforts, and whenever a sample was brought defective in some respect, he discovered, mingled with the yeast cells, other distinct microscopic forms. So skillful did he become in the detection of these various germs that he soon was able to predict the particular flavor of a wine from an examination of the sediment. In "healthy" wines, the foreign forms were absent and yeast cells alone were to be seen.

Although many bacterial species found in spoiled wines were

described by Pasteur, he did not investigate the problem with the completeness that characterized his researches on the production of acetic acid. The chemical changes taking place during the production of wine are more complex than are those involved in vinegar manufacture, and are not completely understood even today. To Pasteur, however, is due the credit for establishing that many diseases of wine are dependent on the activity of foreign organisms, a conclusion which was soon found applicable to the alterations of other beverages and foodstuffs.

In addition to *Mycoderma aceti,* which converts alcohol into acetic acid, there is often present in the casks used in the Orléans process another microscopic organism, called *Mycoderma vini,* which prevents the formation of vinegar by converting the alcohol into carbon dioxide and water. The casks may also contain minute worms, "vinegar worms," which spread over the surface of the mixture and prevent *Mycoderma aceti* from obtaining the oxygen required for the conversion of alcohol to acetic acid. Pasteur showed that mild heating was sufficient to kill the "vinegar worms" and that by seeding the wine-vinegar mixture with large amounts of a pure pellicle of the mother of vinegar, it was possible to overcome the undesirable competitors and establish *Mycoderma aceti* over the whole surface.

Like wine and vinegar, beer was then likely to undergo spontaneous alterations, to become acid, and even putrid, especially during the summer. Pasteur demonstrated that these alterations were always caused by microscopic organisms, and he described his findings in a book published in 1877 under the title *Etudes sur la bière, ses maladies, causes qui les provoquent. Procédés pour la rendre inaltérable, avec une théorie nouvelle de la fermentation.*[3] It is entertaining to compare the wealth of loving detail covered in the studies on wine with the austerity of the discussions in the book on beer. Little is said of brewing in this

[3] A translation of this book was made with the author's sanction under the title *Studies on Fermentation: the Diseases of Beer, their Causes, and the Means of Preventing Them.* (London, Macmillan, 1879.)

book. The middle section has no immediate bearing on beer or on its making. It deals with the mechanism of alcoholic fermentation, the origin, distribution and transformation of microorganisms — all theoretical problems preoccupying Pasteur's mind at this stage of his life. The first chapter shows that the diseases of beer are always due to the development of microscopic organisms foreign to a good fermentation, and the last chapter tells how to prevent their occurrences. In other words, the brewers were told how to keep beer from becoming bad, not how to make it good. The reason was simple: Pasteur did not like beer. He had undertaken a study of the brewing industry after the war of 1870 merely to produce a *bière de la revanche* which would compete with the German product. Duclaux reports that Pasteur was amazed to see his friend Bertin recognize between different brands subtle shades of taste that were unnoticeable to him. Nevertheless, he did succeed in developing practical techniques for the control of beer diseases.[4]

Following his first observations in French breweries, Pasteur decided to visit one of the famous London establishments to confirm his findings and spread his gospel further. The account which he has left of his visit to the Whitbread brewery in London reveals the strength of his convictions, and the courage with which he submitted them to practical tests.

"In September, 1871, I was allowed to visit one of the large London breweries. As no one there was familiar with the microscopic study of yeast, I asked to perform it in the presence of the managers. My first test dealt with some porter yeast, obtained from the outflow of the fermentation vats. One of the disease microorganisms was found to be very abundant in it. . . . I concluded therefore that the porter was probably unsatisfactory, and in fact I was told that it had been necessary to obtain that very

[4] Pasteur's contribution to the understanding of fermentation processes were so important in placing the brewing industry on a rational basis that they stimulated the Danish brewer Jacobsen to organize in the Carlsberg brewery a research laboratory devoted to the scientific aspects of fermentation. The Carlsberg Laboratory soon became one of the greatest world centers of biochemical research.

same day a new sample of yeast from another London brewery. I examined the latter under the microscope and found it purer than the old sample.

"I then asked to study the yeasts of other beers in course of fermentation, in particular of ale and pale ale. Here is the drawing which I made then. One recognizes again the filaments of spoiled beer. It was of interest to study the beers which had been produced just before the ones of which I had examined the yeasts.

"I was given two kinds, both in casks. . . . One was slightly cloudy; on examining a drop of it, I immediately recognized three or four disease filaments in the microscopic field. The other was almost clear but not brilliant; it contained approximately one filament per field. These findings made me bold enough to state in the presence of the master brewer, who had been called in, that these beers would rapidly spoil . . . and that they must already be somewhat defective in taste, on which point every one agreed although after long hesitation. I attributed this hesitation to the natural reserve of a manufacturer whom one compels to declare that his merchandise is not beyond reproach. . . .

"The English brewers . . . confessed that they had in their establishment a large batch of beer that had completely spoiled in the casks in less than two weeks. . . . I examined a sample of it under the microscope without at first detecting the disease organisms; however, presuming that the beer had become clear as a result of being kept still, and that the disease organisms had become inert and settled in the bottom of the casks, I examined the deposit—which turned out to consist exclusively of the disease organisms, without even being mixed with the alcoholic yeast. . . .

"When I returned to the same brewery less than a week later, I learned that the managers had made haste to acquire a microscope and to change all the yeasts which were in operation at the time of my first visit."

Time and time again, Pasteur reiterated his views concerning the origin of the alterations of fermenting fluids, and he summarized in the following terse statements how his investigations had led him to a practical solution of the problem.

"We have shown that the changes that occur in the beer yeast, in the wort, and in the beer itself, are due to the presence of microscopic organisms of a nature totally different from those belonging to the yeast proper. These organisms, by the products resulting from their multiplication in the wort, in the beer yeast, and in the beer, alter the properties of the latter and militate against its preservation.

"We have further demonstrated that the organisms responsible for such alterations, these disease ferments, do not appear spontaneously, but that whenever they are present in the wort or in the beer it is because they have been brought from without, either by dust in the air, or by the vessels, or by the raw materials which the brewer employs.

"We also know that these disease ferments perish in malt wort raised to the temperature of boiling, and, as a necessary consequence of this fact, we have seen that malt wort exposed to pure (sterile) air does not undergo any sort of fermentation after having been boiled.

"As all the disease germs of wort and beer are destroyed in the copper vessels in which the wort is heated, and as the introduction of pure yeast from a pure beer cannot introduce into the latter any foreign ferment of a detrimental nature, it follows that it ought to be possible to prepare beers incapable of developing any mischievious foreign ferments whatever. This can be done provided that the wort coming from the copper vessels is protected from ordinary air . . . and fermented with pure yeast, and that the beer is placed in vessels carefully freed from ferments at the end of the fermentation."

As microorganisms can spoil wine, vinegar and beer, it is essential to avoid introducing them during and after the manufacturing process, or to prevent their multiplication, or to kill them after they have been introduced. The introduction of foreign microorganisms in the finished product can be minimized by an intelligent and rigorous control of the technological operations, but cannot be prevented completely. The problem therefore is to inhibit the further development of these organisms after they have been introduced into the fermented fluid. To this end, Pasteur first tried to add a variety of antiseptics, especially hypophosphites and bisulfites which are without too objectionable an odor and a taste in dilute solutions, and which are converted to nontoxic sulfates or phosphates by oxidation. However, the results were mediocre or negative and, after much hesitation, he considered the possibility of using heat as a sterilizing agent.

There was much to be feared from the damaging effect of heat on fermented fluids, especially on wine, for one does not need to be a connoisseur to realize that "cooked wine" is no longer real wine. Fortunately, Pasteur's knowledge of the susceptibility of microorganisms to heat suggested to him that the problem was not as hopeless as appeared at first sight. He knew that wine is always slightly acidic and that heat is a much more effective disinfectant under acid than under neutral conditions; indeed, a temperature as low as 55° C. proved sufficient to improve the keeping qualities of ordinary wine. On the basis of his prior investigations of the effect of air on the aging of wine, Pasteur further postulated that heat might not have any significantly deleterious effect on the bouquet of wine if applied only after the oxygen originally present in the bottle had become exhausted, and this presumption proved true. These considerations led to the process of partial sterilization, which soon became known the world over under the name of "pasteurization," and which was found applicable to wine, beer, cider, vinegar, milk and countless other perishable beverages, foods and organic products.

It was characteristic of Pasteur that he did not remain satisfied with formulating the theoretical basis of the process of heat sterilization, but took an active interest in designing industrial equipment adapted to the heating of fluids in large volumes and at low cost. His treatises on vinegar, wine and beer are illustrated with drawings and photographs of this type of equipment, and describe in detail the operations involved in the process. The word "pasteurization" is, indeed, a symbol of his scientific life; it recalls the part he played in establishing the theoretical basis of the germ theory, and the phenomenal effort that he devoted to making it useful to his fellow men. It reminds us also of his oft-repeated statement: "There are no such things as pure and applied science – there are only science, and the applications of science."

The pasteurization process was soon attacked from many different angles. First to be overcome was the natural hesitation of those who feared that heating would spoil the qualities of fine wines. With the organizing genius that he exhibited so many times, Pasteur established an experimental cellar in which samples of heated and unheated wines of different origins and qualities were kept for periods of several years. At regular intervals of time, an official commission of winetasters compared the products and published reports, which were uniformly favorable to pasteurization. He also arranged to have heated and unheated wines used in comparison on ships of the French Navy during long sea voyages, and thus obtained additional confirmation of the superiority of the pasteurized products. He published in agricultural journals practical descriptions of his process and its merits. To carry still more conviction, he reported in dialogue form (as reported on pages 71 and 72) the visit received from the Mayor of Volnay who had come to him an unbeliever and had left converted. He even quoted with pride that his process was used with success in faraway California:

"Across continents and oceans, I extend my most sincere thanks to this honest wine maker from California whose name I am sorry not to know.

"It is inspiring to hear from the citizen of a country where the grapevine did not exist twenty years ago, that, to credit a French discovery, he has experimented at one stroke on 100,000 liters of wine. These men go forward with giant steps, while we timidly place one foot in front of the other, often more inclined to disparage than to honor a good deed."

While Pasteur had to struggle to establish the practicability and safety of partial sterilization by heat, at the same time the accusation was leveled against him that the process was not new: that Appert, in particular, had shown that wines could be warmed without altering their taste. As soon as he learned of these old experiments, he acknowledged their importance, but pointed out that there were theoretical and practical differences between Appert's method and his own. Having also been told that the heating of wine had been practiced for a long time at Mèze in the South of France, he hastened there to investigate the matter. After verifying the fact, he made clear that his process differed markedly also from the one practiced at Mèze. "They do heat the wine at Mèze, but it is to age it more speedily. For this purpose, they warm it in contact with the air, for a long time, so as to bring about changes in taste, which sometimes exceed the limit, and which it is then necessary to correct. These gropings about in the dark show that the wine merchants of Mèze do not have any clear idea of what they are about, and have not read my book. It would be to their interest to do so, for I give the theory of their practice. Moreover, what does this long and dangerous warming in contact with the air have in common with the rapid heating to 50° C., protected from the air, that I recommend?"

There were other claims of priority which led to bitter public controversies. Pasteur should have been wise enough to trust to the judgment of time, but profound faith is always a little intolerant, and such faith was his. These polemics are of no interest today, except as they reveal the fundamental weakness of empirical practices in comparison with those based on rational theory. The heating of wine had been practiced sporadically from all antiquity, and some vintners knew that it could be done under certain conditions without spoiling the flavor of the product. But it was Pasteur who first provided a rational basis for the empirical procedure, by establishing that certain alterations were caused by contaminating microorganisms, and that these organisms could be inhibited by heat. His theoretical studies led to standardized and dependable techniques for the preservation, not only of wine, but also of other perishable fluids.

The general body of knowledge and of techniques on which the germ theory of fermentation was based had long been available, and the only surprising thing is that the scientific world had refused to accept the obvious interpretation of the known facts. Intent on their mission to prove that all physiological phenomena could be interpreted in terms of physicochemical reactions, the opponents of the theory did not wish to consider life in the processes which they studied, and consequently they failed to recognize it in the form of yeast or of other ferments. For a similar reason, Pasteur had failed a few years earlier to recognize in synthetic malic acid a mixture analogous to the one that he himself had separated into left and right components, because he did not believe that optically active compounds could be synthesized from inactive precursors in the laboratory. The mind can be a piercing searchlight which reveals many of the hidden mysteries of the world, but unfortunately, it often causes such a glare that it prevents the eyes from seeing the natural objects which should serve as guideposts in following the ways of nature.

Much perspicacity, intellectual courage, and forcefulness were needed to overpower the formidable physicochemical philosophy of the time. In fact, it is to this day a source of wonder that Pasteur, then still a young man known to but a few chemists, dared to challenge Liebig on his own ground, and managed within a few years to impose the vitalistic theory of fermentation upon the scientific and lay public. That he dared is evidence of his fighting temperament and of his faith in the correctness of his intuitive judgment, for he had no proof of the living nature of yeast when he took his stand on the side of Cagniard de la Tour, Schwann, Kützing and Turpin. That he succeeded so rapidly was due to his skill as an experimenter and to the vigor of his fighting campaign.

As will be recalled, all of his former training and research experience had been in the fields of physical and organic chemistry. And yet, within a few weeks after his first contact with the fermentation problem, we find him borrowing the experimental approach and the techniques of the biologist, using the microscope not only to investigate chemical substances, but even more to unravel the nature of the agents involved in the fermentation processes. Just as the study of fermentation and putrefaction was then considered the province of the chemist, so knowledge of microbial life was the specialty of a few botanists interested in the description of the microscopic forms as biological curiosities, with at best only a vague awareness of their chemical activities. It was therefore an extraordinary intellectual feat that Pasteur should have been able to adopt immediately the biological point of view without being inhibited by the fear and inertia that investigators have to overcome in passing from one laboratory discipline to another. The intellectual vigor required by this attitude may not be obvious to the modern man who, through education and publicity, has been made almost hyperconscious of the ubiquitous presence of living germs in the world around him, and of their role as agents of fermentation, putrefaction, rotting and disease. In 1857, however, the chemist who adopted the vitalistic theory of fermentation had to face the same odds that would today confront a telephone engineer interested in developing the use of telepathy for the transmission of thought.

As soon as Pasteur became convinced that living microorganisms were the primary cause of fermentation and putrefaction, he devised means for recognizing and studying them and showed that one could control their activities almost at will. Thus, countless experiments and improved industrial practices emerged as the fruits of the germ theory, whereas Liebig's theory had no operational value whatever. Even when dignified with the name of "catalytic theory" by Berzelius, the view that organic substances in decomposition imparted to the sugar molecules an agitation which converted them into alcohol, or lactic acid, or butyric acid did not lead to any new experiment and was of no help in the technology of fermentation. In fact, the contrast between the bareness of the chemical theory and the wealth of theoretical and practical consequences which were derived from the germ theory assured the rapid growth of the latter as soon as it found in Pasteur a determined leader.

It has been said that Berzelius's and Liebig's theories came closer than did the vitalistic theory to the ultimate understanding of fermentation phenomena. Indeed, in 1897, as will be recounted later, Büchner succeeded in extracting from yeast a lifeless juice capable of converting sugar into alcohol. Berzelius's catalytic theory turned out to be fundamentally correct, and Büchner's yeast juice was one step nearer the ultimate cause of fermentation than was Pasteur's living yeast. It is also true, however, that in 1860 no progress could be made in the understanding of fermentation until the chemical activities of microorganisms had been recognized and until techniques had been worked out for the study and control of microbial life. In their activities, furthermore, men are not governed by concern with ultimate truths, but rather by practical common sense. Liebig could argue that to invoke vitalism was to take a backward step; his contemporaries believed Pasteur, because the emphasis on the living nature of yeast and ferments was productive of useful results. Men may propound many kinds of philosophies in their discourse, but in general, they act pragmatically. Throughout his life, Pasteur was amazingly pragmatic in his operations; for him, a theory was right which was useful in action. In 1860, the germ theory was more useful than the chemical theory because it was better adapted to the discovery of new scientific facts and to the improvement of industrial practices.

There is something pathetic in Liebig's last attempts at defending his views against those of Pasteur. After so many years, he had not been able to contribute one positive finding to substantiate the mechanism of fermentation that he advocated. In his long memoir of 1869, he could report only his inability to obtain fermentation and multiplication of yeast in a synthetic medium free of organic nitrogen. It is probable that his failure came from the fact that he attempted to repeat Pasteur's experiment with a strain of brewer's yeast having such exacting nutritional requirements that the microorganisms were not satisfied with the synthetic media then available. As he did not believe in the possibility of growing yeast in the absence of albuminoid matter, Liebig did not make the effort required to solve the technical problems which would have permitted him to duplicate Pasteur's results. He also stated in the same memoir that the bacterium, *Mycoderma aceti,* which Pasteur claimed to be responsible for the production of acetic acid, was not present in the German vinegar works; for, he said, "on wood shavings which had been used for twenty-five years in a large vinegar factory in Munich, there was no visible trace of Mycoderma, even when observed under the microscope." Liebig was wrong; he failed to see the Mycoderma because he did not want to see them. This great man, whose vision, learning and energy had founded the science of biochemistry, presents with particular acuity the tragic spectacle of a brilliant mind become slave of preconceived ideas and blinded by them.

Pasteur replied to Liebig's memoir by a short note in which he scorned to carry on the argument but instead went straight to the facts. Liebig had questioned the validity of two of his claims. Pasteur challenged him to submit the matter to a commission of scientists before whom these facts would be put to an objective test. He offered to prepare, in an exclusively mineral medium, as much yeast as Liebig could reasonably demand, and to demonstrate the presence of the *Mycoderma aceti* on all the beech shavings of the Munich vinegar factory. He also suggested that the Munich manufacturer bring to boiling temperature the vats containing the wood shavings and then reintroduce alcohol into them. Under these conditions, he affirmed, no vinegar would be produced because the bacteria would have been killed by heating. Liebig ignored the challenge, and never replied, either because he was convinced, or, more likely, because he was overpowered by the greater vigor of his opponent. The germ theory of fermentation immediately gained widespread acceptance. Microbiology, which heretofore had been the odd occupation of a few botanists, became within a few years one of the most rapidly growing sections of biological sciences. A new haunt of life had been discovered and its exploration and exploitation gave rise to one of the boom periods of biology.

The theory states that, in a large measure, the transformations of organic matter are carried out through the agency of microorganisms. For each type of transformation, there exists one or several types of organisms specialized in the performance of the chemical reactions involved in this transformation. Each microbial type is characterized not only by its behavior as a chemical agent, but also by the fact that it demands highly selective conditions for optimum growth and activity. By taking advantage of this selectivity in requirements, man can become master of microbial life, favoring one form by providing conditions that are optimum for its multiplication and activities, repressing others by creating an environment unfavorable to them. He can even modify somewhat the chemical reactions which accompany microbial life by modifying the conditions under which microorganisms grow and carry on their chemical activities. Thus, fermentation and putrefaction are no longer vague and uncontrollable transformations, indeterminate in their cause and origin, taking place in haphazard manner under the influence of ill-defined organic matter; they are predictable phenomena due to the existence and activity of specific microbial agents that can be domesticated to function according to the needs and wishes of man. Such was the central theme that Pasteur was to develop during the rest of his scientific career. With him began the domestication of microbial life.

CHAPTER VI

Spontaneous Generation and the Role of Germs in the Economy of Nature

Omne vivum ex vivo.

— HARVEY

Why then, asked the Sirian, do you quote this Aristotle in Greek?

It is, the learned man replied, because it is wiser to quote that which one does not understand at all, in the language that one comprehends least.

— VOLTAIRE

IT IS common observation that all dead plants and animals undergo decomposition, again to become part of the envelope of soil, water and atmosphere at the surface of the earth. Should any component of organic life remain undestroyed and be allowed to accumulate, it would soon cover the world, and imprison in its inert mass the chemical elements essential to the continuity of life. Of this, however, there is no danger. Substances of animal or plant origin never accumulate in nature, for any organic product which finds its way into soil or water undergoes, sooner or later, a chain of alterations which break it down stepwise, into simpler and simpler compounds — water, carbon dioxide, hydrogen, ammonia, elementary nitrogen, mineral salts. It is in this fashion that after death the chemical elements are returned to nature for the support of new life. "All are of the dust, and all turn to dust again."

The eternal movement from life, through organic matter, and back into life, has inspired the psalms and songs of poets, and scientists have long known that it is essential to the maintenance of life on the surface of the earth. Before the microbiological era, however, the cycle of organic matter was surrounded with mystery, as appears from a note found after Lavoisier's death among his unpublished manuscripts. "Plants extract from the air that surrounds them, from water and in general from the mineral kingdom, all the substances necessary to their organization.

"Animals feed either on plants or on other animals which themselves have fed on plants, so that the substances of which they are constituted originate, in final analysis, from air or from the mineral kingdom.

"Finally fermentation, putrefaction and combustion endlessly return to the atmosphere and to the mineral kingdom the principles which plants and animals had borrowed from them. What is the mechanism through which Nature brings about this marvelous circulation of matter between the three kingdoms?"

During the first six decades of the nineteenth century, chemists had described with ever-increasing detail the chemical transformations by which the chemical constituents derived from the air and from the mineral kingdom become the substances of which plants and animals are made; but the mechanism through which organized matter was returned to nature after death was as little understood in 1860 as in Lavoisier's time.

However, once the demonstration had been made that fermentation and putrefaction were caused by living microorganisms, it became apparent that many of the other transformations of organic matter might also result from the activities of microbial life. Pasteur saw immediately the large implications of the new point of view, and presented his interpretation of Lavoisier's theme in a letter written to the Minister of Public Education in April 1862. The vision of a cosmic cycle of organic matter, eternally carried out by infinitely small microorganisms, appeared to him with dramatic quality. What Lavoisier had said in a few words, in the disciplined language of the Enlightenment, Pasteur elaborated with the vehemence of the prophets. Even in science, the redundance of the romantic period had replaced the classic restraint.

> We know that the substances extracted from plants ferment when they are abandoned to themselves, and disappear little by little in contact with the air. We know that the cadavers of animals undergo putrefaction and that soon only their skeletons remain. This destruction of dead organic matter is one of the necessities of the perpetuation of life.
>
> If the remnants of dead plants and animals were not destroyed, the surface of the earth would soon be encumbered with organic matter, and life would become impossible because the cycle of transformation . . . could no longer be closed.
>
> It is necessary that the fibrin of our muscles, the albumin of our blood, the gelatin of our bones, the urea of our urines, the ligneous matter of plants, the sugar of their fruits, the starch of their seeds . . . be slowly resolved to the state of water, ammonia and carbon dioxide so that the elementary principles of these complex organic substances be taken up again by plants, elaborated anew, to serve as food for new living beings similar to those that gave birth to them, and so on *ad infinitum* to the end of the centuries.

Pasteur outlined in the same letter his creed concerning the all-important role played by microbial agents in the economy of nature and in the causation of disease. Without denying that ordinary chemical forces may slowly attack organic matter, he affirmed that decomposition was chiefly caused by "microscopic living beings, endowed with special properties for the dissociation of complex organic substances, or for their slow combustion with fixation of oxygen." Thus, he pointed out, "When the sweet juice of a plant or a fruit is abandoned to itself, air brings to it yeast which transforms the sugar into alcohol and carbon dioxide; then other microbial agents intervene which oxidize the alcohol to acetic acid, and still others which complete the process of oxidation to carbon dioxide, thereby returning practically all of the carbon originally present in the sugar back to the atmosphere where it becomes once more available for the growth of plants. It is by interrupting the oxidation of sugar either at the alcohol or at the acetic acid level that man has established empirically the industries which give him wine, beer, or vinegar . . . And thus it becomes obvious . . . how pure science . . . cannot advance one step without giving rise, sooner or later, to industrial applications." Pasteur also saw clearly that "the study of germs offers so many connections with the diseases of animals and plants, that it certainly constitutes a first step in the . . . serious investigation of putrid and contagious diseases," and announced in these words the program to which he was to devote the remainder of his scientific life: "To pursue, by rigorous experimentation, the study of what is, in my opinion, the immense physiological role of the infinitely small in the general economy of nature."

One fundamental question had to be answered before "rigorous experimentation" in this field became possible. Where did these "infinitely small" come from? Did they originate from parents identical to themselves? Or did they arise *de novo* whenever conditions were favorable for their existence? Thus Pasteur was compelled to consider the problem of the origin of microbial life and to become involved in the controversy on spontaneous generation.

Under a thousand symbols, men of all religions and philosophies have sung and portrayed the repeated emergence of life from inanimate matter. There is a poetic fascination in the ancient creed that life is always arising anew from matter, as Aphrodite came out of the foam of the sea. Men have also long believed, and indeed many still believe, that vitality is an indestructible property; that all living things are composed of organic parts, in themselves eternal, and capable of entering into the most diverse compositions to recreate life. Spontaneous generation, according to a view which was still prevalent in the nineteenth century, was the recombination of some of these eternal fundamental units of life set free by the prior death of another living thing.

Throughout the ancient civilizations, the Middle Ages and the Renaissance, it was widely believed that plants and animals could be generated *de novo* under certain peculiar circumstances; that eels arose without parents from the ooze of rivers and bees from the entrails of dead bulls. Weird formula for the creation of life found their way into learned textbooks, and as late as the sixteenth century the celebrated physiologist van Helmont affirmed that one could create mice at will by putting in a container some dirty linen together with a few grains of wheat or a piece of cheese.

Slowly, men ceased to believe in the spontaneous generation of grubs, maggots, tapeworms and mice but it remained the general opinion that the microbial life which crowds fermenting and putrefying fluids is a product of the alteration of organic materials and results from some form of spontaneous generation. And, indeed, it was often observed that countless bacteria made their appearance in a flask of broth or milk, previously heated to destroy living forms, whenever the broth or milk went bad. These bacteria were so small and seemed to be so simple in structure that they appeared to be at the threshold of life. Was it not possible therefore that they came into being out of inanimate organic matter? Were they not primitive enough to be excluded from the law *Omne vivum ex vivo?* Experimentation on this problem began around 1750 and was pursued over a century. Some experimenters, like the Irish Catholic priest Needham, claimed that they could bring about at will the creation of living microscopic agents in infusions which had been sterilized by heat, while others, following the Abbé Spallanzani, maintained that life could never be spontaneously generated from dead matter. Naturalists, philosophers and wits kept the debate alive by contributing to it observations and experiments, as well as religious and philosophical arguments. In the article "God" of the *Dictionnaire Philosophique,* the skeptical Voltaire amused himself at the thought that Father Needham should claim the ability to create life, while atheists, on the other hand, "should deny a Creator and yet attribute to themselves the power of creating eels." Despite laborious experimentation and even more strenuous arguments, the problem remained without definite conclusion, each of the adversaries showing clearly that the others were wrong in some details, but not succeeding in proving that he himself was right on all points.

It was known that putrescible organic infusions in which life had been destroyed by prolonged heating at high temperature often remained unspoiled, and that microscopic life usually did not develop in them, as long as they were protected from contact with air. The mere admission of air, however, was sufficient to cause the fluids to enter upon fermentation and putrefaction, and to bring about the appearance of a great variety of microorganisms within a few days. The believers in spontaneous generation saw in these changes evidence that oxygen was necessary to initiate the generation of life. Their adversaries claimed, on the contrary, that air merely introduced into the organic fluids the living germs of fermentation and putrefaction. To prove the latter thesis, Franz Schülze in 1836 and Theodor Schwann in 1837 passed the air through caustic potash or concentrated sulfuric acid, or heated it to a very high temperature, in order to destroy the hypothetical living germs before admitting it to the organic fluid. And indeed, under these conditions, the fluids often remained unaltered, the microscopic agents failed to appear.

In 1854, and again in 1859 and 1861, an interesting modification of these experiments was introduced by Heinrich Schröder and Theodor von Dusch. Instead of heating the air or drawing it through sulfuric acid before permitting it to enter the infusion, these authors simply filtered it through cotton plugs, as had been done a few years earlier by the chemist Loewel. He had found that air could be deprived of its ability to induce the crystallization of supersaturated solutions if it passed through a long tube filled with cotton wool. Similarly, Schröder and von Dusch found that meat boiled in water and continuously receiving fresh air filtered through the cotton tube remained unchanged and devoid of unpleasant odor and of microbial population for long periods of time.

These observations clearly suggested that the germs of fermentation and of putrefaction were introduced from the air into the organic infusions and did not generate there *de novo*. Unfortunately, whatever the precautions used to destroy or exclude these germs, the experiments now and then failed, microbial life appeared, and fermentation and putrefaction set in. These failures shook the faith of many, indeed, often of the experimenters themselves. Thus, Schröder and von Dusch had particular difficulty in protecting milk and egg yolk from putrefaction and concluded that the germs of putrefaction came from the substances themselves and were derived from animal tissues. As mentioned in an earlier chapter, even Helmholtz had found it necessary to reach a similar conclusion. The possibility that spontaneous generation occurred in the case of putrefaction clearly left the door open for the possibility of its occurrence in all other microbial activities.

This confused state of affairs was brought to a crisis in 1858 when Félix Archimède Pouchet read before the Paris Academy of Sciences a paper in which he claimed to have produced spontaneous generation at will by admitting air to the sterilized putrescible matter under carefully chosen conditions. Pouchet was director of the Museum of Natural History in Rouen, and an honored member of many learned societies. He was, according to Tyndall, "ardent, laborious, learned, full not only of scientific but of metaphysical fervour," and had reached his conviction by meditating over the nature of life: "When by meditation, it became evident to me that spontaneous generation was one of the means employed by nature for the reproduction of living things, I applied myself to discover the methods by which this takes place." Conviction based on philosophical premises was a dangerous start on the subject of spontaneous generation, which was so full of experimental pitfalls.

Pouchet's experiment consisted in taking a flask of boiling water, which he hermetically sealed and then plunged upside down into a basin of mercury. When the water had become quite cold, he opened the flask under mercury and introduced half a liter of oxygen, and also a small quantity of hay infusion previously exposed for a long time to a very high temperature. These precautions, he believed, were adequate to plug every loophole against the admission of living organisms into his flask, and yet microbial growth regularly appeared within a few days in the hay infusion.

The following year, Pouchet presented his views and findings in an elaborate work of 700 pages entitled *Hétérogénie,* which he regarded as the final demonstration that life could originate anew from inanimate solutions. Although criticized by some of the leading French physiologists, Pouchet's claims produced a great sensation in scientific circles. In the hope of encouraging experiments that would dissipate the confusion, the Academy instituted in 1860 the Alhumbert Prize with the following program: "To attempt, by carefully conducted experiments, to throw new light on the question of the so-called spontaneous generations."

As early as 1859, the year of publication of the *Origin of Species,* Pasteur had entered into the thick of the fight over the origin of life. It has been suggested that he immediately took sides against those who claimed to have demonstrated spontaneous generation because, as a devout Catholic, he could not accept the thought of a new creation of life. This interpretation is certainly unwarranted. A few years before, Pasteur himself had attempted to create life by the action of asymmetrical physical and chemical forces, but his studies on fermentation had more recently led him to emphasize the specific nature of the fermentative reactions — a concept incompatible with the haphazard appearance of microorganisms which seemed to be a consequence of the doctrine of spontaneous generation. At that time, the specificity of living species had already become associated with the idea of the continuity of the germ cell, and it would have been very astonishing had this relationship failed to operate among the "infinitely small." The idea of specificity, born of the work on fermentation, involved the concept of hereditary characters, which in turn led to the belief in an ordinary kind of generation.

Biot and Dumas strongly dissuaded Pasteur from attacking the problem of spontaneous generation, as they feared that he would lose valuable time on this question, which appeared to them outside the range of scientific inquiry. Pasteur was convinced, however, that the solution of the problem was essential to the successful development of his views on the role of microorganisms in the economy of nature. He could not be deterred from joining the controversy and singled out Pouchet's momentary triumph as the first goal for his attacks.

Pasteur's experiments on the problem of spontaneous generation have achieved great and deserved fame; but even more interesting than their performance is the broad strategy of his attack on the subject. He immediately acknowledged the possibility that in our midst, or somewhere in the Universe, life may still be creating itself. This possibility cannot and should not be denied. But what can be done is to evaluate the claims of those who pretend to have seen life emerge anew, or to have brought about conditions under which its spontaneous generation is possible. To this task, he set himself and devoted an immense amount of labor.

It was generally agreed that an organic infusion, even though subjected to prolonged heating, would always undergo fermentation or putrefaction shortly after ordinary air was admitted to it. Was this due to the fact that natural air was an essential factor for the new emergence of life, as the proponents of spontaneous generation claimed, or was it merely that air contained preformed and viable germs? This question suggested to Pasteur countless experiments with all types of organic fluids: yeast extract, broth, milk, urine, blood—and with air heated and filtered under all kinds of conditions. He borrowed the techniques and procedures developed by his predecessors in the problem, but, by paying infinite attention to minute technical details, he finally succeeded in arranging tests that always gave the desired result. He observed, for example, that the mercury used at the time of readmitting the air into the heated flasks always contained dust and living germs on its surface. By eliminating mercury from the experimental technique, he eliminated at the same time a source of contamination of the fluids and air. He also recognized that if milk, egg yolk and meat did putrefy even after heating at 100° C., it was not, as Helmholtz, Schröder and many others had believed, because putrefaction was a process inherent in these nitrogenous materials, but only because the germs which they contained had to be heated at somewhat higher temperatures to be inactivated. This greater resistance was due in part, but not wholly, to the fact that germs are less readily destroyed by heat at the neutral reaction of milk or egg yolk than at acid reactions. For example, a concoction of yeast which is slightly acidic is easily sterilized by a short boiling under normal conditions, but needs to be heated at 110° C. if neutralized by the addition of sodium carbonate; it then behaves like milk. All organic fluids heated at a sufficiently high temperature remained sterile when air sterilized by heat or by filtration was admitted to them. As these same heated fluids quickly showed living things in the presence of ordinary air, it appeared obvious that the germs of life came from the air.

These new experiments, however reproducible, left open the possibility that life demands the presence of a (chemical) "vital principle," present in organic fluids and in natural air, but destroyed by prolonged heating and by filtration. This objection, so intangible and yet difficult to ignore, stimulated great discussion in the scientific atmosphere of the 1860's and particularly in Pasteur's laboratory. Even those of his friends who had advised him to keep away from the spontaneous generation controversy now took a lively interest in it. The great Jean Baptiste Dumas, who had become one of the majestic officials of the Empire, now and then came down from his heights to watch the details of the scientific debate, and to encourage the Pasteur camp. More involved in the actual proceedings was Balard, who followed Pasteur's career with as much personal interest as during the crystallographic period, and whose important contribution to the new controversy has been brought to light by Duclaux: "Balard loved science. It was sufficient to see him in a laboratory managing a piece of apparatus, or carrying out a reaction, to know that he was a chemist to his finger tips. But he had a certain natural indolence, and he was thenceforth satisfied with his share of glory.

In place of the scientific work that he would have been able to do, he preferred that which he found done in the laboratories he frequented. . . . There he wished to see all the details, and we told him everything, first, because he had an open mind and a generous soul, and also because it would have been difficult to conceal anything from him: he put into his interrogations at the same time so much shrewdness and simplicity! When anyone showed him a nice demonstration he admired it with all his heart! And then, one was sometimes rewarded for this confidence: he would suggest an idea or reveal a method. . . . All the experiments on spontaneous generation transported Balard with delight, and the laboratory became animated with his expansive joy as soon as he entered."

It was in the course of one of these visits that Balard suggested an experimental procedure — the use of the swanneck flask — which was extensively exploited in Pasteur's laboratory to demonstrate the possibility of maintaining heated organic infusions in the presence of natural air, without causing, thereby, the appearance of microscopic life.

After introducing into the flask a fermentable fluid, the experimenter would draw the neck of the flask into the form of a sinuous S-tube (hence the name "swanneck" flask). The liquid was then boiled for a few minutes, the vapor forcing out the air through the orifice of the neck. On slow cooling, the outside air returned to the flask, unheated and unfiltered. As the neck remained open, the air inside the flask communicated freely with the unfiltered and unheated atmosphere outside, and there was constant gaseous exchange; yet the fluid in the flask remained sterile indefinitely. It was obvious, therefore, that failure of life to develop could not be due to any deficiency in the air. That the infusion itself was still capable of supporting life could also be readily shown, for it was sufficient to allow dust to fall in by breaking the neck of the flask to see microscopic life appear, first at the spot directly under the opening.

Why, then, did the flasks remain sterile after air was admitted through the swanneck? Why did the germs fail to enter with it?

It was because the air was washed in the moisture which remained in the curves of the neck after heating had been interrupted. In the beginning, when the entrance of the air was rapid, the purifying action of this washing was increased by the temperature of the liquid, still high enough to kill the germs that came in contact with it. Later, the wet walls of the neck held fast the germs of the air as they passed through the opening. In fact, when the flask was shaken in such a manner as to introduce into the curve of the neck a little drop of the infusion, this drop became clouded even when the open end had been previously closed so that nothing new would enter from the outside. Further, if the drop was mixed with the rest of the liquid, the latter became infected just as if the neck had been broken off.

Pasteur had already obtained direct evidence that germs of life are present in the air by concentrating the fine particles suspended in the atmosphere and observing them under the microscope. He had aspirated air through a tube in which was inserted a plug of guncotton which acted as a filter and intercepted the aerial germs. When at the end of the experiment, the guncotton plug was dissolved by placing it in a tube containing a mixture of alcohol and ether, the insoluble dust separated from the solvent and settled in the bottom of the tube. Under the microscope, the sediment was found to contain many small round or oval bodies, indistinguishable from the spores of minute plants or the eggs of animalcules; the number of these bodies varied depending upon the nature of the atmosphere and in particular upon the height above the ground at which the aspirating apparatus had been placed. The dust recovered from the alcohol and ether solution always brought about a rapid growth of microorganisms when it was introduced into heated organic infusions, despite all precautions taken to admit only air sterilized by heat. It was thus clear that the fine invisible dust floating in the air contained germs which could initiate life in heated organic fluids.

To this evidence, Pouchet and his supporters raised the objection that there could not possibly be *enough* germs in the air to account for the generation of life by the techniques used in the laboratory. They quoted experiments of Gay-Lussac which seemed to show that the smallest bubble of oxygen was sufficient to produce putrefaction. If, urged Pouchet, decomposition were due to the germs present in a minute bubble, then air would be heavily laden with living forms and be "as dense as iron."

There were few physiologists who took Pouchet's remarks seriously, and Pasteur could have afforded to ignore them. This, however, was not his bent. By temperament, he could not leave unanswered any opposition to what he believed to be the truth. Moreover, there were other reasons which made it imperative to study the distribution of microorganisms in nature and thus provide an answer to Pouchet's objections. Pouchet had gained much following among those who welcomed the claim that life could be created at will from inanimate matter; thus, the doctrine of spontaneous generation, firmly rooted in philosophical convictions, could derive enough nourishment to survive even without strong scientific support. More important, knowledge of the quantitative distribution of microorganisms in the atmosphere was an indispensable basis for the development of the germ theory, and Pasteur realized that Pouchet's objection could best be answered by gaining more accurate information concerning the presence of microorganisms in the atmosphere. In fact, the results of filtration of air through guncotton had already suggested that great differences existed between the number and type of germs in the atmosphere of different localities.

It was while attempting to answer Pouchet's objections that Pasteur carried out some of his most famous experiments of the spontaneous generation controversy, experiments which established that the germs of putrefaction and fermentation are quite unevenly distributed. The necks of a number of flasks containing yeast extract and sugar were drawn out in the flame to a fine opening, so that they could be easily sealed when desired. The liquid was then boiled to destroy living things and to drive out the air, which was displaced by the current of water vapor. The flasks, sealed by melting the glass with a blowpipe while the steam was escaping, were thus practically empty of air and their content remained sterile as long as they were kept sealed. Pasteur took these flasks to the places where he wished to make a study of the air, and broke the necks with a long pair of pincers. This was done with the most exacting precaution, the necks and the pincers being passed through the flame of an alcohol lamp in order to kill all the germs deposited on them, and the flasks being kept throughout the operation as high as possible above his head in order to avoid contamination of the air by the dust from his clothing. When the necks were broken, there was a hissing sound; this was the air entering. The flasks were then resealed in the flame and carried to an incubator.

Even though each flask received at least one third of a liter of external air during the operation, there were always some that remained sterile, demonstrating that germs do not occur everywhere. The tests revealed that aerial germs are most abundant in low places, especially near cultivated earth. Their numbers are smaller in air allowed to remain still for long periods of time — as in the cellars of the Paris Observatoire — and also in the mountains away from cultivated and inhabited land. Indeed, most of the flasks that Pasteur opened in the midst of the Swiss glaciers remained sterile, evidence of the cleanness of the atmosphere at these high altitudes; a further proof, also, that pure unheated air is unable to cause alteration of organic fluids if it does not contain the living germs of fermentation.

These experiments produced an enormous sensation in the scientific and lay public by virtue of their very simplicity, but they did not convince the advocates of the theory of spontaneous generation.

The controversy had now reached beyond the scientific arena into that diffuse periphery where religious, philosophical and political doctrines were then confusing so many aspects of French intellectual life. Pasteur's findings seemed to support the Biblical story of Creation, and were in apparent conflict with advanced political philosophy. Writers and publicists took sides in the polemic, not on the basis of factual evidence, but only under the influence of emotional and prejudiced beliefs. Despite his indignation at the war being conducted against him in the scientific bodies as well as in the daily press, Pasteur managed to control his temper for a few months. It is possible that he could not devise at that time any new experimental approach capable of adding weight to the evidence already accumulated and, consequently, he judged it wiser to wait for a tactical error of his adversaries that would expose them to his blows.

In 1863, Pouchet reported that, in collaboration with two other naturalists, Joly and Musset, he had attempted without success to duplicate Pasteur's findings concerning the distribution of germs in the air. At different altitudes in the Pyrenees, up to the edge of the Maladetta glacier at 10,000 feet elevation, the three naturalists had collected air in sterile flasks containing heated hay infusion. Instead of obtaining the results reported by Pasteur, they found that "wherever a liter of air was collected and brought into contact with an organic fluid, in a flask hermetically sealed, the fluid soon revealed the development of living germs." All Pasteur's precautions were said to have been observed, except that the necks of the flasks had been cut with a file, and their contents shaken before being sealed again. To Pasteur, this absolute conflict with his own results appeared as the long-awaited opening for the riposte, a situation where there was no question of theoretical discussion or of philosophical argument. He trusted completely in his technique and had no respect for that of his opponents. To settle the matter, he demanded that a commission be appointed by the Academy of Sciences to repeat the experiments carried out with such apparently incompatible results by the two groups of workers.

By then, the Academy was on Pasteur's side. In 1862, it had granted him the Prix Alhumbert for his *Mémoire sur les corpuscles organisés qui existent dans l'atmosphère* . . . In 1863, the influential physiologist Flourens had dismissed the paper of Pouchet, Joly and Musset with a terse and scornful statement: "M. Pasteur's experiments are decisive. If spontaneous generation is a reality, what is needed to bring about the development of animalcules? Air and putrescible fluids. Now, M. Pasteur succeeds in putting together air and putrescible fluids and nothing happens. Generation therefore does not take place. To doubt any longer is to fail to understand the question."

The commission demanded by Pasteur was appointed and threw down the gauntlet with the following statement: "It is always possible in certain places to take a considerable quantity of air that has not been subjected to any physical or chemical change, and yet such air is insufficient to produce any alteration whatsoever in the most putrescible fluid." Pouchet and his coworkers took the bait. They answered the challenge by declaring that the statement was erroneous and they promised to supply the proof, adding: "If a single one of our flasks remains unaltered, we shall loyally acknowledge our defeat." Nevertheless, on two different occasions and for reasons that need not detain us, they refused to agree to the terms of the test organized by the commission and finally withdrew from the contest. Pasteur, on the contrary, arrived with his assistants laden with apparatus and ready for the test, which was carried out in Chevreul's laboratory in the Museum of Natural History. He first demonstrated three flasks which he had opened on the Montanvert in 1860 and which had remained sterile ever since. One was opened and its air analyzed, revealing a normal content of 21 per cent of oxygen. The second was opened and exhibited countless microorganisms within three days. The third flask was left untouched and was subsequently exhibited at the Academy of Sciences. Pasteur then prepared a new series of sixty flasks before the commission. In each was placed a third of a liter of yeast water. The neck was narrowed and the fluid boiled for two minutes; fifty-six out of the sixty flasks were sealed in the flame. In four the necks were drawn out, bent downwards, and left open. Of the fifty-six sealed flasks, nineteen were opened in the amphitheater of the Museum of Natural History, with the result that fourteen remained sterile and five became infected; nineteen were opened on the highest part of the dome of the amphitheater and there thirteen of them remained sterile. The third set of eighteen flasks was exposed in the open air under some poplar trees and only two of them remained sterile. The four open flasks with swannecks remained sterile. A strongly worded official report, published in the *Comptes Rendus de l'Académie des Sciences,* recorded Pasteur's triumph and, as many then believed, closed the polemic.

We must, at this point, anticipate by a few years the further development of the controversy to point out that, despite the spectacular success of Pasteur's experiments in Chevreul's laboratory, and also despite the eminence and integrity of the scientists who witnessed the tests and acted as referees for the Academy, the judgment of the commission was based on insufficient evidence. Even the most distinguished academicians must bend before the superior court of Time, and we know today that they were hasty in deciding the issue without more thorough appraisal of Pouchet's claims. In reality, both Pasteur and his opponents were right as to what they had observed in their respective experiments, although Pouchet was wrong in his interpretation of the findings. But before such a tribunal, nerve as well as right were indispensable for securing justice, and Pouchet was overawed by the conviction of his opponent.

The facts are these: Pasteur had used yeast infusion as the putrescible material in his experiments, whereas Pouchet had used hay infusion. Yeast infusion is easy to sterilize by heat, hay infusion excessively difficult. The heat treatment applied by Pasteur would have been insufficient to sterilize the latter. Consequently the heat applied by Pouchet, which was the same as that used by Pasteur, failed to sterilize the hay infusion that he employed, thus accounting for the fact that growth usually developed in his flasks whereas Pasteur's flasks containing yeast infusion remained sterile under the same conditions.

The want of self-confidence exhibited by Pouchet in this extraordinary trial, and the judgment in default given by the academic tribunal, delayed knowledge of the whole truth by some years, for it was not until 1876 that Pasteur's triumph was again called into question. There is, perhaps, a moral to be drawn from this story. It is the subtle danger that arises from the assumption by an official body, however distinguished, of responsibilities beyond its real competence. The authoritative pronouncement of the Academy protected Pasteur for a time by throttling renewed investigation, especially in France. Fortunately it could not protect him from attacks by scientific foes owning no allegiance to the august body whose sanction he had so successfully invoked. As we shall see, it was to overcome the claims of spontaneous generation made in England a decade later, by Bastian, that Pasteur was compelled to recognize the limitations of the experimental techniques which he had used in his controversy with Pouchet, and to establish his claims on a more definite basis.

By 1864, Pasteur's triumph appeared complete. He had assembled his results on the origin and distribution of germs in the essay, *Sur les corpuscules organisés qui existent dans l'atmosphère. Examen de la doctrine des générations spontanées,* which was published in 1861, and of which Tyndall wrote: "Clearness, strength and caution, with consummate experimental skill for their minister, were rarely more strikingly displayed than in this imperishable essay." It was indeed, the inauguration of a new epoch in bacteriology.

In 1862, at the age of 40, Pasteur had been elected a member of the Paris Academy of Sciences. As his varied scientific activities did not particularly fit him for any of the specialized sections in the Academy, he had been nominated in the mineralogical section, on the basis of his early studies in crystallography and also of his formal training in chemistry and physics. There is no question, however, that it was the spectacular character of the studies on fermentation and spontaneous generation that had placed him in the forefront of French science.

His fame had now reached beyond scientific circles and the polemic on spontaneous generation had become one of the lively topics of discussion in social gatherings. Although Pasteur himself was careful to limit the debate to the factual evidence for or against the *de novo* emergence of life, he was approved by many and blamed by others, as the defender of a religious cause. A priest spoke of converting unbelievers through the proved non-existence of spontaneous generation, and, on the other side, the celebrated novelist, Edmond About, took up Pouchet's cause with no better scientific understanding. "M. Pasteur preached at the Sorbonne amidst a concert of applause which must have gladdened the angels." The lecture to which About referred had been an enormous triumph. On April 7, 1864, at one of the "scientific evenings" of the Sorbonne, before a brilliant public which counted social celebrities in addition to professors and students, Pasteur had outlined the history of the controversy, the technical aspects of his experiments, their significance and their limitations. Presenting to his audience the swanneck flasks in which heated infusions had remained sterile in contact with natural air, he had formulated his conclusion in these words of singular beauty:

"And, therefore, gentlemen, I could point to that liquid and say to you, I have taken my drop of water from the immensity of creation, and I have taken it full of the elements appropriated to the development of inferior beings. And I wait, I watch, I question it! – begging it to recommence for me the beautiful spectacle of the first creation. But it is dumb, dumb since these experiments were begun several years ago; it is dumb because I have kept it from the only thing man does not know how to produce: from the germs which float in the air, from Life, for Life is a germ and a germ is Life. Never will the doctrine of spontaneous generation recover from the mortal blow of this simple experiment."

It has not recovered yet; it may never do so. Today, after almost a century, the fluids in these very same flasks stand unaltered, witness to the fact that man can protect organic matter from the destructive action of living forces, but has not yet learned the secret of organizing matter into Life.

But, despite Pasteur's scientific and official triumphs, his opponents had not been convinced or entirely silenced. In 1864, Pouchet brought out a new and larger edition of his book, in which he reiterated his belief in spontaneous generation. Here and there, other workers also published a few experiments with results in conflict with Pasteur's teachings. It is not essential to follow these minor skirmishes as there was to arise in England, a few years later, a more formidable challenger to demonstrate that the decision of an official academy was not sufficient to exterminate the hydra of spontaneous generation. In 1872, Henry Charlton Bastian published in London an immense tome of 1115 pages entitled *The Beginning of Life: Being Some Account of the Nature, Modes of Origin and Transformation of Lower Organisms,* in which the theory of spontaneous generation was again reintroduced in its most extreme form.

Bastian contributed one new fundamental observation to the problem. He found that whereas acid urine heated at high temperature remained clear and apparently sterile when kept from contact with ordinary air, it became clouded and swarmed with living bacteria within ten hours after being neutralized with a little sterile potash. According to Bastian, this established the fact that spontaneous generation of life was possible but that Pasteur had failed to provide the complex physicochemical conditions necessary for its occurrence. Bastian's techniques were crude and most of his claims worthless, the results of clumsy experimentation. He was correct in stating, however, that urine heated at 110° C. could still give rise to microbial life following the addition of sterile alkali. If they did not arise *de novo,* where did the germs of this life come from? The manner in which this problem was solved deserves some consideration, not only because it settled, at least for the time being, the problem of spontaneous generation, but also because it led Pasteur and his students in France — and Tyndall in England — to work out some of the most useful techniques of bacteriological science.

Bastian had dissolved heated potash in distilled water, unaware of the fact that the most limpid water can carry living germs. While investigating this problem Pasteur and his assistant Joubert recognized that water from deep wells, which had undergone a slow filtration in sandy soil, was often essentially or even completely free of germs. This observation soon led Chamberland to devise the porcelain bacteriological filters now so widely used in bacteriological laboratories and formerly as household objects.

It soon became obvious that there must have been still other sources of contamination in Bastian's experiments. During the studies on silkworm diseases, which we shall consider in a later chapter, Pasteur had become aware of the fact that bacteria can exist in dormant phases which are more resistant to heat than the active vegetative forms. These resistant forms were extensively studied by John Tyndall and especially Ferdinand Cohn, who named them "bacterial spores." Pasteur surmised that the samples of urine studied by Bastian were contaminated with a few of these bacterial spores which had survived the heating process, but were unable to germinate and give rise to visible microbial growth until the slight acidity of urine had been neutralized by the addition of alkali. Fortunately, heating under pressure at 120° C. was found sufficient to destroy bacterial spores and effect thereby complete sterilization of urine and other organic fluids. Thus was introduced into bacteriological technique, and into many practical operations of public health and technology, the use of the autoclave to effect sterilization with superheated steam. It was found on the other hand that although 120° C. was sufficient for sterilization in the presence of water vapor most forms of life were much more resistant to dry heat. The discovery of this fact led to the elaboration of many bacteriological procedures — such as the use of ovens reaching 160° C. for dry heat sterilization, and the practice of passing test tubes, flasks, and pipettes, through the naked flame for the inoculation and transfer of microbial cultures.

While Pasteur and his school were explaining Bastian's results and combatting his interpretations, the physicist John Tyndall had taken up the torch against spontaneous generation in England. In the course of his studies on the relation of radiant heat to gases, Tyndall had been greatly struck by the difficulty of removing from the atmosphere the invisible particles of dust that float in it. This interest in dust led him to drift progressively into the discussion on spontaneous generation and to recognize that dust particles carry living organisms. His experiments on the subject, published in 1876 and 1877, provided powerful support for Pasteur's views. He presented them again, along with brilliant essays and lectures, in his book *Essays on the Floating Matter of the Air in Relation to Putrefaction and Infection,* which, when published in 1881, played a role almost equal to that of Pasteur's writings in accomplishing the final downfall of the doctrine of spontaneous generation.

Tyndall prepared experimental chambers, the interior surfaces of which had been coated with glycerine. The closed chambers were left untouched for several days until a beam of light passing through lateral windows showed that all floating matter of the air had settled and become fixed on the glycerine surfaces. In Tyndall's terminology, the air was then "optically empty." Under these conditions, all sorts of sterilized organic fluids, urine, broth, vegetable infusions, could be exposed to the air in the chamber and yet remain unaltered for months. In other words, optically empty air was also sterile air. Thus it became certain that the power of the atmosphere to generate bacterial life goes hand in hand with its ability to scatter light and therefore with its content in dust, and that many of the microscopic particles which float in the air consist of microorganisms, or carry them.

Tyndall, who had been trained as a physicist, displayed like Pasteur great biological inventive imagination in all these experiments. It seems worth while to digress for a moment and mention here the circumstances under which he worked out the technique of practical sterilization by discontinuous heating, known today as "Tyndallization." He had been much impressed by the enormous resistance to heat exhibited by the spores of the hay bacillus, an organism universally present in hay infusions. Knowing that vegetative bacteria are easily killed by boiling and that a certain latent period is required before the heat-resistant spores return to the vegetative state in which they again become heat susceptible, he devised a process of sterilization which he first described in 1877 in a letter to Huxley: "Before the latent period of any of the germs has been completed (say a few hours after the preparation of the infusion), I subject it for a brief interval to a temperature which may be under that of boiling water. Such softened and vivified germs as are on the point of passing into active life are thereby killed; others not yet softened remain intact. I repeat this process well within the interval necessary for the most advanced of those others to finish their period of latency. The number of undestroyed germs is further diminished by this second heating. After a number of repetitions which varies with the characters of the germs, the infusion however obstinate is completely sterilized." Boiling for one minute on five successive occasions could render an infusion sterile, whereas one single continuous boiling for one hour might not.

It is likely that Pasteur's and Tyndall's triumph over the upholders of the doctrine of spontaneous generation was not as universal and as complete as now appears through the perspective of three quarters of a century. There must have been many scientists who, while accepting the wide distribution of microbes in the atmosphere, could not dismiss the belief that elementary microscopic life does now and then arise *de novo* from putrefying matter. Indeed all over the world, there are today experimenters watching with undying hope for some evidence that matter can organize itself in forms stimulating the characteristics of life. Man will never give up, and probably should not relinquish, his efforts to evoke out of the chaotic inertia of inanimate matter the dynamic and orderly sequence of living processes.

It was an unexpected event which revealed to Pasteur that, deep in the hearts of some of his most illustrious colleagues, the "chimera" of spontaneous generation was still breathing.

The French physiologist Claude Bernard died in the fall of 1877. Although it appears likely that his social intercourse with Pasteur never went beyond official meetings at the academies, the two men had enjoyed friendly scientific relations and had written in flattering terms of each other. There had been rumors that Bernard had devoted much of his last months of activity, in his country home of Saint-Julien, to the problem of alcoholic fermentation. In fact, sketchy notes outlining a few crude experiments dealing with the fermentation of grapes were found in the bottom of a drawer after his death, and they were immediately published by Bernard's friends, the chemist Berthelot and the physiologists Bert and d'Arsonval. Bernard stated in these notes that, contrary to Pasteur's views, fermentation could occur independently of living processes and he seemed to imply that yeast might arise as a result of fermentation instead of being the cause of it. "Yeast is produced only in those extracts of grape in which the protoplasmic function exists. It does not occur in the very young juice. It no longer occurs in the juices which have rotted, in which the plasmatic power has been killed. . . ."

Pasteur recognized in those obscure lines a disguised reappearance of the doctrine of spontaneous generation, and he expressed his dismay in a communication to the Academy: "On reading these opinions of Bernard, I experience both surprise and sorrow; surprise because the rigorous mind which I used to admire in him is completely absent in this physiological mysticism; sorrow, because our illustrious colleague seems to have forgotten the demonstrations which I have presented in the past. Have I not, for example, carefully described as early as 1872, and more particularly in my *Studies on Beer* in 1876, a technique to extract grape juice from the inside of a berry, and to expose this juice in contact with pure air, and have I not shown that, under these conditions, yeast does not appear and ordinary alcoholic fermentation does not take place? . . .

"It has also been painful for me to realize that all this was taking place under the auspices of our eminent colleague M. Berthelot."

Pasteur could not leave unanswered the veiled hint that yeast might after all originate from grape juice. Even though Bernard's posthumous statement could hardly be construed as anything more than a vague suggestion, the immense prestige of its author was enough to give new life to the lingering doctrine of spontaneous generation. For this reason, he immediately decided to demonstrate once more that alcoholic fermentation was dependent upon the prior introduction of yeast cells into the grape juice, and this project naturally led him into the question of the origin of yeast under the natural conditions of wine making.

The wine maker does not need to add yeast to start wine fermentation. The wild yeasts which are present in large numbers on the grapes and their stems are mixed with the grape juice at the time of pressing and they initiate lively fermentation shortly after the grape juice has been loaded into the vats. Depending upon the localities and the type of grapes, the wild yeasts differ in shape and physiological characteristics as well as in the flavor which they impart to the liquids they ferment. The same kind of yeasts reappear at the appointed place in the vineyard every year, ready to start the fermentation of the new grape crop. It would seem, indeed, as if Providence had provided yeast to complete in a natural process the evolution of the ripe grape into fermented wine. Yeast almost appears as a normal component of the grape crop, and unfermented grape juice as the crippled and mutilated fruit of the sunny vineyards. Where do these wild yeasts come from, and what are the workings of the time clock by which Providence brings them to the grape at the right time?

Yeast cells are present in significant numbers on the plant only at the period when the grape ripens; then they progressively decrease in number on the stems remaining after the harvest, and finally disappear during the winter. Pasteur's discovery that ripe grape carries its own supply of yeast explained the rapid course of fermentation under the practical conditions of wine making. It accounted also for some of the laboratory observations that had been quoted to support the theory of spontaneous generation, particularly for the fact that Bernard had observed the formation of some alcohol in the clear juice extracted from crushed grapes. All these conclusions had been anticipated in the *Studies on Beer* but, as Bernard had ignored or forgotten them, Pasteur resolved to repeat his earlier experiments on a larger and more convincing scale. In addition to its scientific interest, this episode has the merit of illustrating Pasteur's working methods, his ardor in returning to already conquered positions when they were threatened, and the suddenness with which he took decisions when he judged that an important issue was at stake. His plan was formulated the very day following the posthumous publication of Bernard's manuscript: "Without too much care for expense," he wrote, "I ordered in all haste several hothouses with the intention of transporting them to the Jura, where I possess a vineyard some dozens of square meters in size. There was not a moment to lose. And this is why.

"I have shown, in a chapter of my *Studies on Beer,* that the germs of yeast are not yet present on the grape berry in the state of verjuice, which, in the Jura, is at the end of July. We are, I said to myself, at a time of the year when, thanks to a delay in growth due to a cold rainy season, the grapes are just in this state in the Arbois country. By taking this moment to cover some vine with hothouses almost hermetically closed, I would have, in October at grape harvest time, vines bearing ripe grapes without any yeasts on the surface. These grapes, being crushed with the precautions necessary to exclude yeast, will be able neither to ferment nor to make wine. I shall give myself the pleasure of taking them to Paris, of presenting them to the Academy, and of offering some clusters to those of my confreres who still believe in the spontaneous generation of yeast.

"The fourth of August, 1878, my hothouses were finished and ready to be installed. . . . During and after their installation, I searched with care to see if yeasts were really absent from the clusters in the state of verjuice, as I had found hitherto to be the case. The result was what I expected; in a great number of experiments I determined that the verjuice of the vines around Arbois, and notably that of the vines covered by the hothouses, bore no trace of yeast at the beginning of the month of August, 1878.

"For fear that an inadequate sealing of the hothouses would allow the yeasts to reach the clusters, I decided to cover a certain number on each vine with cotton wrappings previously heated to a temperature of about 150° C. . . .

"Toward the tenth of October, the grapes in the hothouses were ripe; one could clearly distinguish the seeds through their skin and they were as sweet in taste as the majority of the grapes grown outside; the only difference was that the grapes under the cotton, normally black, were scarcely colored, rather violaceous than black, and that the white grapes had not the golden yellow tint of white grapes exposed to the sun. Nevertheless, I repeat, the maturity of both left nothing to be desired.

"On the tenth of October, I made my first experiment on the grapes of the uncovered clusters and on those covered with the cotton, comparing them with some which had grown outside. The result, I may say, surpassed my expectation. . . . Today, after a multitude of trials, I am just where I started, that is to say, it has been impossible for me to obtain one *single time* the alcoholic yeast fermentation from clusters covered with cotton.

"A comparative experiment naturally suggested itself. The hothouses had been set up in the period during which the germs are absent from the stems and clusters, whereas the experiments which I have just described took place from the tenth to the thirty-first of October during the period when the germs were present on the plant. It was then to be expected that if I exposed hothouse clusters from which the cotton had been removed on the branches of vines in the open, these clusters . . . would now ferment under the influence of the yeasts which they could not fail to receive in their new location. This was precisely the result that I obtained."

In the presence of these results, nothing was left of Bernard's inconclusive experiments. Once more, spontaneous generation had been ruled out of existence. Pasteur had to fight a few more oratorical battles in the Paris Academy of Medicine against the last articulate upholders of the doctrine, but after 1880 little more was heard of them – except the lonely voice of Bastian, who continued to proclaim his faith until the time of his death in 1910.

It is unrewarding for a philosopher to demonstrate his thesis with too much thoroughness and too convincingly. His ideas soon become part of the intellectual household of humanity, and the genius and labors which had to be expended in establishing them either are forgotten, or their memory becomes somewhat boring. For this reason, one often reads and hears that Pasteur and Tyndall wasted much talent and energy in a useless fight, for the belief in spontaneous generation was dying a natural death when they took arms against it. In reality, they had to overcome not only the teachings of the most eminent physiologists of the day, but also emotional prejudices based on philosophical convictions.

Neither Pasteur nor Tyndall ever devised an experiment which could prove that spontaneous generation does not occur; they had to be satisfied with discovering in each claim of its occurrence experimental fallacies that rendered the claim invalid. It was this ever-renewed necessity of discovering sources of error in the techniques of the defenders of spontaneous generation that made it necessary to reopen the debate time and time again, thus giving the impression of endless and wasteful repetition. The problem was clearly stated by Pasteur before the Academy of Medicine in March 1875, at the occasion of a debate during which Poggiale had spoken disdainfully of his experiments on spontaneous generation. "Every source of error plays in the hands of my opponents. For me, affirming as I do that there are no spontaneous fermentations, I am bound to eliminate every cause of error, every perturbing influence. Whereas I can maintain my results only by means of the most irreproachable technique, their claims profit by every inadequate experiment."

In addition to settling the controversy on spontaneous generation, Pasteur's and Tyndall's effort served to establish the new science of bacteriology on a solid technical basis. Exacting procedures had to be devised to prevent the introduction of foreign germs from the outside into the system under study, and also to destroy germs already present in it. Because of this necessity, the fundamental techniques of aseptic manipulation and of sterilization were worked out between 1860 and 1880. Incidental to the controversy also, there were discovered many facts concerning the distribution of microorganisms in our surroundings, in air and in water. It was also found that the blood and urine of normal animals and of man are free from microbes and can be preserved without exhibiting putrefying changes if collected with suitable aseptic precautions. All these observations constituted the concrete basis on which would be built the natural history of microbial life and, as we shall see, Pasteur saw in them many analogies which helped him to formulate the germ theory of disease and the laws of epidemiology. The controversy on spontaneous generation was the exacting school at which bacteriology became aware of its problems and learned its methodology.

However, it must be emphasized that what had been settled was not a theory of the origin of life. Nothing had been learned of the conditions under which life had first appeared, and no one knows even today whether it is still emerging anew from inanimate matter. Only the simple fact had been established that microbial life would not appear in an organic medium that had been adequately sterilized, and subsequently handled to exclude outside contamination. The germ theory is not a philosophical theory of life, but merely a body of factual observations which allows a series of practical operations. It teaches that fermentation, decomposition, putrefaction, are caused by living microorganisms, ubiquitous in nature; that bacteria are not begotten by the decomposing fluid, but come into it from outside; that sterile liquid, exposed to sterile air, will remain sterile forever.

It was this concept that Pasteur exposed in his Sorbonne lecture before an amphitheater overflowing with a fashionable audience come to hear from him a statement concerning the origin, nature and meaning of life. But wisely he refrained from philosophizing. He did not deny that spontaneous generation was a possibility; he merely affirmed that it had never been shown to occur.

The words which he pronounced on that occasion constitute the permanent rock on which were built whole sections of biological sciences:

"There is no known circumstance in which it can be affirmed that microscopic beings came into the world without germs, without parents similar to themselves. Those who affirm it have been duped by illusions, by ill-conducted experiments, by errors that they either did not perceive, or did not know how to avoid."

CHAPTER VII

The Biochemical Unity of Life

> As in religion we are warned to show our faith by works, so in philosophy by the same rule the system should be judged of by its fruits, and pronounced frivolous if it be barren; more especially if, in place of fruits of grape and olive, it bear thorns and briars of dispute and contention.
>
> — FRANCIS BACON

IN 1860 Pasteur recognized that the microorganisms responsible for the butyric fermentation and for putrefaction can grow in the absence of oxygen. As we have seen, this discovery was a landmark in the history of biological sciences. It revealed a new and unexpected haunt of life and it served as a powerful beacon to search into some of the most intimate mechanisms of the chemistry of living processes. And yet Pasteur's discovery was immediately belittled. Some questioned the validity of his observations, although few took the trouble to attempt to duplicate them. Many sneered at the wording of his descriptions because, impressed by the motility of the organisms that he had seen, he had referred to them as "infusoria" to suggest their animal nature. It is true that Pasteur's lack of familiarity with the terminology of the naturalist often rendered him somewhat inaccurate in the description of biological phenomena.[1] But he had the genius to

[1] Pasteur was aware of this limitation, but did not worry about it as he considered that the knowledge of microorganisms was still too imperfect to justify formal systems of classification. "It was on purpose that I used vague words: mucors, torula, bacteria, vibrios. This is not arbitrary. What would be arbitrary would be to adopt definite rules of nomenclature for organisms that can be differentiated only by characteristics of which we do not know the true significance."

The word "microbe" was introduced in 1878 by a surgeon, Sédillot, in

reach beyond the channel of specialized knowledge into the vast horizons of general biological laws, as illustrated by the following defiant words. "Whether the progress of science makes of this vibrio a plant or an animal is immaterial; it is a living being, which is motile, which lives without air, and which is a ferment." Thus, Pasteur entered biology, not through the narrow doors of classification and nomenclature, but by the broad stream of physiology and function. It is one of the most remarkable facts of his career that, although trained as a pure chemist, he attacked biological problems by adopting, straightaway, the view that the chemical activities of living agents are expressions of their physiological processes. He steadfastly maintained this attitude even when it brought him into conflict with the chemists and physiologists of his time who looked with mistrust upon any attempts to explain the chemical happenings of life in terms of vital forces.

So rapidly did Pasteur adopt the physiological attitude, in its most extreme form, that it appears of interest to document the evolution of his concepts with a few dates.

It was on February 25, 1861, that he reported for the first time the existence of the butyric acid organisms and their anaerobic nature and suggested at the same time a possible causal relation between life without air and fermentation. On April 12 of the same year, he mentioned before the Société Chimique that yeast ferments most efficiently in the absence of air whereas it grows most abundantly in its presence. These observations led him to suggest again that there exists a correlation between life without oxygen and the ability to cause fermentation. He also stated at the same time that, under anaerobic conditions, yeast respires with the oxygen borrowed from the fermentable substance. He presented the new theory of fermentation in more precise terms on June 17, 1861. "In addition to the living beings so far known

---

the course of a discussion at the Paris Academy of Medicine, to designate any organism so small as to be visible only under the microscope. Pasteur himself rarely used it, but preferred the expression "microorganisms."

Nevertheless he suggested in 1882 that the science of microbial life be designated "microbie" or "microbiologie," words which he properly regarded as less restricted in meaning than "bactériologie."

which . . . can respire and feed only by assimilating free oxygen, there appears to exist a class of beings capable of living without air by obtaining oxygen from certain organic substances which undergo a slow and progressive decomposition during the process of their utilization. This latter class of organized beings constitutes the ferments, similar in all particulars to those of the former class, assimilating in the same manner carbon, nitrogen and phosphates, requiring oxygen like them, but differing from them in the ability to use the oxygen removed from unstable organic combinations instead of free oxygen gas for their respiration." The same view was again presented in the form of a more general biological law on March 9, 1863. "We are thus led to relate the fact of nutrition accompanied by fermentation, to that of nutrition without consumption of free oxygen. There lies, certainly, the mystery of all true fermentations, and perhaps of many normal and abnormal physiological processes of living beings."

It is clear that Pasteur arrived very early at a well-defined concept of the relation of fermentation to metabolic processes, and at a realization that this concept had very broad implications for the understanding of the chemical processes of life. Although this generalization is perhaps the most original and profound thought of his long career, he never devoted much time to the subject, and his contributions to it were only by-products of other preoccupations, particularly of his studies on the technological aspects of the fermentation industries. Most of his fundamental thoughts on the physiological aspects of fermentation were published in the *Studies on Beer*. This book, intended to serve as a guide to the brewing industry, illustrates in a striking manner the struggle for the control of Pasteur's scientific life that went on beneath the apparently logical flow of his work: the everlasting conflict between his desire to contribute to the solution of the practical problems of his environment, and his emotional and intellectual urge to deal with some of the great theoretical problems of life.

* * *

It was then a common belief that many molds and other microorganisms can become transformed into yeast when submerged in a sugar solution, and thus give rise to alcoholic fermentation. Pasteur himself long remained under the impression that the vinegar organism (*Mycoderma aceti*), which oxidizes alcohol to acetic acid in the presence of air, can also behave as yeast and produce alcohol from sugar under anaerobic conditions. As these beliefs were in apparent conflict with one of the fundamental tenets of the germ theory of fermentation, namely the concept of specificity, Pasteur devoted many ingenious experiments to prove or disprove their validity, and arrived at the conclusion that they were erroneous. Yeasts, he pointed out, are ubiquitous in the air, and were often introduced by accident into the sugar solutions along with the other microbial species under study. It was therefore necessary to exclude this possibility of error, and after succeeding in eliminating it by elaborate precautions, he stated with pride, "Never again did I see any yeast or an active alcoholic fermentation follow upon the submersion of the flowers of vinegar. . . . At a time when belief in the transformation of species is so easily adopted, perhaps because it dispenses with rigorous accuracy in experimentation, it is not without interest to note that, in the course of my researches on the culture of microscopic plants in a state of purity, I once had reason to believe in the transformation of one organism into another, of Mycoderma into yeast. I was then in error: I did not know how to avoid the very cause of illusion . . . which the confidence in my theory of germs had so often enabled me to discover in the observations of others." There remained, however, one case of apparent transformation of a mold into yeast, accompanied by alcoholic fermentation, which seemed to be confirmed by experiment.

In 1857, Bail had asserted that *Mucor mucedo*, a mold commonly present in horse manure, induced a typical alcoholic fermentation if grown out of contact with air by immersion in a sugar solution. Instead of the long mycelial filaments that are characteristic of the mold growing in the presence of air, there were then produced chains of round or oblong cells which Bail had taken for those of brewer's yeast. Pasteur confirmed Bail's claim and established that alcohol and bubbles of carbon dioxide were produced out of the sugar in the absence of air. But instead of interpreting these phenomena as a change of the Mucor into yeast, he recognized in them a manifestation of his physiological theory of fermentation. He found that the short cells of Mucor which fermented in the bottom of sugar solutions immediately recovered their typical mycelial morphology when allowed to grow again in the presence of air. Under these conditions, they destroyed the sugar by complete oxidation instead of converting it into alcohol. In other words *Mucor mucedo,* which looked and behaved like yeast under anaerobic conditions, resumed the aspect and behavior of a mold when in contact with oxygen. Thus, in the case of this microorganism at least, there existed a striking correlation between morphological characteristics and biochemical behavior. The phenomena observed by Bail were not due to a transformation of species, but represented a transformation in cell form as a result of adaptation to a new life. The alteration in form coincided with a change of functions and Bail's phenomenon was merely the expression of a functional plasticity of the cell, allowing it to become adapted to a new environment.

Pasteur asked himself whether the counterpart of this situation might not occur in the case of true yeast. In agreement with his preconceived idea, he found that indeed the morphological and physiological characteristics of yeast were also influenced by the conditions of growth. Yeast grew slowly and fermentation took a long time in the total absence of air but the amount of sugar transformed into carbon dioxide and alcohol per unit of yeast was then extremely high. For example, 0.5–0.7 gm. of yeast was sufficient to transform 100 gm. of sugar into alcohol in the absence of air, a ratio of 1 to 150 or 1 to 200. On the other hand, as the amount of air admitted during fermentation was increased, the development of yeast became more rapid and more abundant, and the ratio of weight of sugar fermented to weight of yeast became smaller. When an excess of oxygen was provided throughout the process, hardly any alcohol was formed, although the development of yeast was very abundant and the ratio of sugar consumed to yeast produced fell to 4 or 5. Some alteration of the morphological characteristics of the yeast also occurred concomitantly with these dramatic changes in physiological behavior.

The correlation between the morphology of the fungus Mucor and the conditions of growth had served as a guide to recognize a correlation of a more fundamental nature, namely the dependence of biochemical behavior upon the availability of oxygen. Once before in his life, Pasteur had given an example of this change in emphasis, from the morphological to the functional level, in order to achieve a broader interpretation of observed phenomena. As will be remembered, it was the recognition of morphological differences between hemihedric crystals of tartaric acid that had led him to postulate a relation between crystal morphology, molecular structure, and ability to rotate the plane of polarized light. With the progress of his crystallographic studies he had later become less concerned with crystal shape, and had looked upon optical activity as a more direct expression of molecular structure. Similarly, the alteration in morphology of Mucor had now made him aware of a more fundamental fact, namely that fermentation was the result of life without oxygen.

These examples help in understanding Pasteur's attitude toward morphological studies for the investigation of natural phenomena. Many students of the history of bacteriology, and even his disciple Duclaux, have asserted that Pasteur was completely indifferent to considerations of morphology, and some have seen in this fact an indication that he had little interest in biology. This interpretation appears unjustified. In all phases of his scientific life, Pasteur observed and described morphological characteristics as carefully as his training and natural gifts permitted him. However, because he always had a specific goal in mind and because this goal was in all cases the understanding or the control of a function, he used morphology only as a guide to the discovery of functional relationships. Although he never studied morphological characteristics for their own sake, he used morphology wherever it provided information useful for the description of a system. For him, experimental techniques and procedures of observation were never an end unto themselves, but only tools, to be used for the solution of a problem and to be abandoned as soon as more effective ones became available.

Pasteur had postulated in 1861 that fermentation was the method used by yeast to derive energy from sugar under anaerobic conditions. In 1872 he restated these views in more precise terms. "Under ordinary conditions, the heat (energy) necessary for development comes from the oxidation of foodstuffs (except in the case of utilization of solar light). In fermentation, it comes from the decomposition of the fermentable matter. The ratio of the weight of fermentable matter decomposed to the weight of yeast produced will be higher or lower depending upon the extent of action of free oxygen. The maximum will correspond to life with participation of free oxygen."

This theory did not explain the known fact that access of a small amount of oxygen often increases the rate of production of alcohol by yeast, and that consequently oxygen must play a certain role in the fermentation process. To account for this apparent discrepancy between theory and fact, Pasteur postulated that respiration in the presence of oxygen permits the accumulation of reserve materials which are utilized under anaerobic conditions. Oxygen is beneficial because "the energy that it communicates to the life of the cell is later used up progressively." This prophetic view, for which evidence would not be forthcoming until half a century later, was suggested to Pasteur by morphological considerations. He described with great detail the appearance of youthfulness and of improved health and vigor in yeast exposed to oxygen. "In order to multiply in a fermentable solution deprived of oxygen, yeast cells must be young, full of life and health, under the influence of the vital activity which they owe to free oxygen and which perhaps they have stored. . . . When the cells are older, they generate bizarre and monstrous forms. Still older, they remain inert in a medium free of oxygen. It is not that these old cells are dead; for they can be rejuvenated in the same fluid after it has been aerated."

The calling of other problems, of duties which he considered more pressing, prevented Pasteur from pursuing very far the demonstration of these affirmations. Moreover, theoretical knowledge and experimental techniques were not adequate to permit at that time a convincing demonstration of his theory of fermentation. Realizing this fact, he predicted in a visionary statement that final evidence would have to come from consideration of thermodynamic relationships, a science which was yet in its infancy. "The theory of fermentation . . . will be established . . . on the day when science has advanced far enough to relate the quantity of heat resulting from the oxidation of sugar in the presence of oxygen to the quantity of heat removed by yeast during fermentation."

Although the understanding of the relation of oxygen to fermentation had been derived from the study of yeast, Pasteur was convinced that, with proper modifications, the new knowledge would be valid for all living cells. He stated, for example, that the truly anaerobic forms such as the butyric ferment "differ from yeast only by virtue of the fact that they are capable of living independently of oxygen in a regular and prolonged manner." Lechartier and Bellamy had shown in 1869 that the plant cells of ripe fruits transform a part of the sugar that they contain into alcohol if the fruits in question are preserved in an atmosphere of carbon dioxide. This observation suggested that alcohol production from sugar was a general property of plant protoplasm functioning in the absence of free oxygen. In a similar vein, Pasteur observed that, whereas plums kept in an open container took up oxygen and became soft and sweet, they remained firm, lost sugar, and produced alcohol if placed in an atmosphere of carbon dioxide. Time and time again he restated his belief that: "Fermentation should be possible in all types of cells. . . . Fermentation by yeast is only a particular case of a very general phenomenon. All living beings are ferments under certain conditions of their life . . ."

The relation between fermentation, respiration and availability of oxygen is not limited to microbial and plant cells. "Similarly, in animal economy, oxygen gives to cells an activity from which they derive, when removed from the presence of this gas, the faculty to act in the manner of ferments."

Pasteur never engaged in experiments with animal tissues. It appears of peculiar interest, therefore, to quote the *a priori* views which he expressed concerning the metabolism of muscle:

"(*a*) An active muscle produces a volume of carbon dioxide larger than the volume of oxygen consumed during the same time . . . this fact is not surprising according to the new theory, since the carbon dioxide which is produced results from fermentation processes which bear no necessary relation to the quantity of oxygen consumed. (*b*) One knows that muscle can contract in inert gases . . . and that carbon dioxide is then produced. This fact is a necessary consequence of the life continued by the cells under anaerobic conditions, following the initial stimulus which they have received from the oxygen. . . . (*c*) Muscles become acid following death and asphyxia. This is readily understandable if . . . fermentation processes go on after death in the cells, which function then as anaerobic systems."

Thus, Pasteur arrived at the conclusion that all living cells, whatever their own specializations and peculiarities, derive their energy from the same fundamental chemical reactions. By selecting yeast and muscle to illustrate this law, he anticipated modern biochemistry, not only in one of its most far-reaching conclusions, but also in its methodology, for the study of yeast and muscle physiology has provided much of our understanding of the chemistry of metabolic processes.

Surprisingly enough, these large implications of Pasteur's views on the essential biochemical unity of life did not impress his contemporaries and are not mentioned even by Duclaux, who of all disciples assimilated most completely the spirit of the master's discoveries. It is the more surprising that they did not immediately become integrated into the physiological thinking of the time, because they fitted so well into the prevalent desire to explain physiological phenomena in terms of physicochemical reactions. Indeed, Liebig, with his great generalizing mind, had prophesied the new era in biochemistry in his last memoir of 1869, entitled *On Fermentation and the Source of Muscular Energy*. By the irony of fate, it was Pasteur who first expressed in clear chemical terms the analogy between the metabolism of yeast and the workings of muscle, thus giving reality to the prophetic views of his great German rival.

Pasteur reiterated time and time again that, as far as it had been observed, fermentation was a manifestation of the life of yeast, causally connected with its metabolism and growth; but most contemporary physiologists believed that it was caused by simple chemical forces and they refused to explain the phenomenon in terms of vital action. It is now clear that the physiological view of fermentation held by Pasteur, and the chemical theories defended by his opponents, were not incompatible but indeed were necessary for the completion of each other. And yet, because Pasteur was convinced that fermentation could be more profitably considered as a function of life than as a chemical reaction, and because his opponents refused to meet him on this ground for reasons of scientific philosophy, there arose a battle of words in which many of the most vigorous minds of the nineteenth century took part. As will become obvious on reading some of the statements made by the leading contestants, there was no real justification for this controversy. Minor adjustments would have sufficed to compose the differences between the proponents of the two theories, merely the willingness to recognize that all natural phenomena can be profitably investigated at different levels of integration. Nevertheless, this great debate is of historical interest in recalling the struggles out of which modern physiology has evolved. It also illustrates how slow and painful is the maturation of a scientific concept that appears simple to the following generation; and reveals that, like other men, scientists become deaf and blind to any argument or evidence that does not fit into the thought pattern which circumstances have led them to follow. Truth is many-faceted, and the facet which one happens to see from any given angle and at any given time is often different from, but not necessarily incompatible with, that which appears to one approaching from a different direction. For this reason, as Goethe said: "History must from time to time be rewritten, not because many new facts have been discovered, but because new aspects come into view, because the participant in the progress of an age is led to standpoints from which the past can be regarded and judged in a novel manner."

Two decades earlier, the sarcasm and haughty edicts of Berzelius and Liebig had silenced the voice of those who believed in the living nature of yeast. In the 1860's Pasteur's fighting temperament had made many scientists somewhat fearful of denying in public the vitalistic theory of fermentation. In science as in politics, however, it is easier to silence than to convince the opponents of a doctrine. That the vitalistic theory was still held in ill-repute was revealed to Pasteur when Marcellin Berthelot published, in 1877, the fragmentary and sibylline notes in which Claude Bernard had expressed his belief that alcoholic fermentation could occur in the absence of living cells. Pasteur was certainly right in regarding these notes as tentative projects and thoughts that Bernard had never intended to publish, and he accused Berthelot of utilizing the authority of the illustrious physiologist without the latter's consent. Although he was then heavily engaged in the study of anthrax, he did not hesitate to undertake new experiments proving that Bernard's assertions were based on faulty observations and he led a vigorous attack against Bernard and Berthelot in the Academy. Thus began a weird controversy, in which one of the main protagonists was in the grave and appeared only in the form of a few posthumous notes.

Throughout his meditations on the mechanism of physiological processes, Claude Bernard had attempted to find a compromise between two sets of facts that he believed characteristic of life. On the one hand, all physiological phenomena proceed according to the same physicochemical laws which govern other natural events. On the other hand, it is equally obvious that each living being has its own characteristic potentiality of development which, prearranged in the ovum, is an expression of the properties of the species and appears to depend upon forces which operate at a higher level of organization. It is this dual concept that Bernard, lacking an explanation, nevertheless felicitously expressed in a famous statement: "Admitting that vital processes rest upon physicochemical activities, which is the truth, the essence of the problem is not thereby cleared up; for it is no chance encounter of physicochemical phenomena that constructs each being according to a pre-existing plan, and produces the admirable subordination and the harmonious concert of organic activity. There is an arrangement in the living being, a kind of regulated activity, which must never be neglected, because it is in truth the most striking characteristic of living beings."

This concept greatly influenced Bernard's views of the phenomena of alcoholic fermentation. He regarded the growth and development of yeast as a result of synthetic processes regulated by the property of organization which is characteristic solely of life. As to the processes of organic destruction, they were explained by simple physicochemical laws. According to Bernard, it was this purely destructive aspect of the activities of the cell which presided over the return of dead matter to nature, and which, in the case of yeast, was responsible for the breakdown of sugar into alcohol and carbon dioxide. Despite their philosophical dressing, these views were essentially a modernized return to Berzelius's and Liebig's earlier concepts. The chemical theory of fermentation had recently received some support from the discovery of enzymes, those complex components of living cells which catalyze certain organic reactions. Berthelot had demonstrated the existence in yeast of a soluble agent, the enzyme invertase, which was capable of splitting cane sugar and which retained its activity even after being extracted in a soluble form free of yeast cells. Was it not possible that yeast could also produce another enzyme capable of converting sugar into alcohol, precisely as invertase broke down cane sugar?

With this possibility in mind, Bernard planned experiments to demonstrate the production of alcohol and carbon dioxide without the intervention of yeast, by the natural play of forces exterior to the cell. Having the preconceived notion that fermentation was the result of cellular disintegration, he undertook to find the alcoholic enzyme in grapes which had begun to decay. He crushed ripe grapes and observed in the clear juice the production of alcohol within forty-eight hours, in the absence of yeast as he believed. In other experiments, it is true, he did find yeast globules in his fermenting juice but, confident that he had not introduced them from the outside, he came to consider the possibilities that yeast might be a consequence, and not the origin, of the whole process. So crude were his experiments that they serve only to demonstrate how unwillingly experimenters, however great, will subjugate the shaping of their concepts to the hard reality of facts.

With the most exacting technique, Pasteur disposed of Bernard's claims within a few weeks, and demonstrated once more the dependence of alcoholic fermentation upon the presence of living yeast. But his triumph was incomplete, for it was at the level of biochemical doctrine, and not at the level of fact, that Berthelot attempted to defend Bernard's point of view. Berthelot argued that, in final analysis, all biological processes are the results of chemical reactions. Since several activities of living cells had been shown to be caused by enzymes capable of remaining active in solution, away from the cells which had produced them, it was likely that production of alcohol could also occur in the absence of living yeast cells. To this plausible hypothesis Pasteur could rightly answer that the theory that fermentation is correlative with the life of yeast did account for all the known facts, and was in conflict with none. He was extremely careful to define his position in these exact terms — so careful, indeed, that his attitude suggested the skill of an attorney who knows how to use statements expressing the letter of the law, even though the cause which he defends is in obvious conflict with common sense. In this strange debate, during which the two opponents used polished academic language to accuse each other of bad faith, the facts were on Pasteur's side, but history was soon to show that it was Berthelot who was speaking the voice of future common sense.

Twenty years after the controversy, a new fact became available, which established at one stroke the theory that Liebig, Berthelot, Bernard and their followers had attempted to uphold by means of logic, analogy and bad experiments. In 1897, Büchner extracted from yeast a soluble fraction, which he called "zymase," and which was capable of producing alcohol from sugar in the absence of formed, living yeast cells. It is unfortunate for the dignity of the scientific method that this epoch-making achievement was not the outcome of an orderly intellectual process, but the unplanned result of an accident. Hans and Eduard Büchner were attempting to break up yeast by grinding it with sand in order to obtain a preparation to be used for therapeutic purposes. The yeast juice thus obtained was to be employed for animal experiments but underwent alteration rapidly. As the ordinary antiseptics were found unsuitable to prevent the growth of bacteria, Büchner added sugar in high concentration to the juice as a preservative. To his great surprise, an evolution of carbon dioxide accompanied by production of some alcohol took place immediately and it was the marked action of the juice upon the added sugar which revealed to him that fermentation was proceeding in the absence of living yeast cells.

Despite the fact that the discovery of zymase was the result of an accident, chance appears as the main actor only in the last scene of this great drama. In reality, physiologists and chemists had long worked hard trying to release the alcohol-forming enzyme from yeast, beginning in 1846 with Lüdersdorff. Roux has reported that Pasteur himself had attempted to grind, freeze and plasmolyze the yeast cells with this purpose in mind, but all in vain; and many were those who, before and after Pasteur, also failed. Büchner was the fortunate heir of a long tradition of experience, which had made him aware of the problem and its difficulties. As in the case of so many discoveries, the new phenomenon was brought to light, apparently by chance, as the result of an investigation directed to other ends, but fortunately fell under the eye of an observer endowed with the genius that enabled him to realize its importance and give to it the true interpretation.

Büchner's discovery inaugurated a new era in the study of alcoholic fermentation, and from it has evolved the analysis of physiological phenomena in terms of the multiple individual steps of intermediate metabolism. Proof of the existence of zymase constituted a violent setback for the physiological theory of Pasteur, and biochemists came to regard cells as bags of enzymes performing this or that reaction without much regard to their significance for the life of the organism. Within a decade, however, a new point of view began to manifest itself; it was forcefully expressed by the German physiologist Rubner: "The doctrine of enzymes and their action must be brought into relationship with living processes . . . modern literature offers no explanation of the part played by sugar fermentation; this is regarded merely as a result of ferment action. Our knowledge from the biological standpoint cannot be satisfied by this statement; the life of an organism cannot consist only in the production of a ferment causing decomposition."

It is obvious today that the chemical processes of the cell subserve functions, and that there always exists a relation between the chemical changes observed within a living organism and the varied activities of this organism. It is equally certain on the other hand that all chemical changes which go on within living cells are carried out by means of a machinery, consisting largely of enzymes, and powered by energy-yielding reactions, which can operate outside the cell and function independently of life. The physiological and the chemical theories of fermentation and metabolism are therefore both right, and both are essential to the description of living processes. In fact, even the combination of the two theories may not be sufficient to explain life, for life is more than its mechanisms and functions. It is characterized, as

Claude Bernard always emphasized, by the integration and organization which "produce the admirable subordination and the harmonious concert of organic activity." To define the nature and mechanisms of this mysterious organization will be the exciting venture of the generations to come.

From our vantage point in time, it is now possible to recapture the general trend of the long series of experiments and debates from which has evolved the modern doctrine of fermentation chemistry. Lavoisier, Gay-Lussac and Dumas had utilized to the best of their talent quantitative chemical techniques to establish an approximate balance sheet describing the conversion of sugar into alcohol and carbon dioxide. Their equation, however, did not provide any clue as to the nature of the forces involved in the reactions. When Pasteur approached the problem in 1856, two conflicting views confronted each other. The belief that yeast was a small living plant, and fermentation an expression of its life, had been ably presented twenty years earlier by Cagniard de la Tour, Schwann, Kützing and Turpin. Their views had been silenced by Berzelius and Liebig, who taught that the disruption of the sugar molecule was brought about by contact with unstable organic substances. Pasteur's experiments established beyond doubt that, under the usual conditions of fermentation, the production of alcohol was the result of the life of yeast in the absence of oxygen; fermentation was the process by which yeast derived from sugar the energy that it needed for growth under anaerobic conditions.

Berthelot, who like Liebig had originally held that fermentation was caused by dead protein and that yeast acted by virtue of the protein it contained and not as a living agent, had now recast his views to take into account the facts demonstrated by Pasteur. The following statement, which Berthelot published in 1860, defined in the clearest possible terms the role of the recently discovered enzymes in the chemical activities of the cells.

"From the work of M. Cagniard Latour, and even more from that of M. Pasteur, it has been proved that yeast consists of a mycodermic plant. I consider that this plant does not act on sugar through a physiological action but merely by means of ferments which it is able to secrete, just as germinating barley secretes diastase, almonds secrete emulsin, the animal pancreas secretes pancreatin, and the stomach secretes pepsin. . . . Insoluble ferments, on the other hand, remain attached to the tissues and cannot be separated from them.

"In short, in the cases enumerated above which refer to soluble ferments, it is clearly seen that the living being itself is not the ferment but the producer of it. Soluble ferments, once they have been produced, function independently of any vital act; this function does not necessarily show any correlation with any physiological phenomenon. I lay stress on these words so as to leave nothing equivocal in my manner of picturing the action of soluble ferments. It is evident, moreover, that each ferment can be formed preferentially, if not exclusively, by one or another plant or animal; this organized being produces and increases the corresponding ferment in the same way as it produces and increases the other chemically defined substances of which it is composed. Hence the success of M. Pasteur's very important experiments on the sowing of ferments, or rather, in my opinion, of the organized beings which secrete the actual ferments."

Liebig, also, had now come to regard yeast as a small living plant. Despite the gross factual errors that they contain, his long and confused memoirs of 1869 remain of interest in showing his generalizing and philosophical mind attempting to arrive at a reconciliation between the vitalistic and the chemical theories of fermentation.

"I admit," he wrote, "that yeast consists of plant cells which come into existence and multiply in a liquid containing sugar and an albuminoid substance. The yeast is necessary for fermentation in order that there may be formed in its tissues, by means of the albuminoid substance and the sugar, a certain unstable combination, which alone is capable of undergoing disruption. . . .

"It appears possible that the only correlation between the

physiological act and the phenomenon of fermentation is the production by the living yeast cell of a substance which, because of peculiar properties similar to the one exerted by emulsin on salicin or amygdalin, brings about the decomposition of sugar into other organic molecules. According to this view, the physiological act would be necessary for the production of this substance, but would bear no other relation to fermentation." Pasteur expressed complete agreement with this statement. It is unfortunate, therefore, that Liebig spoiled the chance of complete understanding by refusing to admit that one could bring about fermentation by growing yeast in a synthetic medium free of extraneous albuminoid matter, a refusal that he based on the failure of his own attempts, as well as on irrelevant and trivial arguments – "If it were possible to produce or to multiply yeast by adding ammonia to the fermenting fluid, industry would soon have taken advantage of this fact . . . but so far, nothing has been changed in the manufacture of beer."

Shortly thereafter, Traube also attempted to reformulate the mechanisms of fermentation by incorporating all the known facts within a single theory, in particular Pasteur's recent discoveries on the role of oxygen in the process: "The protoplasm of plant cells is itself, or contains, a chemical ferment which produces the alcoholic fermentation of sugar; the effectiveness of this ferment appears to depend upon the presence of the living cell for no one, so far, has succeeded in extracting it in an active form. In the presence of air, this ferment oxidizes sugar by fixing oxygen onto it; when protected from air, the ferment decomposes sugar by transporting oxygen from one group of atoms of the sugar molecule to another group, thus giving on one side a product of reduction (alcohol), and on the other a product of oxidation (carbon dioxide)." With this statement, again, Pasteur agreed, but in this case also he could not come to terms with Traube, who denied that yeast could grow and ferment with ammonium salts as sole source of nitrogen.

Thus, despite the lack of direct experimental proof, the view that alcoholic fermentation was due to the chemical action of some substance elaborated by the cell, and not directly to the vital processes of the cell, found supporters even among those who regarded yeast as a plant. Claude Bernard expressed this point of view in the following statement to his disciples: "Pasteur's experiments are correct, but he has seen only one side of the question. . . . The formation of alcohol is a very general phenomenon. It is necessary to banish from fermentation the vitality of cells. I do not believe in it." On October 20, 1877, in the last laboratory notes written before his death, he again wrote that fermentation "is not life without air, for in air, as well as protected from it, alcohol can be formed without yeast. . . . Alcohol can be produced by a soluble ferment in the absence of life."

While Pasteur's assertion that the production of alcohol was a manifestation of life remained uncontroverted by experience until 1897, it is clear that all leading physiologists of the time regarded fermentation as caused by enzymes which, theoretically at least, could act independently of the life of the cell that produced them. Pasteur himself had not ignored the possible role of enzymes. As already mentioned, he had even tried to separate the soluble alcoholic ferment from living yeast. The statement of the problem appears in the most clear-cut terms as early as 1860 in his first memoir on alcoholic fermentation.

". . . If I am asked what is the nature of the chemical act whereby the sugar is decomposed and what is its real cause, I reply that I am completely ignorant of it.

"Ought we to say that yeast feeds on sugar and excretes alcohol and carbonic acid? Or should we rather maintain that yeast produces some substance of the nature of a pepsin, which acts upon the sugar and then disappears, for no such substance is found in fermented liquids? I have nothing to reply to these hypotheses. I neither admit them nor reject them; I wish only to restrain myself from going beyond the facts. And the facts tell me simply that all true fermentations are correlative with physiological phenomena."

Again in 1878, he stated: "I would not be at all surprised if I were shown that the cells of yeast can produce a soluble alcoholic ferment," but he also pointed out, "Enzymes are always the products of life, and consequently the statement that fermentation is caused by an enzyme does not contribute to our further understanding of the problem as long as no one has succeeded in separating the fermentation enzyme in an active form, free of living cells." Because of this constant and overbearing emphasis on the aspects of the problem which had already been proved, and of this contempt for speculations which did not lead to experiment, Pasteur's attitude will appear to some narrow and unphilosophical. At the same time, however, these very limitations made him the most effective worker of all the participants in the debate. To all theories and discussions he could reply by the facts that he had established and that became, indeed, the basis of all subsequent discoveries. "One is in agreement with me if one accepts that, (1) true fermentation depends upon . . . microscopic organisms, (2) these organisms do not have a spontaneous origin, (3) life in the absence of oxygen is concomitant with fermentation."

Pasteur's reply to Liebig in 1871 reveals with pungency an attitude characteristic of the mind of the experimenter, who proceeds from one limited scope to another, in contrast with that of the more speculative mind which attempts to arrive at an all-embracing concept by a broad intellectual process. "If you agree with me that fermentation is correlative with the life and nutrition of yeast, we agree on the fundamental issue. If this agreement exists between us, let us concern ourselves, if you wish, with the intimate cause of fermentation, but let us recognize that this is a problem far different from the first. Science proceeds by successive answers to questions more and more subtle, coming nearer and nearer to the very essence of phenomena."

It is perhaps, fitting to conclude the account of this celebrated controversy, to which so many of the most vigorous minds of the nineteenth century contributed their genius, and also their share of human frailties, by quoting from a letter written to Duclaux by Liebig, in 1872, one year before his death.

> I have often thought in my long practical career and at my age, how much pains and how many researches are necessary to probe to the depths a rather complicated phenomenon. The greatest difficulty comes from the fact that we are too much accustomed to attribute to a single cause that which is the product of several, and the majority of our controversies come from that.
>
> I would be much pained if M. Pasteur took in a disparaging sense the observations in my last work on fermentation. He appears to have forgotten that I have only attempted to support with facts a theory which I evolved more than thirty years ago, and which he had attacked. I was, I believe, in the right in defending it. There are very few men whom I esteem more than M. Pasteur, and he may be assured that I would not dream of attacking his reputation, which is so great and has been so justly acquired. I have assigned a chemical cause to a chemical phenomenon, and that is all I have attempted to do.

CHAPTER VIII

The Diseases of Silkworms

> Let no man look for much progress in the sciences — especially in the practical part of them — unless natural philosophy be carried on and applied to particular sciences, and particular sciences be carried back again to natural philosophy.
>
> — FRANCIS BACON

THE FIRST triumphs of microbiology in the control of epidemics came out of the genius and labors of two men, Agostino Bassi and Louis Pasteur, both of whom were untrained in medical or veterinary sciences, and both of whom first approached the problems of pathology by studying the diseases of silkworms.

Although Bassi's findings exerted no detectable influence on Pasteur's later work, it is only fair, for the sake of historical justice, to salute in the romantic person of the great Italian the dawn of the science of infectious diseases. Agostino Bassi was not a trained scientist, but a public servant in Lodi — with such a love of scientific pursuits that he sacrificed to them not only the physical comforts of life but his eyesight, which he ruined by countless hours at the microscope.

A disease known as *mal del segno* was then causing extensive damage to the silkworm industry in Lombardy. Bassi demonstrated that the disease was infectious and could be transmitted by inoculation, by contact, and by infected food. He traced it to a parasitic fungus, called after him *Botrytis bassiana,* which invaded the tissues during the life of the worm and covered its dead body with a peculiar white effervescence containing the fungal spores. An exact understanding of the etiology of the disease and of its mode of dissemination allowed Bassi to work out methods to prevent its spread through the silkworm nurseries. After twenty years of arduous labor, he published in 1836, under the title *Del mal del segno* . . . an extensive account of his theoretical and practical findings. Then, forced to give up microscopic investigations by the onset of blindness, he began to work and write on agricultural subjects. He continued, however, to develop the view that contagion is caused by living parasites, and applied his theory to the infectious diseases of man, and to the related problems of antisepsis, therapy and epidemiology. Although unable to see the bacterial agents of disease because of blindness, Bassi envisioned from his studies on the *mal del segno* the bacteriological era which was to revolutionize medicine two decades after his death.

Toward the middle of the nineteenth century a mysterious disease began to attack the French silkworm nurseries. It reached disastrous proportions first in the southern districts. In 1853, silkworm eggs could no longer be produced in France, but had to be imported from Lombardy; then the disease spread to Italy, Spain and Austria. Dealers procuring eggs for the silkworm breeders had to go farther and farther east in an attempt to secure healthy products; but the disease followed them, invading in turn Greece, Turkey, the Caucasus — finally China and even Japan. By 1865, the silkworm industry was near ruin in France, and also, to a lesser degree, in the rest of Western Europe.

Before describing the manifestations of the disease and the studies that led to its control, it may be useful to describe in Duclaux's words the techniques by which the silkworm is commercially raised on the leaves of the mulberry tree.

> Everybody knows, at least in a general way, the principal phenomena of the life of the silkworm: its birth from an egg, whose resemblance to certain plant seeds has led to its being given the name of "seed," its four "molts," or changes of skin — during which the worm ceases to eat, remains motionless, seems to sleep upon its litter, and clothes itself, under

its old skin, with a new supple and elastic skin, which allows it to undergo further development. The fourth of these molts is followed after two or three days by a period of extreme voracity during which the worm rapidly increases in volume and acquires its maximum size: this is called the "*grande gorge.*" This period ended, the worm eats no more, moves about uneasily, and if sprigs of heather on which it can ascend are present, it finds thereon a suitable place to spin its *cocoon,* a kind of silky prison which permits it to undergo in peace its transformation first into a chrysalis, and then into a moth. In this cocoon, the body of the worm, emptied of all the silky matter, contracts and covers itself with a resistant tunic in the interior of which all the tissues seem to fuse into a pulp of homogeneous appearance. It is in the midst of this magma that, little by little, the tissues of the moth are formed and become differentiated.

The moth has only a rudimentary digestive canal, for it no longer has any need of eating: the worm has eaten for it. It has wings, but, in our domestic races, it makes no use of them. It is destined only for the reproduction of the species, and the sex union takes place as soon as the moth comes out of the cocoon. The female then lays a considerable number of eggs, which may reach six hundred to eight hundred. In the races that we call annual, which are the most sought after, this "seed" does not hatch until the following year and is delayed until the reawakening of vegetation, the spring of the following year.

It is only when the grower wishes to induce the laying of eggs that he awaits this coming-forth from the cocoon, in which case the transformation of the worm into a moth requires about fifteen days. By adding thereto the thirty-five or forty days required for the *culture* of the worm, and the time necessary for the laying of the eggs, we see that the complete evolution of the silkworm, from the egg to the egg, is about two months. The period of industrial life is sensibly shorter. When the grower wishes to use only the cocoons, he must not wait until the moth, in coming forth, has opened them and thereby rendered them unfit for spinning. They are smothered five or six days after they have climbed the twigs of heather. That is to say, the cocoons are put into a steam bath, to kill the chrysalids by heat. In this case, scarcely six weeks separate the time of egg-hatching from the time when the cocoons are carried to market, from the time the silk grower sows to the time when he reaps. As, in former times, the harvest was almost certain and quite lucrative, the *Time of the Silkworm* was a time of festival and of joy, in spite of the fatigues which it imposed, and, in gratitude, the mulberry tree had received the name of *arbre d'or,* from the populations who derived their livelihood from it.

The disease that was now afflicting the French silkworm nurseries was different from that studied by Bassi. It was usually characterized by the existence, within the worm and especially upon its skin, of very small spots resembling grains of black pepper, and for this reason it was often referred to as "pébrine." Instead of growing in the usual uniform and rapid manner, from one molt to the other, the worms with pébrine became arrested at different stages of development. Many died in the first stages, and those which passed the fourth molt successfully could not complete their development, but instead faded away and gave insignificant yields. However, it often happened that the worms showed the spots without being sick; and contrariwise, within a diseased group those worms which were not spotted did not necessarily give good cocoons or eggs.

In addition to pébrine there were at the same time other forms of disease, known under the names "flacherie," *morts-flats* and *gattine,* which riddled the French silk economy. They all had much in common and were considered probably different aspects of the same illness. Most frequently after the fourth molt, during the period of voracity called the *grande gorge,* the diseased worms were seen to be indifferent to the provender, crawling over the leaf without attacking it, even avoiding it, and giving the appearance of seeking a quiet corner in which to die. When dead, they generally softened and rotted, but sometimes remained firm and hard, so that one had to touch them to be certain that they were dead. When attacked more slowly by the disease, the worm climbed the heather, but with difficulty, slowly spun its cocoon, sometimes left it unfinished and died without changing into a chrysalis or a moth.

The nature of pébrine was as mysterious as its origin. It had long been known that there existed in the diseased worms and moths peculiar microscopic structures, designated as "corpuscles"; but these corpuscles could also be found in apparently healthy moths. There were several facts, nevertheless, which pointed to the relation of the corpuscles to the disease. In 1849 an Italian biologist, Osimo, had described them in the eggs of silkworms, and another Italian, Vittadini, had later stated that their numbers increased in proportion as the eggs approached the period of hatching. Convinced of the relation of the presence of corpuscles to the disease, Osimo had advised as early as 1859 that the eggs and chrysalids be examined microscopically with the aim of rejecting the stocks found to be too corpuscular. This suggestion had been tested by still another Italian, Cantoni, who, having cultivated the eggs coming from noncorpuscular moths, had seen the worms develop corpuscular elements during the culture; this proved, Cantoni concluded, that the microscopic examination of moths was as worthless as the countless other remedies that had been advocated, and found wanting.

At the request of Jean Baptiste Dumas, who came from one of the afflicted regions, the Minister of Agriculture appointed a mission for the study of pébrine. With an extraordinary foresight Dumas asked Pasteur to take charge of it. Although Pasteur knew nothing of silkworms or of their diseases, he accepted the challenge under circumstances that have been described by Duclaux:

"I still remember the day when Pasteur, returning to the laboratory, said to me with some emotion in his voice, 'Do you know what M. Dumas has just asked me to do? He wants me to go south and study the disease of silkworms.' I do not recall my reply; probably it was that which he had made himself to his illustrious master [Dumas]: 'Is there then a disease of silkworms? And are there countries ruined by it?' This tragedy took place so far from Paris! And then, also, we were so far from Paris, in the laboratory!" To Pasteur's remark that he was totally unfamiliar with the subject, Dumas had replied one day: "So much the better! For ideas, you will have only those which shall come to you as a result of your own observations!"

Pasteur accepted Dumas's request, in part because of his great devotion to his master. It is probable also that he welcomed the opportunity to approach the field of experimental pathology, as is suggested by a sentence in his letter of acceptance: "The subject . . . may even come within the range of my present studies." He had long foreseen that his work on fermentation would be of significance in the study of the physiological and pathological processes of man and animals. But he was aware of his lack of familiarity with biological problems and Dumas's insistence helped him to face an experience that he both desired and dreaded.

It is through his studies on the diseases of silkworms that Pasteur came into contact with the complexities of the infectious processes. Surprisingly enough, he approached the problem unready to accept the idea that pébrine was caused by a parasitic agent foreign to the tissues of the worm. Instead, he retained for two years the belief that the disease was primarily a physiological disturbance, and that the corpuscles were only secondary manifestations of it, products of disintegration of tissues. The intellectual struggle and blundering steps which led him to the concept that a foreign parasite was the primary etiological agent can be recaptured from two sources of information. One is the brilliant analysis by Duclaux of the mental activities of his master during this period of their common labors. The other is found in the official documents prepared by Pasteur himself. As he was on a governmental mission and working under the public eye on a problem of great practical urgency, he had to make his results immediately available through official channels and his progress reports were naturally colored by the theoretical views he held of the nature of the disease. "In this phase of his researches," Duclaux points out, "he had not the right to keep the Olympian silence with which he loved to surround himself until the day when his work appeared to him ripe for publication. Under normal circumstances, he said not a word about it, even in the laboratory, where his assistants saw only the exterior and the skeleton of his experiments, without any of the life which animated them. Here, on the contrary, he was under obligation as soon as he had found out something, to speak and to excite the public judgment, as well as that of industrial practice, on all his laboratory discoveries."

The struggle against error, always imminent in these studies on silkworm diseases, is of peculiar interest because it provides a well-documented example of the workings of a scientific mind. As Pasteur himself said: "It is not without utility to show to the man of the world, and to the practical man, at what cost the scientist conquers principles, even the simplest and the most modest in appearance." Usually, the public sees only the finished result of the scientific effort, but remains unaware of the atmosphere of confusion, tentative gropings, frustration and heartbreaking discouragement in which the scientist often labors while trying to extract, from the entrails of nature, the products and laws which appear so simple and orderly when they finally reach textbooks and newspapers.

Pasteur arrived at Alais in early June and established primitive headquarters in a silkworm nursery. He immediately familiarized himself with the black spots of pébrine, and with the appearance of the corpuscles — which were easy to find in the diseased worms and moths throughout the Alais district. As he began a systematic comparison of the appearance and behavior of different cultures (broods) of worms, there came his way an observation which suggested that the disease might well be independent of the presence of corpuscles. He found in a certain nursery two different cultures, one of which had completed its development and had ascended the heather, while the other had just come out of the fourth molt. The first one appeared healthy and behaved normally; on the contrary the worms of the second ate little, did not grow and gave a poor harvest of cocoons. To his great surprise Pasteur found that whereas the corpuscles were abundant in the chrysalids and in the moths which had done well, they were scarce in the worms of the bad culture. Further observation of the bad culture revealed that the number of worms containing corpuscles increased as the culture advanced; even more corpuscles were present in the chrysalids, and finally not a single one of the moths was free of them.

There appeared to be an obvious and inevitable conclusion to be drawn from these observations: namely that pébrine was fundamentally a physiological disturbance which weakened the worms, independent of the presence of the corpuscles, and that the latter constituted only a secondary and accidental expression of the disease, probably a breakdown product of tissues. Whatever their origin, however, the corpuscles could be used as an index of the disease; and this led Pasteur to adopt again the method of egg selection that had been advocated by Osimo and found ineffective by Cantoni. He arrived at this conclusion within two weeks after his arrival at Alais, and recommended the egg-selection method in the following terms: "The technique consists in isolating each couple, male and female, at the moment of egg-laying. After the mating, the female, set apart, will be allowed to lay her eggs; she should then be opened as well as the male, in order to search for the corpuscles. If they are absent both from male and female, this laying should be preserved, as it will give eggs absolutely pure which should be bred the following year with particular care." He described the same technique later in greater detail. "As soon as the moths have left their cocoons and mated, they should be separated and each female placed on a little square of linen where she will lay her eggs. The moth is afterwards pinned up in a corner of the same square of linen, where it gradually dries up; later on, in autumn or even in winter, the withered moth is moistened in a little water, pounded in a mortar, and the paste examined with a microscope. If the least trace of corpuscles appears, the linen is burnt, together with the seed which would have perpetuated the disease."

As soon as the silkworm season was over, Pasteur moved back to Paris to resume his duties at the Ecole Normale. Early in February, 1866, he started again for Alais, accompanied by two assistants, Gernez and Maillot, and soon followed by Duclaux. After a short time, the party settled at Pont Gisquet, a lonely house at the foot of a mountain, and a laboratory was soon arranged in an empty orangery.

Multiple tragedies afflicted Pasteur's life during these few months. In 1865 he had lost his father and one of his daughters, Camille, then two years old. Another daughter, Cécile, died of typhoid fever at the age of twelve during May 1866. As the weight of these sorrows and the burden of the immense responsibilities which he had undertaken were leaving a mark on his health, Madame Pasteur, accompanied by their last surviving daughter, Marie-Louise, came to join the hard-working group at Pont Gisquet.

Comparative cultures of worms were immediately begun, with the eggs obtained the preceding summer from different pairs of moths which had exhibited corpuscles in varying degrees. Preliminary tests with small lots fed on the leaves of mulberry trees cultivated in hothouses were followed by the natural cultures in May and June. The results of these cultures revealed a number of facts of great importance.

First, it became obvious that the larger the number of corpuscles found in the parents of a given batch of eggs, the smaller was the yield of cocoons given by these eggs. There was no doubt, therefore, that the corpuscles bore a direct relation to the disease, even though it was not yet established that they were the cause of it. Very striking also was the fact that certain eggs laid by corpuscular moths remained capable of yielding acceptable cocoons, particularly eggs imported from Japan and those of a few sturdy French races. However, all the moths which originated from these cocoons were strongly corpuscular and the following generation of worms was, therefore, unsuited to the production of eggs and silk. This explained why selection of the seed from the gross appearance of the cocoon had given such unfortunate results and why it was necessary to know the extent of its contamination with corpuscles to judge of the value of an egg. Thus, additional evidence was obtained for the necessity of selection based on microscopic study. Finally, it was recognized that, even in cultures derived from very corpuscular insects, one could find here and there a few noncorpuscular moths which in their turn would produce healthy eggs. This observation meant that it was possible to recover a sound and productive stock from an infected nursery.

Needless to say, many skeptics sneered at the suggestion that the microscope could ever become an effective tool in the control of the disease. The microscope, they felt, might have its place in the hands of chemists, but how could one expect a practical silkworm grower to use such a complicated instrument? "There is in my laboratory," answered Pasteur, "a little girl eight years of age who has learned to use it without difficulty." This little girl was his daughter, Marie-Louise, for at Pont Gisquet everyone had joined in the common task, and had become experienced in the art of growing silkworms.

By June 1866 Pasteur was in a position to send to the Minister who had organized his mission in the South a statement that embodies the most important facts concerning the practical control of pébrine:

"In the past, the evil had been sought in the worm and even in the seeds, but my observations prove that it develops chiefly in the chrysalis, especially in the mature chrysalis, at the moment of the moth's formation, on the eve of the function of reproduction. The microscope then detects its presence with certainty, even when the seed and the worm seem very healthy. The practical result is this: You have a full nursery; it has been successful or it has not; you wish to know whether to smother the cocoons or whether to keep them for reproduction. Nothing is simpler. You hasten the development of about one hundred moths by raising the temperature, and you examine these moths through the microscope.

"The evidence of the disease is then so easy to detect that a woman or a child can do it. If the producer is a peasant, unable to carry out this study, he can do this: instead of throwing away the moths after they have laid their eggs, he can bottle them in brandy and send them to a testing office or to some experienced person who will determine the value of the seed for the following year."

From then on, the egg-selection method was used systematically in Pasteur's laboratory, and permitted the selection of healthy eggs.

Why had Cantoni failed to obtain useful results with the same method? A similar question occurs again and again throughout Pasteur's scientific career. In many circumstances, he developed reproducible and practical techniques that in other hands failed, or gave such erratic results as to be considered worthless. His experimental achievements appear so unusual in their complete success that there has been a tendency to explain them away in the name of luck, but the explanation is in reality quite simple. Pasteur was a master experimenter with an uncanny sense of the details relevant to the success of his tests. It was the exacting conscience with which he respected the most minute details of his operations, and his intense concentration while at work, that gave him an apparently intuitive awareness of all the facts significant for the test, and permitted him always to duplicate his experimental conditions. In many cases, he lacked complete understanding of the reasons for the success of the procedures that he used, but always he knew how to make them work again, if they had once worked in his hands.

Although Pasteur had been adventurous enough to suggest the use of the egg-selection method on the basis of very sketchy observations, he was well aware at that time of the inadequacy of the evidence available to prove that the corpuscles were a constant index of the existence of the disease in the worms. Where his predecessors had been satisfied with this uncertain state of affairs, he now decided to undertake comparative cultural experiments upon healthy and diseased eggs, in the hope of arriving at a more accurate knowledge of the relation of corpuscles to the disease. To this end he used the egg-selection method to secure seed originating from chrysalids containing varying numbers of corpuscles. So widespread was the disease in the Alais region, where he was located, that more than eight days of constant microscopic work were necessary to find among many hundred moths two or three pairs free of corpuscles.

In a few preliminary experiments, Pasteur attempted to determine whether healthy worms would contract the disease when fed food contaminated with corpuscles, but the results had been equivocal; some of the worms had remained healthy, and others had died without exhibiting the corpuscles. A later experiment, carried out by his assistant Gernez, had been more instructive. All the worms fed ordinary mulberry leaves, or leaves moistened with clean water, had yielded beautiful cocoons free of corpuscles. The worms fed with leaves contaminated with the debris of corpuscular moths had given only few cocoons, all very corpuscular, even when the contaminated leaves were first introduced after the third molting. Finally, the worms fed the contaminated leaves only after the fourth molt gave a normal number of cocoons, but most of these were corpuscular.

Pasteur's assistants had become convinced that the corpuscles were the cause of the disease and they now believed that he also had reached the same conclusion. A new experiment of Gernez's appeared particularly convincing in this regard. It showed that worms issued from healthy eggs could give healthy cocoons, whereas the introduction of corpuscles either prevented the worms from reaching the cocoon stage if severe infection occurred early enough, or gave rise to corpuscular cocoons if the infection was delayed. When this experiment was reported to the Academy in November 1866, Gernez was much surprised to see Pasteur describe it merely as another evidence of the effectiveness of the egg-selection method, without even mentioning that the results suggested the contagious nature of the disease and the corpuscle as its cause.

Duclaux claims that Pasteur had, in fact, failed to derive from the results of Gernez's experiments the conclusion that his disciples had inferred. And yet, that he had given earnest thought to the possibility of infection is clear from a report that he presented two months later, in January 1867. To the question "Is the disease parasitic?" he answered in the negative, for the following reasons:

"(*a*) There are many circumstances in which the disease precedes the corpuscles and therefore appears constitutional in nature.

"(*b*) The feeding of mulberry leaves contaminated with corpuscular material, either in the form of dust from a silkworm nursery, or of ground-up moths or worms loaded with corpuscles, often kills the worms very rapidly without giving them the corpuscles.

"(*c*) I have not been able to discover a mode of reproduction of the corpuscles, and its manner of appearance makes it resemble a product of transformation of the tissues of the worms."

Although these reasons indeed appeared sufficient to justify doubts concerning the role of the corpuscle as parasitic cause of the disease, they represented an erroneous interpretation of experimental findings. The first two points were invalidated by the fact, recognized by Pasteur himself the following year, that there was not one disease but two (or perhaps still more) occurring simultaneously in the same nursery and often in the same worm. Empirically, the silkworm growers had recognized this by using those different names – pébrine, *morts-flats*, flacherie, *gattine* – depending upon the symptoms of the diseased worms. No one, however, had yet extended these practical observations and postulated the existence of several different causal agents.

It will be recalled that, during the first weeks of his studies, in 1865, Pasteur had seen instances of diseased worms in which the corpuscles had appeared only in the late phases of the disease. It is almost certain, on the basis of present knowledge, that these worms had suffered first from the disease flacherie, and had only later become infected with the corpuscles of pébrine. Similarly, on several occasions when Pasteur had fed to healthy worms mulberry leaves contaminated with corpuscular material taken from worms known to have pébrine, he had unwittingly introduced at the same time the causative agent of flacherie. As the latter disease often ran a much more rapid course than pébrine, many worms had become sick, or even died, before the corpuscles had had time to multiply to detectable numbers. Thus, the first two arguments used by Pasteur to rule out the parasitic nature of the corpuscles were invalid because his observations dealt with patterns of disease other than of those of pébrine alone.

The third reason given by Pasteur, namely his failure to recognize processes of reproduction of the corpuscles, was a penalty for his lack of knowledge of microscopic morphology. Pébrine, it is now known, is caused by a protozoan parasite (*Nosema bombycis*) which invades practically all the tissues of the embryo, larva, pupa and adult silkworm, and destroys the invaded cells. Pasteur's studies on fermentation and spontaneous generation had familiarized him with the morphology of yeasts and bacteria, which reproduce respectively by budding and binary fission (cleavage along the short axis), but the corpuscles of the protozoan *Nosema bombycis* undergo a more complex morphological evolution, penetrating the tissue cells of the worm and becoming at a certain stage almost invisible before they divide again into distinct and sharply contoured corpuscles. Pasteur, who was a masterful observer, had detected under the microscope many of these morphological details, and had had them reproduced in a number of drawings to illustrate his memoirs. He described with precision the slow, progressive differentiation of the corpuscles out of tissue substances. However, being totally unfamiliar with protozoology, he failed at first to place the proper interpretation on his findings, and seeing the corpuscles appear as it were *de novo* in the midst of the tissues of the diseased worms, he concluded that the pébrine corpuscles were not independent elements but were the products of pathological transformation of diseased cells.

The scientific method is usually regarded as an orderly, logical process evolving from a correct interpretation of accurate findings to inescapable conclusions. It would seem that Nature had amused herself in this instance by leading Pasteur to a practical solution of the first problem of infectious disease that he tackled, through the peculiar pathway of complex observations and erroneous interpretations. He, who had made himself the champion of the role of microorganisms in nature, denied for two years that the corpuscles were living parasitic agents, and stated that "pébrine is a physiological hereditary disease."

During the 1866 season, Pasteur had prepared by egg selection large amounts of healthy eggs. He had used them in his own cultures and had also distributed them to many breeders for tests under practical conditions. Many of the results had fulfilled his expectations. However, it soon became obvious that certain batches issued from moths free of corpuscles gave disastrous results, the worms dying rapidly with the symptoms of flacherie. Out of sixteen broods of worms which he had raised, and which presented an excellent appearance, the sixteenth almost perished entirely immediately after the first molt. "In a brood of a hundred worms," wrote Pasteur, "I picked up fifteen or twenty dead ones every day—black, rotting with extraordinary rapidity. . . . They were soft and flaccid like an empty bladder. I looked in vain for corpuscles; there was not a trace of them."

The outbreaks followed a well-defined pattern. The new disease attacked *all* the worms issuing from certain batches of eggs—even though these eggs had been distributed to different breeders, who had raised them under various conditions of place, time, climate and culture. The worms were attacked *at the same age*, as if they had all brought with them an inescapable germ of destruction. The disease, clearly, came from the eggs, and not from the environment. Many of the worms dying of flacherie remained free of corpuscles and failed to exhibit the spots characteristic of pébrine.

Pasteur became more and more anxious as he realized the gravity of the situation—but, Duclaux says, "he kept us so remote from his thought that we could not explain his uneasiness until the day when he appeared before us almost in tears, and, dropping discouraged into a chair, said: 'Nothing is accomplished; there are two diseases!' "

We have already described, in anticipation of this phase of Pasteur's studies, some of the characteristics of flacherie. Obviously, the disease had been frequently associated with pébrine and had not been readily differentiated from it. As soon as pébrine could be eliminated by growing worms issued from noncorpuscular moths, differential diagnosis of the two diseases became possible, and the road was opened for the study of flacherie.

The etiology of flacherie is much more complex than that of pébrine, and Pasteur never succeeded in formulating a complete picture of it; indeed the cause of the disease is not clear even today. Instead of describing piecemeal the many detailed observations made on the subject in the Pont Gisquet laboratory from 1866 to 1870, we shall summarize the point of view finally reached by Pasteur, although this involves the risk of presenting in terms of misleading simplicity a problem which must have appeared hopelessly confusing to the experimenters.

Under normal conditions, the intestinal contents of the healthy silkworms are almost free of microorganisms. On the contrary, the digestive tracts of diseased worms contain immense numbers of bacteria of various types, spore-bearing bacilli and streptococci appearing to predominate among them. Contamination of the mulberry leaves with the excrement of diseased worms causes the appearance of the disease in healthy worms; the disease is, therefore, contagious. Eggs derived from infected moths give rise to infected cultures, whatever the conditions under which they are raised, indicating either that the infection is carried in the egg or that certain batches of eggs exhibit a peculiar susceptibility to it.

There was another puzzling observation. The disease now and then appeared spontaneously in a nursery, especially when technical errors had been made in the handling of the mulberry leaves, or in controlling the temperature or aeration of the rooms. This, Pasteur believed, was because the disease agent was commonly distributed on the leaves; for he found that it was sufficient to let bruised mulberry leaves stand in high humidity at summer temperature to witness the development of bacteria similar to those seen in the intestines of worms with flacherie, and to produce the disease in healthy worms fed on them. Pasteur concluded from these observations that the silkworms normally ingested a few bacteria with the leaves, but in numbers too small to establish a pathological state; at times, however, high temperature or excessive humidity or poor ventilation in the nursery allowed unusual multiplication of the bacteria on the leaves, and perhaps also caused a decrease in the physiological resistance of the worms. Under these conditions, Pasteur felt the bacteria gained the upper hand, and multiplied without restraint in the digestive organs; then the disease broke out.

Recent information suggests that the etiology of flacherie is even more complex, and that the primary cause of the disease belongs to the class of submicroscopic agents known as filtrable viruses, too small to be seen by ordinary microscopy. It is not unlikely that this hypothetical virus of flacherie can cause a disease so mild that it escapes detection, but capable of rendering the silkworms more susceptible to a variety of bacteria eaten along with mulberry leaves; these bacteria might be the spore-bearing bacilli, or the streptococci seen by Pasteur in the intestines of the worms. In fact, the symptoms of the disease appear to vary, depending upon the nature of these bacterial invaders; and this variation probably accounts for the several different names under which the second disease of silkworms is known, *gattine* being the form in which *Streptococcus bombycis* predominates, whereas *Bacillus bombycis* occurs in the true flacherie. Several other mixed infections involving both a filtrable virus and a bacterium have been recognized in men and in animals during recent years. In the influenza pandemic of 1917–1918, for example, the primary cause of the infection was probably the influenza virus, which, alone, causes only a fairly mild disease — but at that time the virus infection was complicated in many cases by a superimposed bacterial infection, which greatly increased its severity and modified its character.

Pasteur did not, and could not, recognize and identify the multiple factors that condition flacherie. However, thanks to the constant and penetrating supervision which he exercised over all the phases of the work in the experimental nursery at Pont Gisquet, he gained a thorough knowledge of the manifestations of the disease and of the factors which aggravated its course, and he soon succeeded in developing techniques to avoid its spread.

Having become an expert breeder of silkworms, Pasteur could detect subtle differences in their behavior during the course of their development. In 1866, he had noted that in one of the cultures entirely free of pébrine, the worms had exhibited a peculiar behavior at the time of climbing up the heather to undergo transformation; they had appeared to him sluggish and unhealthy. Following this lead, he secured silkworm cultures in which flacherie was prevailing, obtained from them cocoons free of pébrine corpuscles, and confirmed that the eggs derived from these cocoons gave rise to cultures which failed almost entirely, especially in the fourth molt, with the characteristic symptoms of flacherie.

On the basis of these observations, he formulated practical rules to prevent the development of the new disease. He emphasized "the imperious necessity of never using, for the egg laying—whatever may be the external appearance of the moths or the results of their microscopic examination—broods which have shown any languishing worms from the fourth molt to the cocoon, or which have experienced a noticeable mortality at this period of the culture, due to the disease of the *morts-flats*." Later, he also advocated a microscopic technique to detect the presence of bacterial infection in the moths used for the production of seed. The method consisted "in extracting with the point of a scalpel a small portion of the digestive cavity of a moth, then mixing it with a little water and examining it with a microscope. If the moths do not contain the characteristic microorganism, the strain from which they came may be considered as suitable for seeding."

Thus, as in the case of pébrine, a practical solution had been found for the prevention of flacherie even before the cause of the disease had been thoroughly worked out.

* * *

Most of the investigations on silkworm diseases were carried out near Alais, one of the most important centers of the industry. From the beginning and through the five years of his campaign, Pasteur established direct contact with the practical breeders, studying their problems at close range, concerning himself with the broad interests of the industry, and submitting his own views and methods to the acid test of application in the field. As we have seen, Pont Gisquet was not merely a research laboratory, it was an operating silkworm nursery where everybody, including Madame Pasteur and the little Marie-Louise, engaged in the raising of the worms and in the collection and selection of the eggs. Countless experiments and microscopic examinations; painstaking control and watch over the trial cultures; worry over the ever-present threat of mice, which preferred silkworms to the most succulent baits; the feverish harvest of mulberry leaves when rain was threatening—all these occupations left, to those who participated in the work, the memory of laborious days, but also that of one of the happiest periods in the scientific life of the master.

Small lots of the seed, selected at Pont Gisquet, were tested in the laboratory and the balance distributed among producers who sent reports on the results of their cultures. This co-operative enterprise soon led to an enormous volume of correspondence, which was handled by Pasteur himself. He spent his evenings dictating to his wife replies to distant collaborators, polemical articles for the trade journals, scientific articles for the academies and, finally, his book on the diseases of silkworms.

There were then no typewriters, telephones, or efficient offices and secretaries. A photograph of Pasteur dictating a scientific note to his wife in a garden, with a large sun hat in the background, calls forth a scene of olive trees and the brilliant skies of Provence, with cicadas humming their endless chant in the cool of the evening. It must have been good to work at Pont Gisquet, with an orangery for laboratory, and trees and water for office furniture.

After the 1867 season, when techniques had been worked out to control the spread of pébrine and flacherie, it became more urgent to prove the practical character of the new method of silkworm breeding. In this work, Pasteur showed the qualities of a chief of industry who watches everything, lets no detail escape him, wishes to know and control all operations, and who, at the same time, keeps up personal relations with his clientele, asking both those who are satisfied and those who are not the reasons for their opinions. To complete his apostolate, he became a practical silkworm breeder; he traveled to the Alps and to the Pyrenees to supervise the installation of his process in the nurseries of growers who had implored his aid.

There was, of course, much opposition to the new method, and to the personality of its discoverer. That a chemist should invade a purely biological industry and try to modify ancestral practices appeared nonsense to some, and wounded the susceptibilities and prejudices of others. In addition to professional jealousies, there were the fears of the dealers in silkworm eggs, whose financial interests were threatened. Slanderous reports of the activities of the Pont Gisquet laboratory began to circulate through the peasant population and reached the newspapers. During June 1868, for example, Madame Pasteur received from her father a letter in which he expressed concern over their welfare. "It is being reported here that the failure of Pasteur's process has excited the population of your neighborhood so much that he has had to flee from Alais, pursued by infuriated inhabitants throwing stones at him."

Pasteur responded to these attacks with his usual vigor. Every letter was acknowledged, whether friendly or threatening, every article answered with facts and also with passion. Addressing one who had questioned the value of the egg-selection program, Pasteur concluded his argument with the edifying remark: "*Monsieur le Marquis* . . . , you do not know the first word of my investigations, of their results, of the principles which they have established, and of their practical implications. Most of them you have not read . . . and the others, you did not understand."

However, words and arguments were not sufficient to convince the unbelievers. Sure of his facts, Pasteur engaged in bold predictions of the outcome of the cultures issued from eggs which he had selected, or which he had subjected to microscopic examination. In February 1867, for example, he sent to the trade journal *Jean-Jean* a prognosis, to be opened only at the end of the season, of results on certain batches of eggs, and the prognosis turned out to be true.

In 1868, he wrote to the Mayor of Callas, who had submitted two samples of eggs for examination: "These two batches will fail completely, whatever the skill of the breeders and the importance of their establishment."

The Silk Commission of Lyons, while interested in the selection method, had expressed some reserve as to its dependability, and had asked Pasteur in March 1869 for a little guaranteed healthy seed. He sent it, as well as other sample lots of which he predicted the future fate:

> 1. One lot of healthy seed, which will succeed.
> 2. One lot of seed, which will perish exclusively from the corpuscle disease known as pébrine.
> 3. One lot of seed, which will perish exclusively from flacherie.
> 4. One lot of seed, which will perish partly from corpuscle disease and partly from flacherie.
>
> It seems to me that the comparison between the results of those different lots will do more to enlighten the Commission on the certainty of the principles I have established than could a mere sample of healthy seed.

A few months later the Commission acknowledged the correctness of Pasteur's predictions.

In April 1869, the Minister of Agriculture asked Pasteur to submit a report on three lots of eggs that Mademoiselle Amat, a celebrated silkworm breeder, was distributing throughout the country. Pasteur's answer came four days later:

> . . . *Monsieur le Ministre* . . . , these three samples of seed are worthless. . . . They will in every instance succumb to corpuscle disease. If my seeding process had been employed, it would not have required ten minutes to dis-

> cover that Mademoiselle Amat's cocoons, though excellent for spinning purposes, were absolutely unfit for reproduction. . . .
>
> I shall be much obliged, *Monsieur le Ministre,* if you will kindly inform the Prefect of the Corrèze of the forecasts which I now communicate to you, and if you will ask him to report to you the results of Mademoiselle Amat's three lots.
>
> For my part, I feel so sure of what I now affirm, that I shall not even trouble to test, by hatching them, the samples which you have sent me. I have thrown them into the river. . . .

Marshal Vaillant, Minister of the Emperor's household, finally conceived of a test that would establish the effectiveness of Pasteur's method and silence his opponents. The Prince Imperial owned, near Triesta, an estate called Villa Vicentina, where pébrine and flacherie had completely ruined the culture of silkworms. In October 1869 the Marshal requested Pasteur to send selected seed and invited him to spend several months on the property. There he could supervise the raising of silkworms according to his methods, and at the same time recuperate from the attack of paralysis which had struck him the year before. Pasteur accepted the invitation, but instead of resting, completed his book on silkworm diseases, which was ready for publication in April 1870. He spent the spring organizing the culture of his selected eggs, on the imperial property and in neighboring farms. The results fulfilled all expectations and the property paid a profit for the first time in ten years. The egg-selection method was gaining ever-wider recognition, and soon came to be applied on a large scale in Italy and Austria.

The study on silkworm diseases constituted for Pasteur an initiation into the problem of infectious diseases. Instead of the accuracy of laboratory procedures he encountered the variability and unpredictability of behavior in animal life, for silkworms differ in their response to disease as do other animals. In the case of flacherie, for example, the time of death after infection might vary from twelve hours to three weeks, and some of the worms invariably escaped death. Pasteur repeatedly emphasized that the receptivity to infection of different individuals of the same species is of paramount importance in deciding the course and outcome of the disease. He also realized that the susceptibility of the worms was not solely conditioned by their inherited characteristics, but depended in part upon the conditions under which they lived. Excessive heat or humidity, inadequate aeration and stormy weather were all factors which he considered inimical to general physiological health of the worms, and capable of decreasing their resistance to infection. Similarly, atmospheric conditions and poor handling could cause the spoilage of the mulberry leaves and render them unfit as food. Even without attempting to analyze the role of these factors, he learned to raise the worms under sanitary conditions by giving them enough space so that they would not infect each other, by isolating the diverse lots in separate baskets, by exposing them to the open air — all practices which, in his mind, improved their well-being and protected them against contagion. He devoted much thought to the engineering and architectural planning of the nurseries to provide hygienic conditions. As ever interested in the most minute details, he quoted that in China the woman in charge of the nursery, the "mother of silkworms," was instructed to regulate the temperature of the room according to her own feelings of warmth or cold when dressed in a traditional costume.

Time and time again, he discussed the matter of the influence of environmental factors on susceptibility, on the receptivity of the "terrain" for the invading agent of disease. So deep was his concern with the physiological factors that condition infection that he once wrote, "If I had to undertake new studies on silkworms, I would investigate conditions for increasing their vigor, a problem of which one knows nothing. This would certainly lead to techniques for protecting them against accidental diseases."

He also kept constantly in mind the part played by the contaminated leaves, equipment and dust in spreading the infection to the worms coming in contact with them. By thinking about these problems, he discovered through direct experience many of the laws and practices of epidemiology – a knowledge that served him in good stead when he began to deal with the diseases of higher animals and man, a few years later.

He was aware that this had been his apprenticeship into the study of pathological problems, and he was wont to tell those who later came to work in his laboratory: "Read the studies on the silkworms; it will be, I believe, a good preparation for the investigations that we are about to undertake."

CHAPTER IX

The Germ Theory of Disease

> A whole flock in the fields perishes from the disease of one.
>
> — JUVENAL

IT WAS in April 1877, that Pasteur published, in collaboration with Joubert, his first paper on anthrax, twenty years after having presented the Manifesto of the germ theory in the memoir on lactic acid fermentation. One might be tempted to infer that these years had been necessary for the slow evolution and maturation, step by step and in a logical orderly manner, of the concept that microorganisms participate in the various processes of life and death. In reality, Pasteur had envisaged the role of microorganisms in disease as soon as he had become familiar with the problem of fermentation, and had stated early his intention of applying himself to the study of contagion.

While discussing the breakdown of plant and animal tissues by fermentation and putrefaction, he had written as early as 1859: "Everything indicates that contagious diseases owe their existence to similar causes." In 1860, he ventured the prediction that his studies on spontaneous generation and on the origin of microorganisms "would prepare the road for a serious investigation of the origin of various diseases." After having demonstrated that microorganisms are present in the dust of the air and vary in type and number depending upon the location, the time and the atmospheric conditions, he suggested in 1861: "It would be interesting to carry out frequent microscopic analysis of the dust floating in the air at the different seasons, and in different localities. The understanding of the phenomena of contagion, especially during the periods of epidemic diseases, would have much to gain from such studies."

In a letter of April 1862 to the Minister of Education, quoted on page 161, he clearly indicated that contagious diseases were on his program of study; and again, in March 1863, he wrote to Colonel Fave, Aide-de-Camp to the Emperor: "I find myself prepared to attack the great mystery of the putrid diseases, which I cannot dismiss from my mind, although I am fully aware of its difficulties and dangers."

When in September 1867 he appealed to the Emperor for a new and larger laboratory, he emphasized the application of his studies on fermentation and putrefaction to the problem of disease, pointing out that the handling of experimental animals, living and dead, would require adequate working facilities. Interestingly enough, he singled out in his request the subject of anthrax, although ten years were to elapse before his first experimental studies of this problem.

The experience with silkworm diseases had greatly sharpened his awareness of the problems of epidemiology, and allowed him to recognize in apparently unrelated observations many lessons that were applicable to the understanding of the spread of disease. "In Paris, during the month of July when the fruit trade is active, there must be large numbers of yeasts floating in the air of the streets. If fermentations were diseases, one could speak of epidemics of fermentation."

As we shall remember, Pasteur had found that wild yeasts become abundant in vineyards and on grapes only at the time of the harvest. Guided by this knowledge, he had succeeded in allowing grapes to ripen without coming into contact with yeast, by covering the vines early in the season with portable glass houses. The spores of the mold Mucor, on the other hand, were present in the vineyard throughout the year and therefore always contaminated the grapes despite the protection of the glass houses. It was from these simple facts that he casually formulated, in the following prophetic words, a statement that reads like a preview of the epidemiological laws of human diseases: "Can we fail to observe that the further we penetrate into the experimental study of germs, the more we perceive unexpected lights and ideas leading to the knowledge of the causes of contagious diseases! Is it not worth noting that in this vineyard of Arbois . . . every particle of soil was capable of inducing alcoholic fermentation, whereas the soil of the greenhouses was inactive in this respect. And why? Because I had taken the precaution of covering this soil with glass at the proper time. The death, if I may use this expression, of a grape berry falling on the ground of any vineyard, is always accompanied by the multiplication on the grape of the yeast cells; on the contrary, this kind of death is impossible in the corner of soil protected by my greenhouses. These few cubic meters of air, these few square meters of soil, were then in the midst of a zone of universal contamination, and yet they escaped it for several months. But what would be the use of the shelter afforded by the greenhouses in the case of the Mucor infection? None whatever! As the yeast cells reach the grape berries only at a certain time of the year, it is possible to protect the latter by means of a shelter placed at the proper time, just as Europe can be protected from cholera and plague by adequate quarantine measures. But the Mucor parasites are always present in the soil of our fields and vineyards, so that grapevines cannot be protected from them by shelters; similarly, the quarantine measures effective against cholera, yellow fever or plague are of no avail against our common contagious diseases."

Thus, Pasteur had become convinced of the role of microorganisms as agents of disease long before he had had any direct contact with animal pathology. One should not conclude, however, that this prescience of the bacteriological era in medicine was an act of pure divination. Often in the preceding centuries, and especially during Pasteur's own time, natural philosophers and physicians had prophesied in more or less confused words that disease was akin to fermentation and to putrefaction, and that minute living agents were responsible for contagion. The story of the slow process by which men arrived at this concept forms a large part of the history of medicine, and cannot be told here. However, mention of a few of the milestones on this long road will help us to identify the intellectual struggles, and the conflict of theories still in evidence during the second part of the nineteenth century, before the triumph of the germ theory of disease.

"Pest," "cholera," "malaria," "influenza" – in all languages these are words that evoke thoughts of terrifying scourges spreading over the land in a mysterious and inexorable manner. It was an awareness of the transmissibility of disease that led many early societies to formulate quarantine measures, in the hope of preventing contact with the sick or the introduction of the causative agents of epidemics.

Physicians had pondered and argued endlessly on the origin and nature of contagion. Even for those who believed that disease was a visitation by deities intent on punishing human sin and corruption, it remained no less a problem to comprehend how – through what mechanism – it could affect so many men in a similar manner at approximately the same time. Many assumed the prevalence during epidemic periods of certain telluric factors residing in the atmosphere, the soil, the waters, and the foods, which rendered most men susceptible to the scourge, as a drought causes the vegetation to wither, or as excessive exposure to the sun causes men to suffer sunstrokes. Ancient medicine was satisfied with this explanation, and codified it in the Hippocratic writings on "Air, Water and Places." Although science will certainly return to the study of the telluric influences postulated by ancient biology, it is in another direction that European medicine turned in its effort to decipher the riddle of epidemic disease.

It was early suspected that men could transmit a contagious principle to each other by direct emanations or bodily contact, or through the intermediary of clothing or of objects used in common. As early as the first century B.C., Varro and Columella had expressed the idea that disease was caused by invisible living things – "*animalia minuta*" – taken into the body with food or breathed in with air.

The epidemic of syphilis which spread through all of Europe in the late fifteenth and early sixteenth century gave many physicians frequent occasions to observe, often in the form of a personal experience, that a given disease can pass from one individual to another. In this case, the mechanism of transmission was sufficiently self-evident to give to the concept of contagion a definite meaning. It is an interesting coincidence that Fracastoro, who coined the name "syphilis" in the sixteenth century, also formulated the first clear statement that communicable diseases were transmitted by a living agent, a *contagium vivum.* In 1546, he described that contagion could occur by direct contact with the sick person, through the intermediary agency of contaminated objects, and through the air *ad distans.* He regarded the agents of disease as living germs, and expressed the opinion that the seeds of these agents could produce the same disease in all individuals whom they reached. These essentially true statements were unconvincing, because they were not based upon a demonstration of the physical reality of the hypothetical organisms, and confirmation of Fracastoro's theories was long delayed, even after the discovery of bacteria.

Very early, analogies came to be recognized between certain disease processes and the phenomena of putrefaction and of fermentation. Just as contagious diseases were alterations of the normal animal economy communicable from one individual to another, similarly, different types of alterations and decay often appeared to spread through organic matter. Indeed, in the making of bread, a small amount of leaven taken from fermented dough could be used to bring about the rising of new dough, an obviously communicable change. Tenuous as these analogies were, they sufficed to induce many physicians and scholars to think and speak of fermentation, putrefaction, and communicable diseases in almost interchangeable terms, an attitude which was felicitously expressed in 1663 by Robert Boyle in his essay

Offering Some Particulars Relating to the Pathological Part of Physick:

> He that thoroughly understands the nature of ferments and fermentations, shall probably be much better able than he that ignores them, to give a fair account of divers *phenomena* of several diseases (as well fevers as others) which will perhaps be never thoroughly understood, without an insight into the doctrine of fermentation.

In fact, the concepts dealing with fermentation and contagious diseases followed a parallel evolution during the two centuries which followed Boyle's statement. In both cases, two opposing doctrines competed for the explanation of the observed phenomena. According to one, the primary motive force — be it of fermentation, putrefaction or disease — resided in the altered body itself, being either self-generated, or induced by some chemical force which set the process in motion. According to the other doctrine, the process was caused by an independent, living agent, foreign in nature and origin to the body undergoing the alteration, and living in it as a parasite. It is the conflict between these doctrines which gives an internal unity to the story of Pasteur's scientific life. He took an active and decisive part in all phases of the conflict, and succeeded in uniting in a single concept those aspects of microbial life that have a bearing on fermentation, putrefaction and contagion. He was aware of the dramatic quality of this achievement and took pride in the fact that science had had to wait two centuries before Robert Boyle's prophecy became fulfilled in Pasteur's person.

Among those who believed that certain minute living agents could pass from one individual to another, and transfer at the same time a state of disease, there were some who postulated that these carriers of contagion might be the small animalcules which the Dutch microscopist Leeuwenhoek had seen in the tartar of his teeth and in the feces of man and animal. Leeuwenhoek's discoveries, made public by his letters to the Royal Society between 1675 and 1685, had aroused much interest and assured him immediate as well as immortal fame. However, they remained without any real influence on medical thinking for almost two centuries. In fact, it is questionable if any of the students of infection before 1850 succeeded in visualizing how a microbial parasite could attack a large animal host and cause injury to it.

Before the theory of contagion could gain acceptance it was essential that the various diseases be separated as well-defined entities. In the seventeenth century, Sydenham in London taught that there were species of diseases just as there were species of plants, and he gave lucid accounts of the differential diagnosis of contagious diseases such as smallpox, dysentery, plague and scarlet fever. However, Sydenham and his followers differentiated diseases only in terms of symptoms and did not attempt to classify them according to causes. With the growth of the knowledge of pathological anatomy, it became possible to base classification on the characteristics of the pathological lesions, and to begin inquiry as to the causes of disease. The French clinician Bretonneau emphasized in the early nineteenth century that "It is the *nature* of morbid causes rather than their *intensity* which explains the differences in the clinical and pathological pictures presented by diseases." Bretonneau thought, furthermore, that specificity of disease was due to specificity of cause, and that each disease "developed under the influence of a contagious principle, capable of reproduction." And he concluded, "Many inflammations are determined by extrinsic material causes, by real living beings come from the outside or at least foreign to the normal state of the organic structure."

Surprisingly enough, Bretonneau did not even suggest that the causative agents capable of reproduction might be the microscopic organisms already so well known in his days, and he did not get beyond formulating a lucid but purely abstract concept of contagion.

It is a remarkable coincidence that both the germ theory of fermentation and the germ theory of disease passed at exactly the same time from the level of abstract concepts to that of doctrines supported by concrete illustrations of their factual validity. Schwann, Cagniard de la Tour and Kützing had recognized simultaneously and independently in 1835–1837 that yeast is a small living plant, and fermentation a direct expression of its living process. In 1836 also Bassi had demonstrated that a fungus (*Botrytis bassiana*) was the primary cause of a disease of silk-worms. Shortly thereafter (1839) Schönlein found in favus the fungus since known as *Achorion schönleinii*, and another fungus *Trichophyton tonsurans* was shown in 1844 to be the cause of the "Teigne tondante" (*herpes tonsurans*). Within a few years, several species of fungi were discovered as parasites of animal tissues and the parasitic role of these microorganisms achieved wide acceptance through the publication in 1853 of Robin's *Histoire naturelle des végétaux parasites.*

A few pathologists then began to reconsider the origin of contagious diseases in the light of the new knowledge. Prominent among them was Jacob Henle, who in his *Pathologische Untersuchungen* formulated the hypothesis that "the material of contagion is not only organic but living, endowed with individual life and standing to the diseased body in the relation of a parasitic organism." It is interesting to note that Henle was the intimate friend of Schwann (who had recognized in 1837 the living nature of yeast) and the teacher of Koch – who, with Pasteur, was to substantiate the germ theory of disease a few decades later. Henle asserted that the demonstration of the causal role of a given microscopic agent in a given disease would require that the agent be found consistently in the pathological condition, that it be isolated in the pure state, and that the disease be reproduced with it alone. Robert Koch was the first to satisfy in the case of a bacterial disease – namely anthrax – all the criteria required by his teacher, and for this reason the rules so clearly formulated by Henle in 1840 are always referred to as "Koch's postulates."

We have selected from the thought patterns of many centuries some of the shrewd guesses which led a few careful observers to formulate a correct statement of the mechanism of contagion, but it is certain that their views were not in line with the generally held theories. While most physicians were willing to grant that certain skin diseases – such as favus, *herpes tonsurans,* thrush and itch – were produced by minute animals or plants, only a few believed that the important diseases – like cholera, diphtheria, scarlet fever, childbirth fever, syphilis, smallpox – could ever be explained in these terms.

The reaction of scientists to the pandemic of cholera that began to spread over Europe in 1846 puts in a clear light the prevailing confusion concerning the origin of epidemic diseases. Cholera was regarded by some as due to a change in the ponderable or imponderable elements of the air. Others regarded it as the result of a vegetable miasma arising from the soil, or of certain changes in the crust of the earth. Some held it was contagious, others that it came from animalcules existant in the air. From Egypt, the scourge had reached Paris, where its victims numbered more than two hundred daily during October 1865; it was feared that the days of 1832 would be repeated – when the death rate reached twenty-three per thousand population. A French commission, consisting of Claude Bernard, Pasteur, and Sainte-Claire Deville, was appointed in 1865 to study the nature of the epidemic; and Pasteur himself has told how the eminent scientists went into the attics of the Lariboisière Hospital, above a cholera ward, in the hope of identifying in the air a poisonous agent responsible for the disease. "We had opened one of the ventilators communicating with the ward and had fitted to the opening a glass tube surrounded by a refrigerating mixture; we drew the air of the ward into our tube, so as to condense into it as many as we could of the air constituents." All this misconceived effort was of course in vain.

The cholera epidemic, however, stimulated one study which appeared in agreement with the doctrine of *contagium vivum.* Having formed the belief that cholera begins with an infection of the alimentary canal, John Snow in London assumed that water might be the vehicle of transmission, and he verified his hypothesis by collecting exact data on a large number of outbreaks and correlating them with water supplies. At first ignored, his views gained ground following the spectacular Broad Street outbreak in 1854 in London. Within two hundred and fifty yards from the spot where the disease began, there were five hundred deaths from cholera in ten days, at the end of which time the survivors took flight and the street was deserted. With unerring exactitude, Snow traced the outbreak to the contamination of the water of a particular pump in Broad Street, and found in this water evidence of contamination with organic matter. Thus, even though John Snow did not deal with the ultimate cause of cholera, he clearly established that the epidemic was water-borne, and made it evident that some agent capable of surviving outside the body was concerned in its causation.

Now that the concept of microbial parasitism has become so familiar – even though so rarely understood – it is difficult to realize why the medical mind remained impervious to the germ theory until late into the nineteenth century. Physicians probably found it difficult to believe that living things as small as bacteria could cause the profound pathological damage and physiological disturbances characteristic of the severe diseases of animals and man. It was fairly easy to invoke parasitism to explain the invasion of the hair follicles by an insect, as in scabies; or of the skin surface by fungi, as in favus or *herpes tonsurans;* but there was something incongruous in a bacterium of microscopic size challenging and attacking a man or a horse. Moreover, it seemed ridiculous to assume that the specificity of the different disease processes could ever be explained in terms of these microbes, all apparently so similar in the simplicity of their shape and functions. Even today, the bacteriologist looks in vain for morphological or chemical characteristics that might explain why typhoid fever, bacillary dysentery, or food poisoning, for example, can be caused by bacterial species in other respects so alike that they can hardly be differentiated one from the other, or even from other bacteria that are not capable of causing disease. The modern physician is indoctrinated in the belief that certain contagious diseases are caused by microorganisms, but there was no reason for the physician of 1860 to have such a faith, which appeared in many respects inconsistent with common sense.

Under the leadership of Virchow in Germany, pathology was then achieving immense progress by recognizing and describing the alterations that different diseases cause in the various types of tissue cells composing the animal body. Every pathological modification was regarded as a physiological transformation, developing in an organ which could not tolerate it, or at another time than the normal one. The secret of the disease appeared, accordingly, to reside in the anatomy of the tissues. Furthermore the idea that there were organisms coming from the exterior which could impress specific modifications upon tissues was in disagreement with the general current of physiological science. A pleiad of illustrious physiologists – Helmholtz, Du Bois-Reymond, Ludwig, Brücke – had taken position against the existence of a vital force, and were attempting to explain all living processes in terms of physicochemical reactions, just as Liebig was doing in the study of fermentations. The idea of the intervention of living microorganisms could not be received with sympathy and understanding in such an intellectual atmosphere. The germ theory of disease faced the same fundamental hostility which had stood in the way of the germ theory of fermentation.[1] According to Liebig, Virchow, and their followers, the similarity between the causation of fermentation and contagious disease had its seat in the intrinsic properties of fermenting fluids or diseased cells, whereas Pasteur took the view that Boyle's prediction could be fulfilled by another unifying concept, namely the germ theory of fermentation and of disease.

There is no doubt that Pasteur's demonstration, between 1857 and 1876, that the "infinitely small" play an "infinitely great role" in the economy of matter prepared the medical mind to recognize that microorganisms can behave as agents of disease. The proof

[1] Many odd arguments were advanced to discredit the evidence derived from bacteriological science in favor of the germ theory. For example it was claimed that tests carried out in rabbits are not convincing, because "the rabbit is a melancholy animal to whom life is a burden and who only asks to leave it." (Quoted by H. D. Kramer in *Bull. Hist. Med.* 22, p. 33, 1948.)

that fermentation and putrefaction were caused by fungi, yeasts and bacteria revealed a number of relationships which had their counterparts in the phenomena of contagion. It established that the effects of microorganisms could be entirely out of proportion to their size and mass and that they exhibited a remarkable specificity, each microbial type being adapted to the performance of a limited set of biochemical reactions. The microorganisms carried out these reactions as a result of their living processes, they increased in number during the course of the reaction, and thus could be transferred endlessly to new media and induce again the alterations over which they presided.

A few physicians who had retained contact with the evolution of natural sciences were struck by the analogies between fermentation and contagion and saw in them a sufficient basis to account for the origin of disease. In 1850, Davaine (who was then assistant to Rayer in Paris) had seen small rods in the blood of animals dead of anthrax, but had failed to comprehend their nature and importance. Pasteur's brief note on butyric fermentation made Davaine realize that microscopic organisms of a dimension similar to that of the rods present in anthrax blood had the power of producing effects entirely out of proportion to their weight and volume. This gave him the faith that the rods of anthrax might well be capable of causing the death of animals, and led him into the investigations which we shall consider in a later part of this chapter.

Pasteur's studies on spontaneous generation had aroused much interest throughout the scientific world, as we have shown; and his demonstration that different types of living germs are widely distributed in the atmosphere gave a concrete basis to the vague view that agents of disease could be transmitted through the air. Pasteur himself had repeatedly emphasized this possible consequence of his findings, but it was the work of Joseph Lister which first established the medical significance of his teaching.

Lister was the son of a London wine merchant who had made distinguished contributions to the development of the modern microscope. Although he was trained in surgery, he developed, probably under his father's influence, a lively and continued interest in bacteriological problems. Long after he had achieved international fame for his work on antiseptic surgery, he contributed theoretical and technical papers of no mean distinction to the science of bacteriology. Lister was a young surgeon in Glasgow when the impact of Pasteur's studies on the distribution of bacteria in the air convinced him of the role of microorganisms in the varied forms of "putric intoxications" which so commonly followed wounds and surgical interventions. Around 1864, he developed the use of antiseptic techniques in surgery with the object of destroying the microorganisms that he assumed to be responsible for the suppurative processes. Lister's methods, at first criticized and ridiculed – particularly in England – were progressively accepted, and became a powerful factor in transferring the germ theory from the experimental domain to the atmosphere of the clinic. In a most generous manner Lister often acknowledged publicly his intellectual debt to Pasteur, for example in the following letter that he wrote to him from Edinburgh in February 1874:

> MY DEAR SIR:
> Allow me to beg your acceptance of a pamphlet, which I sent by the same post, containing an account of some investigations into the subject which you have done so much to elucidate, the germ theory of fermentative changes. I flatter myself that you may read with some interest what I have written on the organisms which you were the first to describe in your *Mémoire sur la fermentation appelée lactique.*
> I do not know whether the records of British Surgery ever meet your eye. If so, you will have seen from time to time notices of the antiseptic system of treatment, which I have been labouring for the last nine years to bring to perfection.
> Allow me to take this opportunity to tender you my most cordial thanks for having, by your brilliant researches, demonstrated to me the truth of the germ theory of putrefaction, and thus furnished me with the principle upon which alone the antiseptic system can be carried out. Should you at any time visit Edinburgh it would, I believe, give you

sincere gratification to see at our hospital how largely mankind is being benefited by your labours.

I need hardly add that it would afford me the highest gratification to show you how greatly surgery is indebted to you.

Forgive the freedom with which a common love of science inspires me, and

Believe me, with profound respect,

Yours very sincerely,

JOSEPH LISTER

Lister again gave generous recognition to Pasteur in the introduction to his classical paper "On the Antiseptic Principle in the Practice of Surgery": "When it had been shown by the researches of Pasteur that the septic property of the atmosphere depended, not on the oxygen or a gaseous constituent, but on minute organisms suspended in it, which owed their energy to their vitality, it occurred to me that decomposition in the injured part might be avoided without excluding the air, by applying as a dressing some material capable of destroying the life of the floating particles."

It is probable, although less certain than was believed by Pasteur, that his studies on the alterations of vinegar, wine and beer had some influence on medical thought. The very use of the word "diseases" (*maladies*) to describe these alterations rendered more obvious the suggestion that microorganisms might also invade human and animal tissues, as they had already been proved to do in the case of silkworms. In opposition to the point of view expressed by the Paris clinician, Michel Peter, "Disease is in us, of us, by us," Pasteur emphasized that contagion and disease could be the expression of the living processes of foreign microbial parasites, introduced from the outside, descending from parents identical to themselves, and incapable of being generated *de novo*. Time and time again he reiterated with pride his belief that the germ theory of fermentation constituted the solid rock on which had been erected the doctrine of contagious diseases. As early as 1877, at the very beginning of his studies on animal pathology, he contemplated writing a book on the subject and described in a few manuscript notes the outline of the argument that he would have developed.

"If I ever wrote a book entitled *Studies on contagious or transmissible diseases,* . . . it could properly begin by a reproduction of my memoir of 1862 (*Mémoire sur les corpuscles organisés qui existent dans l'atmosphère: Examen de la doctrine des générations spontanées*) and of part of the whole of my memoir of 1860 (*Mémoire sur la fermentation alcoolique*), along with notes which would show at each step that this or that passage has suggested this or that memoir, this or that passage of Cohn, Lister, Billroth, etc. . . .

"Do not forget to emphasize in this book that medicine has been carried into the new avenues: 1. By the facts on putrefaction of 1863 (*Examen du rôle attribué au gaz oxygène atmosphérique dans la destruction des matières animales et végétales après la mort. Recherches sur la putréfaction*). 2. By the fact of butyric fermentation by a vibrio, a vibrio living without air; and the observations which I published on this subject . . . should be reproduced (*Animalcules infusoires vivant sans gaz oxygène libre et déterminant des fermentations. Expériences et vues nouvelles sur la nature des fermentations*). 3. By my notes on wine diseases from 1864 on (*Etudes sur les vins. Des altérations spontanées ou maladies des vins, particulièrement dans le Jura*). Diseases of wines and microorganisms! What a stimulus this must have given to the imagination and intelligence; were it only through the connection between these words *maladies* and *microorganisms.*

"Then in 1867, flacherie and its microorganisms. All this resting on facts inassailable, absolute, which have remained in science. . . . Do not forget to point out that, in the preface of my studies on silkworms, there is mention of contagion, of contagious disease. . . ."

By 1875, the association of microorganisms with disease had received fairly wide acceptance in the medical world. Bacteria had been seen in many types of putrid wounds and other infections. Obermeier had demonstrated in Berlin under Virchow's skeptical eyes the constant presence of spiral-like bacteria (spirochetes) in the blood of patients with relapsing fever. But the mere demonstration that bacteria are present during disease was not proof that they were the cause of it. As revealed by the discussions in the Paris Academy of Medicine, there were still physicians who believed that microorganisms could organize themselves *de novo* out of diseased tissue. The belief in spontaneous generation died hard in medical circles. More numerous were those who believed that bacteria, even introduced from the outside, could gain a foothold only after disease had altered the composition and properties of the tissues. For them, bacterial invasion was only an accidental and secondary consequence of disease, which at best might modify and aggravate the symptoms and pathological changes, but could not be a primary cause. As will be remembered, a similar point of view had been held by Liebig, Helmholtz, Schröder and many others with reference to the role of bacteria in putrefaction.

The germ theory of disease was also condemned in the name of plain common sense. Common sense is the expression of two unrelated mental traits; it is based in part on the recognition of an obvious, direct relationship between certain events, uncomplicated by theories. As such it has a pragmatic value and allows its possessor to behave effectively in ordinary situations. The same expression, "common sense," is also used to express beliefs and opinions which are not the result of personal experience, but are only inherited along with the conventions which make up our everyday life. It was because the germ theory was in conflict with these two forms of common sense that its acceptance was so difficult.

The occurrence of contagious disease was known to be often associated with insalubrious living conditions, and the belief had been transmitted from Hippocratic time that the physical environment decided the health of a community. This point of view was expressed forcefully by Florence Nightingale, the woman who, through her experience in military hospitals during the Crimean War and in India, and by virtue of her fighting temperament, did so much to make of nursing an efficient part of medical care:

"I was brought up by scientific men and ignorant women distinctly to believe that smallpox was a thing of which there was once a specimen in the world, which went on propagating itself in a perpetual chain of descent, just as much as that there was a first dog (or first pair of dogs) and that smallpox would not begin itself any more than a new dog would without there having been a parent dog. Since then I have seen with my eyes and smelled with my nose smallpox growing up in first specimens, either in close rooms or in overcrowded wards, where it could not by any possibility have been 'caught' but must have begun. Nay, more, I have seen diseases begin, grow up and pass into one another. Now dogs do not pass into cats. I have seen, for instance, with a little overcrowding, continued fever grow up, and with a little more, typhoid fever, and with a little more, typhus, and all in the same ward or hut. For diseases, as all experiences show, are adjectives, not noun substantives . . .

"The specific disease doctrine is the grand refuge of weak, uncultured, unstable minds, such as now rule in the medical profession. There are no specific diseases: there are specific disease conditions."

Despite the official and popular hostility to the germ theory, several physicians and veterinarians attempted to prove between 1860 and 1876 that bacteria could by themselves initiate disease in a healthy body. Pasteur followed these efforts with eagerness but, we are told by Roux, "they caused him at the same time pleasure and worry. These experiments by physicians often appeared to him defective, their methods inadequate and the proofs without rigor, more likely to compromise the good cause than to serve it. Soon he could no longer help himself and resolutely decided that he too would attack the problem of anthrax."

This was in 1876. Unknown to him, a young German country

doctor, Robert Koch, had embarked on the same venture the year before; and on April 30, 1876, had presented to Ferdinand Cohn, in the Botanical Institute in Breslau, the complete life history of the anthrax bacillus.

Koch, then thirty-three years old, was practicing medicine in Wollstein in Posen. He had studied in Göttingen and Berlin under distinguished scientists, most notable among them being the chemist Wöhler and the pathologist Henle. From the latter, he had learned the difficulties which stood in the way of establishing the germ theory of disease, and the exacting criteria which had to be met to prove the etiological role of a given bacterium. It was within the rigid framework of this experimental and intellectual discipline that Koch placed himself throughout his laborious studies of contagious diseases. Within a few years after working out the life cycle of the anthrax bacillus, he published his work on the *Etiology of Traumatic Infective Diseases* (1878) and achieved immortal fame by isolating the tubercle bacillus in 1882, and the cholera vibrio in 1883. These spectacular achievements, and the development of experimental and diagnostic techniques which are still universally employed today, soon made him the leader in Berlin of a school to which students from all over the world flocked to learn the methods of the new science of medical bacteriology. We shall not follow the meteoric career or describe the stern personality of the great German master, as he crossed Pasteur's path but a few times in the course of his busy and successful life. For the time being, we shall be content with describing how the work of the two founders of medical bacteriology met in the problem of anthrax, to establish once and for all the germ theory of disease.

The story of the work on anthrax prior to Koch and Pasteur illustrates how great discoveries are prepared by the laborious efforts of the "unknown soldiers" of science. Of these forgotten workers some failed to win the final victory because they could not encompass in a single theme all the elements of the struggle, others because they arrived too early on the scene of combat, at a time when the ground had not yet been sufficiently cleared to permit the marshaling of all the forces necessary for victory. But the part they played in commencing to clear the ground is often as important as the more spectacular achievements of those participating in the last phases of the battle.

Rayer and Davaine in 1845 and Pollender in 1855 had seen, in the blood and the spleen of cattle dead of anthrax, large numbers of microscopic, straight, nonmotile rods. Whereas the two French workers had failed to understand the significance of their observation, Pollender had considered the possibility that the rods might be the contagious elements of anthrax but he did not succeed in ruling out the possibility that they were merely products of putrefaction. Further observations by Brauell in Germany appeared to favor the latter interpretation. Brauell had inoculated the blood of animals dead of anthrax into sheep and horses, and had searched for the appearance of the rods seen by Rayer, Davaine and Pollender. Like his predecessors, he had found them in the blood of many inoculated animals, but often, and especially in blood kept for several days, the rods were different in shape from those described before, and furthermore they appeared actively motile. Pollender also observed that, in certain cases, horses injected with anthrax blood would die without showing any rods whatsoever in their blood. It appeared consequently that the rods were not the real cause of the disease, but only one of its accidental consequences.

Two years later, Delafond pointed out that the motile bacteria seen by Brauell were not characteristic of true anthrax; they started to multiply in the blood only after putrefaction had begun to set in, following the death of the animal, precisely at the time when the anthrax rods described by Rayer, Davaine and Pollender began to disappear. Delafond was convinced of the living nature of the anthrax rods. In the hope of proving it, he let blood stand, expecting to see the rods undergo a complete evolution to what he called the seed stage, but detected only a limited increase in length of the rod structure in the course of several days.

After reading Pasteur's work on the butyric vibrio in 1861,

Davaine became confident that the rods which he had seen in 1845 might, after all, be the cause of the disease. By experimental inoculations of animals and thorough microscopic studies, he arrived at a precise knowledge of the relation of true anthrax to the secondary processes of putrefaction. He based his belief that the rods must be the cause of the disease on the fact that they were constantly present during the disease, that the disease could be transmitted by inoculation, and that there was no anthrax in the absence of the rods.

Many workers, however, still questioned the validity of his conclusions and reported, in agreement with Brauell's findings, that animals sometimes died without exhibiting the presence of the typical rods in their blood. Experiments carried out by two French workers, Leplat and Jaillard, are worth discussing in this respect as they stimulated Pasteur's first observations on anthrax. These two workers had inoculated anthrax blood into a large number of rabbits, but had never found any trace of Davaine's rods, notwithstanding the fact that their animals died; they naturally concluded that the rods were merely an epiphenomenon of the disease. In a discussion of their paper, Pasteur agreed with Davaine that the disease induced experimentally by Leplat and Jaillard was not anthrax, and that the cow from which they had obtained their original material had died of another septic disease. To refute this, Leplat and Jaillard obtained blood from an animal which had unquestionably died of anthrax and which contained myriads of immobile rods similar to Davaine's rods. Rabbits inoculated with this blood died without showing any rods and yet their blood could cause death when injected into other rabbits. Thus, once more it appeared as if Davaine's rods were not the real cause of anthrax. Davaine pointed out again that the disease which killed the rabbits in Leplat and Jaillard's tests differed in its clinical course and pathological characteristics from true anthrax. Pasteur agreed with him after recognizing in the original blood used by Leplat and Jaillard putrefactive bacteria and others similar to the butyric ferment, instead of Davaine's rods.

The physiologist Paul Bert was one of those who long remained unconvinced even after Koch's work. Believing that all living agents could be killed by adequate pressures of oxygen, he exposed the blood of an animal dead of anthrax to the action of compressed oxygen, in order to kill any living form that it might contain. Yet, inoculation of this blood produced disease and death, without the reappearance of bacteria. Therefore, Bert concluded, the bacteria were neither the cause nor the necessary effect of anthrax.

It was in the midst of this confusion that Koch's classical paper appeared, describing in the most exquisite and complete details the life history of the anthrax bacillus and its relation to the disease.

Koch had frequent occasion to observe anthrax in farm animals in the course of his medical duties. Working in a primitive laboratory that he built in his own home, he established the fact that the disease was transmissible from mouse to mouse and produced typical and reproducible lesions in each member of the successive series of mice. He had also the original idea of placing minute particles of spleens freshly removed from infected animals in drops of sterile blood serum or of aqueous humor, and he began to watch, hour after hour, what took place. His technique was simplicity itself, his apparatus homemade. After twenty hours he saw the anthrax rods grow into long filaments, especially at the edge of the cover glass; and, as he watched, he saw round and oval granular bodies appear in the filaments. He realized that they were spores, similar to those described by Ferdinand Cohn in other bacteria; and he recognized that his cultures underwent a cycle including every stage, from Davaine's motionless rod to the fully formed spore. He determined the optimal thermal conditions for spore formation and saw that the spores could again grow into typical anthrax rods. Recognizing that the spores were highly resistant to injurious influences, he grasped at once the significance of this property for the maintenance and spread of infection. He learned to differentiate true anthrax from the septicemic disease which had confused the observation of Brauell,

Leplat and Jaillard. He further established that the hay bacillus (*Bacillus subtilis,* commonly found in hay infusion), an organism very similar to Davaine's rod, and like it capable of producing spores, did not cause anthrax when injected into animals. From all these facts he finally concluded that true anthrax was always induced by only one specific kind of bacillus and he formulated on the basis of this conclusion a number of prophylactic measures aimed at preventing the spread of the disease.

One of Koch's experiments was of particular interest in proving the etiological role of Davaine's rods. He had sown fragments of infected tissues into drops of serum or of aqueous humor of the rabbit, and had allowed this primitive culture to incubate until the bacilli had multiplied to large numbers; then, from this first culture, he had inoculated a new drop of serum. After repeating the process eight times he found to his great satisfaction that the last culture injected into a susceptible healthy mouse was as capable of producing anthrax as blood taken directly from an animal just dead of the disease. Despite their thoroughness and elegance, these experiments still left a loophole for those who believed that there was in the blood something besides the rods, capable of inducing anthrax. Although Koch had transferred his cultures eight times in succession, this was not sufficient to rule out the possibility that some hypothetical component of the blood had been carried over from the original drop and was responsible, instead of the bacteria, for transmitting the infection to the inoculated animal. It was this last debatable point that Pasteur's experiments were designed to settle.

Pasteur knew from his earlier studies on spontaneous generation that the blood of a healthy animal, taken aseptically during life, and added to any kind of nutrient fluid, would not putrefy or give rise to any living microorganism. He felt confident, therefore, that the blood of an anthrax animal handled with aseptic precautions should give cultures containing only the anthrax bacillus. Experiment soon showed this to be the case, and showed also that rapid and abundant growth of the bacillus could be obtained by cultivating it in neutral urine; these cultures could be readily maintained through countless generations by transfers in the same medium. By adding one drop of blood to fifty cubic centimeters (nearly two ounces) of sterile urine, then, after incubation and multiplication of the bacilli, transferring one drop of this culture into a new flask containing fifty cubic centimeters of urine, and repeating this process one hundred times in succession, Pasteur arrived at a culture in which the dilution of the original blood was so great — of the order of 1 part in 100^{100} — that not even one molecule of it was left in the final material. Only the bacteria could escape the dilution, because they continued to multiply with each transfer. And yet, a drop of the hundredth culture killed a guinea pig or a rabbit as rapidly as a drop of the original infected blood, thus demonstrating that the "virulence principle" rested in the bacterium, or was produced by it.

Pasteur devised many other ingenious experiments to secure additional evidence of the etiological role of the anthrax bacillus. He filtered cultures through membranes fine enough to hold back the bacteria and showed that the clear filtrate injected into a rabbit did not make it sick. He allowed flasks of culture to rest undisturbed in places of low and constant temperature, until the bacteria had settled to the bottom; again the clear supernatant fluid was found incapable of establishing the disease in experimental animals, whereas a drop of the deposit, containing the bacterial bodies, killed them with anthrax. These results constituted the strongest possible evidence that the anthrax bacillus itself was responsible for the infection. However, Pasteur took care to point out that there still remained a possibility which had not been explored, namely that the bacilli produced a virus which remained associated with them throughout the culture, and which was the active infective agent. But even this hypothesis did not change the conclusion that the bacilli were living and were the cause of anthrax. The germ theory of disease was now firmly established.

* * *

As soon as it became possible to grow the anthrax bacillus in pure culture, to identify it and to establish with it a reproducible disease in experimental animals, the way was open to elucidate many riddles that had baffled students of the problem during the preceding decades.

Leplat and Jaillard had shown that rabbits inoculated with putrid anthrax blood died quickly without showing rods in their blood. Although Davaine had claimed that the disease so induced differed from true anthrax in length of the incubation period and in many other ways, he had not been able to prove his point. He clearly realized that the most convincing exposition of Leplat and Jaillard's error would be to discover the cause of the disease they had produced, but in this he had failed completely. It was in the blood that Davaine had originally seen the anthrax bacilli, and it was in the blood that he searched obstinately, and in vain, for the cause of the new disease. On the contrary, Pasteur, less bound by tradition and more resourceful as an investigator, soon discovered that Leplat and Jaillard's disease was associated with another type of bacillus present in immense numbers in many tissues but absent or rare in the blood. This bacillus has remained famous in all languages under the name of *vibrion septique* that Pasteur gave it. It was probably the *vibrion septique* that had been responsible for Paul Bert's results. The blood which he had treated with oxygen to kill the rods probably contained the *vibrion septique*, in the resistant spore stage, and thus was capable of causing a special disease in the inoculated animal.

Pasteur found the new organism to be very common in nature, often present as a normal inhabitant of the intestinal canal, where it is harmless until certain circumstances allow it to pass through the intestinal barrier and into other organs. It invades the blood shortly after death and, as the disease which it causes has an extremely rapid course, it often kills animals infected with old anthrax blood before the anthrax bacillus itself has a chance to multiply. Thus were explained all the earlier observations in which animals receiving post-mortem blood from cases of anthrax died without exhibiting a trace of the rods originally seen by Davaine.

Pasteur undertook a physiological study of the *vibrion septique* and he recognized with surprise and pleasure that, like the bacillus of butyric acid fermentation discovered sixteen years before, the new organism was an obligate anaerobe which could be cultivated only in the absence of air. With the assurance that long practice had given him, he derived from this fact important conclusions concerning the physiology of the *vibrion septique.* "It is a ferment, and . . . forms carbon dioxide, hydrogen, and a small amount of hydrogen sulfide which imparts an odor to the mixture. . . . When a post-mortem examination is made on an animal which has died of septicemia, we find tympanites, gas pockets in the cellular tissue of the groin or of the axilla, and frothy bubbles in the fluid which flows when an opening is made in the body. The animal exhales a characteristic odor toward the end of its life. Its parasites, driven out perhaps by this production of hydrogen sulfide, leave the skin, to take refuge at the extremity of its hairs. In short, septicemia may be termed a putrefaction of the living organism."

Except in a few prejudiced minds these studies on the anthrax bacillus and on the *vibrion septique* established the germ theory of disease, once and for all. The difficulties involved in separating the anthrax bacillus from the *vibrion septique,* and in disentangling the two distinct diseases that they cause, must be measured by taking into account the lack of previous experience and the paucity of experimental techniques then available. This historical perspective helps in appreciating the intensity of the travail which preceded the birth of the new theory, and the convincing character of Koch's and Pasteur's achievements.

The causative agents of most other bacterial diseases were discovered and described within two decades after Koch's and Pasteur's studies. It is worth remarking that this triumph, which looms so great in the history of medicine and has such import for the welfare of mankind, was to a large extent achieved through men of very ordinary talent. The harvest comes abundantly, and often without much effort, to those who follow the pioneers. Discovery of the causative agent and of the means of control of a given contagious disease constitutes, in reality, the last stage in centuries of unrecorded labor. It was necessary first for the different diseases to be recognized and separated — a process which required of the early clinicians prolonged observation and much judgment. Then came the intellectual struggle of naturalists and chemists, who conceived the idea of the existence of microorganisms and of the huge potential activities of those tiny forms. Natural philosophers, physicians, and epidemiologists had to have the creative imagination to foresee that the contagious behavior of certain diseases could some day be explained in terms of the activity of minute living agents, traveling from patient to patient through many different channels. Then the experimenters had to establish that "spontaneous generation" does not occur, at least in the ordinary events of life; and that microorganisms do come from parents similar to themselves. Finally before microbiology could become a science, it had to be shown that microbial species are well-defined biological entities, and that each microorganism exhibits dependable specificity in its action, be it as a ferment or as an agent of disease.

In deciding which events were epoch-making in the development of the germ theory, medical scientists and historians of medicine focus attention on those achievements that have the most obvious bearing on the life of man. Thus, the isolation of the tubercle bacillus in 1882 is almost universally regarded as the high point in the unfolding of the science of medical bacteriology; but it is the importance of tuberculosis for man, even more than the distinction of the scientific discovery of its cause, which determines this judgment. The earlier elucidation of the cause of anthrax, a disease of less importance than tuberculosis for man, is usually discussed in less enthusiastic terms. Bassi's demonstration as early as 1836 that a microscopic fungus was the agent of one of the silkworm diseases is dismissed in the form of a few statements. And no student of the history of infection ever mentions the early work done in the field of plant diseases, although all the great debates which presided at the birth of the germ theory of animal diseases were foreshadowed by the discussions on the causation of the epidemics of plant crops.

Early in the nineteenth century, the fungus *Claviceps purpurea* was found to be the cause of ergot, a disease which blackens and elongates the kernels of rye in wet seasons.

In 1813, Knight stated before the Horticultural Society in London that "Rust or mildew . . . of wheat originates in a minute species of parasitical fungus which is propagated like other plants by seeds."

When, in 1845, the potato blight broke out on a disastrous scale in Europe, and in 1846 particularly in Ireland, it brought in its train sufferings and economic upheavals greater than those caused by most human diseases. Potato blight presents a special interest for the historian of the germ theory as many of the debates which enlivened animal pathology during the second half of the nineteenth century had their counterpart a few decades earlier in the establishment of the causation of this disease of plants.

Weather had been very unpleasant shortly before the blight broke out. For several weeks, the atmosphere had been one of continued gloom, with a succession of chilling rains and fog, the sun scarcely ever visible, the temperature several degrees below the average for the previous nineteen yars. The botanist, Dr. Lindley, held the theory that bad weather had caused the potato plants to become saturated with water. They had grown rapidly during the good weather; then when the fogs and the rain came, they absorbed moisture with avidity. As absence of sunshine had checked transpiration, the plants had been unable to get rid of their excess of water and in consequence had contracted a kind of dropsy. According to Lindley, putrefaction was the result of this physiological disease. The Reverend Berkeley, "A gentleman eminent above all other naturalists of the United Kingdom in his knowledge of the habits of fungi," held a different theory and connected the potato disease with the prevalence of a species of mold on the affected tissues. To this, Lindley replied that Berkeley was attaching too much importance to a little growth of mold on the diseased potato plants. He added furthermore that "as soon as living matter lost its force, as soon as diminishing vitality took the place of the customary vigour, all sorts of parasites would acquire power and contend for its destruction. It was so with all plants, and all animals, even man himself. First came feebleness, next incipient decay, then sprang up myriads of creatures whose life could only be maintained by the decomposing bodies of their neighbours. Cold and wet, acting upon the potato when it was enervated by excessive and sudden growth, would cause a rapid diminution of vitality; portions would die and decay, and so prepare the field in which mouldiness could establish itself." [2]

Thus, the professional plant pathologists, represented by the learned Dr. Lindley, believed that the fungus (*Botrytis infestans*) could become established on the potato plant only after the latter had been debilitated by unhealthy conditions, whereas the Reverend Berkeley, while not ignoring the influence of bad weather, saw the fungus as the primary cause of the disease with fog and rain as circumstances which only favored its spread and growth. In this manner the controversies which were to bring Pasteur face to face with the official world of the French Academy of Medicine were rehearsed in the *Gardener's Chronicle,* over the dead body of a potato invaded by the fungus *Botrytis infestans.* This happened thirty years before the beginnings of bacteriological times as recorded in medical histories.

The two decades which followed the work of Koch and Pasteur on anthrax, and which saw the discovery of so many agents of disease, have been called the golden era of bacteriology. But they were in reality only an era of exploitation during which a host of competent but often uninspired workers applied to the problems of contagion the techniques and intellectual approach which had reached maturity in the persons of Pasteur and Koch, after two centuries of scientific efforts.

* * *

[2] Quoted by E. D. Large in *The Advance of the Fungi.* (London: Jonathan Cape, 1940.)

Most of the bacterial agents of disease were discovered by the German school of bacteriology. This was due in part to the mastery by Koch and his disciples of the techniques used in the isolation and identification of microbial cultures. Even more important was the fact that, under the dominating influence of Pasteur, the French school, numerically smaller and far less completely organized, became chiefly concerned with another aspect of the study of infectious disease, namely the problem of immunity. One should not assume, however, that Pasteur had become indifferent to the etiological problems of infection. Although he did not pursue systematically the isolation of pathogenic agents, he contributed to this field many observations that reveal his pioneering mentality.

The causative agents of anthrax, chicken cholera and swine erysipelas were, until 1884, the microorganisms most extensively used by Pasteur for his investigations on immunity. In addition, he found time in the midst of all his studies to visit hospital wards and morgues, where he would arrive accompanied by Roux and Chamberland, carrying culture flasks and sterile pipettes. With precaution that appeared meaningless even to many enlightened physicians of the time, he would take samples of pathological material for microscopical and bacteriological study.

"Childbirth" or "puerperal" fever was then causing immense numbers of deaths in maternity wards. Despite the visionary teachings of Semmelweis in Vienna, and of Oliver Wendell Holmes in Boston, physicians did not regard the disease as contagious, but rather explained it in terms of some mysterious metabolic disorder. Pasteur had observed in the uterus, in the peritoneal cavity and in blood clots of diseased women a microorganism occurring "in rounded granules arranged in the form of chains or string of beads," and he became convinced that it was the most frequent cause of infection among women in confinement.

In March 1879 there took place in the Paris Academy of Medicine a discussion on the cause of puerperal fever. One of the academicians, Hervieux, had engaged in an eloquent discourse, during which he spoke in sneering terms of the role of microorganisms in disease; as we have shown, in 1879 the germ theory was not yet universally accepted in medical circles. Hervieux had contrasted the true "miasm or puerperal fever" with "those microorganisms which are widely distributed in nature, and which, after all, appear fairly inoffensive, since we constantly live in their midst without being thereby disturbed."

Irritated by the vague reference to the "puerperal miasm," Pasteur interrupted the speaker from his place in the audience and retorted with vigor: "The cause of the epidemic is nothing of the kind! It is the doctor and his staff who carry the microbe from a sick woman to a healthy woman!" And when Hervieux retorted that he was convinced that no one would ever find this microbe, Pasteur darted to the blackboard replying, "There it is," and he drew the organisms "shaped like strings of beads" which are now so well known under the name of "streptococcus."

Every occasion was for Pasteur a pretext for microscopic study. Uninformed as he was of medical problems, he had the genius to make observations and establish correlations which, unorthodox at the time, have been vindicated by subsequent developments. Illustrative of his keen judgment of the role of microorganisms in the pathogenesis of disease is the case of the relation of staphylococcus to bone infections. The story has been told by his assistant Duclaux. "I was then suffering from a series of boils. The first thing that Pasteur did when I showed him one of them was to prick it, or rather have it pricked, for he was not fond of operating himself, and to take therefrom a drop of blood in order to make a culture, an undertaking in which he was successful. A second boil gave the same result, and thus the staphylococcus was discovered, so well known since that time. He found the same microbe, made up of little agglomerated granules, in the pus of an infectious osteomyelitis which M. Lannelongue had submitted to him for examination. With a fine audacity, he declared immediately that osteomyelitis and boils are two forms of one and the same disease, and that the osteomyelitis . . . is the boil of the bone. What could be bolder than to liken a grave disease taking place in the depths of the tissues to a superficial abscess generally trifling! To confound internal and external pathology! When he launched this opinion before the Academy of Medicine, I picture to myself the physicians and surgeons present at the meeting, staring at him over their spectacles with surprise and uneasiness. Nevertheless, he was right, and this assertion, daring at the time, was a first victory of the laboratory over the clinic."

In these examples of his mode of attack on medical problems, we see Pasteur applying the methods which were then rapidly becoming standard practice in the bacteriological laboratories of Europe. Of greater interest to illustrate the pioneering and adventurous quality of his genius are the discoveries that he made toward the end of his scientific life, in attempting to bring rabies within the scope of the germ theory of disease.

Rabies was then known as a disease contracted by man or a few species of large animals from the bite of rabid dogs or wolves. In the hope of discovering the causative microorganism Pasteur collected saliva from an infected child and injected it into a rabbit. In agreement with his expectations, he produced a fatal disease, readily transmissible from rabbit to rabbit. For a short time, he held the belief that he had discovered the cause of rabies and described the organism he had isolated in words which show his skill and care in reporting those morphological characteristics that he considered of special interest. "It is an extremely short rod, somewhat constricted in its center, in other words shaped like an 8 . . . Each one of these small microorganisms is surrounded, as can be detected by proper focusing of the microscope, with a sort of aureola that really seems to belong to it . . . it appears that the aureola consists of a mucous substance . . ." The trained bacteriologist will have no difficulty in recognizing the pneumococcus in this accurate description.

Pasteur soon realized, however, that the microorganisms with an aureola isolated from the saliva of the rabid child could also be found in the saliva of normal individuals, and was often absent in other persons suffering from rabies. Moreover, the disease which it caused in rabbits was different from true rabies. It was not the microorganism that he had been looking for, and he turned immediately to other techniques for the solution of his problem.

Bacteriological studies — which must have been very disheartening — failed to reveal the cause of rabies. Attempts were made to cultivate a microorganism in spinal fluid, and even in fresh nerve substance obtained from normal animals, but all in vain. This failure is not surprising, for it is now known that rabies is caused by a filterable virus, which cannot be seen by ordinary microscopy, and which has not yet been cultivated in lifeless bacteriological media. With an uncommon and truly admirable intellectual agility, Pasteur then gave up the in vitro cultural techniques, to the development of which he had contributed so much. Heretofore, he had emphasized the necessity of discovering for each type of microorganism the nutrient medium most selectively adapted to its cultivation. He now conceived the idea of using the susceptible tissues of experimental animals, instead of sterile nutrient solutions, to cultivate the virus of the disease; the concept of selectivity of cultural conditions was thus simply carried over from lifeless media to receptive living cells.

The general symptoms of rabies suggested that the nervous system was attacked during the disease. Indeed, there had been published experiments showing that the infective matter of rabies was present not only in the saliva, but also in the nerve substance of mad animals. On the other hand, infected nerve tissue inserted under the skin of an animal was known to be able to induce rabies. Unfortunately this method of transmission was as uncertain and capricious as transmission through the saliva; rabies did not always appear, and when it did, it was often after a prolonged incubation of several months. Inoculation under the skin, therefore, was ill adapted to the designing of convincing experiments. Someone in the laboratory (probably Roux) suggested depositing the virus in the nerve centers; the proof of its presence and development would then be the appearance of rabies in the inoculated animal. Nerve tissue seemed to be an ideal medium for the virus of rabies, and to fulfill naturally for it the condition of selectivity which was the foundation of the cultural method. As the main problem was to gain access to this tissue under aseptic conditions, the surest way was to attempt to inoculate dogs under the dura mater, by trephining. Roux, who took a leading part in this phase of the work, has left the following account of the circumstances under which the operation was introduced in Pasteur's laboratory: "Ordinarily an experiment once conceived and talked over was carried out without delay. This one, on which we counted so much, was not begun immediately, for Pasteur felt a veritable repugnance toward vivisection. He was present without too much squeamishness at simple operations, such as a subcutaneous inoculation, and yet, if the animal cried a little, he immediately felt pity and lavished on the victim consolation and encouragement which would have been comical had it not been touching. The thought that the skull of a dog was to be perforated was disagreeable to him; he desired intensely that the experiment be made, but he dreaded to see it undertaken. I performed it one day in his absence; the next day, when I told him that the intracranial inoculation presented no difficulty, he was moved with pity for the dog: 'Poor beast. Its brain is certainly badly wounded. It must be paralyzed.' Without replying, I went below to look for the animal and had him brought into the laboratory. Pasteur did not love dogs; but when he saw this one full of life, curiously ferreting about everywhere, he showed the greatest satisfaction and straightaway lavished upon him the kindest words. He felt an infinite liking for this dog which had endured trephining without complaint and had thus relieved him of scruples concerning the operation."

The dog inoculated by trephination developed rabies in fourteen days and all the dogs treated in the same fashion behaved in a similar manner. Now that the cultivation of the virus in the animal body was possible the work could progress at a rapid pace, as in the case of anthrax, fowl cholera and swine erysipelas.

Thus was discovered a technique for the cultivation of an unknown infectious agent in the receptive tissues of a susceptible animal. This technique has permitted the study of those agents of disease which are not cultivable in lifeless media, and has brought them within the fold of the germ theory of disease. The Henle-Koch postulates in their original form could not be applied to the study of filtrable viruses and it is one of the most telling examples of Pasteur's genius that he did not hesitate to free himself of their requirements as soon as they proved unadapted to the solution of his problem. For him, doctrines and techniques were tools to be used only as long as they lent themselves to the formulation and performance of meaningful experiments.

The demonstration that invisible viruses could be handled almost as readily as cultivable bacteria was a great technical feat, and its theoretical and practical consequences have been immense. Even more impressive, perhaps, is the spectacle of Pasteur, then almost sixty years of age and semiparalyzed, attacking with undiminished vigor and energy a technical problem for which his previous experience had not prepared him. Throughout his life the concept of selectivity of chemical and biological reactions had served him as the master key to open the doors through which were revealed many of nature's secrets. From the separation of left- and right-handed crystals of tartaric acid by selective procedures or agents, through the cultivation of yeast and of lactic, acetic and butyric bacteria in chemically defined media, to the differentiation of the anthrax bacillus from the *vibrion septique* by cultivation in vitro, and by infection of experimental animals, he had in the course of twenty-five years applied the concept of selectivity to many different situations. The propagation of the rabies virus in receptive nervous tissue demonstrated that, if used with imagination, the same concept was applicable to still other biological problems. In his hands, the experimental method was not a set of recipes, but a living philosophy adaptable to the ever-changing circumstances of natural phenomena.

CHAPTER X

Mechanisms of Contagion and Disease

> So, naturalists observe, a flea
> Has smaller fleas that on him prey;
> And these have smaller still to bite 'em;
> And so proceed *ad infinitum.*
>
> —SWIFT

THE DEMONSTRATION that microbial agents can be the primary cause of disease left unanswered most of the questions relevant to the mechanisms by which contagion spreads from one individual to another, and by which it manifests itself in the form of characteristic symptoms and pathological alterations. Countless species of microorganisms swarm in the air that we breathe, the foods and fluids that we ingest, the objects that we touch. And yet, few of them become established and multiply in the bodies of plants, animals and man; still fewer are able to cause disease. How do the disease-producing species differ from their innocuous relatives? What weapons do they possess which give them the power to inflict on the invaded host injury more or less profound, symptoms more or less distressing? Why do so many individuals, in plant and animal as well as in human populations, remain apparently unaffected in the course of an epidemic, although they are as fully exposed as their stricken brothers? The course of epidemics is sometimes predictable, more often capricious, never explicable in the simple terms of the mere presence or absence of the causative microorganism. Whence do epidemics originate? What factors determine their growth and their decline, both in space and in time? Why do they subside, as mysteriously and often as abruptly as they began?

These questions may appear abstract and even meaningless to the citizen of a well-policed community, living in a state of peace and economic well-being. However, they have frightening significance for those populations which are victims of the upheavals of war or of social disasters; three decades ago, they were made part of universal consciousness by the impact of the influenza epidemic in 1918–1919.

Before the twentieth century, the riddle of contagion was ever present in the mind of man; the threat of infection and of epidemics introduced a constant element of mystery and terror in the life of the individual and of society. For example, cholera – a disease practically unknown today in the Western world – made several devastating incursions in Europe during Pasteur's lifetime.

For the modern man, cholera is a disease of the East. It suggests the Mohammedan pilgrimages to and from Mecca, and the polluted rivers of Hindustan. Often traveling as a silent member of the caravans, it may suddenly become raging, decimate its fellow travelers, then again quiet down and become, as before, unnoticeable. In the villages as well as in the crowded cities of Asia, along the rivers and caravan trails, cholera appears unexpectedly in a few isolated victims, spreads rapidly through the communities, reaches its maximum in a few weeks, killing half of the persons whom it strikes, then declines to a few sporadic cases before disappearing as mysteriously as it came, for an unpredictable length of time. So much terror and so much mystery in a name!

There had been well-identified cases of cholera in Europe before the nineteenth century, in Nîmes in 1654, in London in 1669 and 1676, in Vienna in 1786. However, the epidemics which occurred after 1817 differed markedly in their "dispersiveness" from these isolated outbreaks. In the wet season of May 1817, cholera appeared in the northern provinces of Bengal, this time affecting not only the untouchables, but other castes and even Europeans. It spread ultimately over almost all India and from there began a world-wide dissemination that reached Russian cities after six years. Another wave of cholera, also coming from India, struck

Russia in 1830 and England in October 1831. Thus, the first pandemic occurred in two parts, one from 1817 to 1823, when the borders of Europe were reached, the other from 1826 to 1838, when the disease spread over most of Western Europe. The end came in 1838 unexpectedly; no one knew the cause of the epidemic or the reason for its sudden termination.

The second pandemic, like the first, came from India. It began around 1840, reached Europe in 1847, remained there some twelve years and again petered out mysteriously. It was during this pandemic that John Snow established the connection of the disease with contaminated water, tracing the London outbreak of 1854 to a particular well in Broad Street. He ascertained that a cesspool drained into this well, and that there had been a case of cholera in the house served by the cesspool.

A third pandemic became manifest in Europe around 1865; after a very irregular course, during which Russia, Austria and Germany were affected, it came to an end in 1875. Despite Snow's discovery, the cause of the disease was still mysterious, as is illustrated by the episode reported on page 241, during which Pasteur, Claude Bernard and Sainte-Claire Deville made their futile attempts at analyzing the gases of the air in the cholera wards of Paris.

The fourth pandemic began in 1881. Traveling with the Moslem pilgrims to Mecca, it reached Egypt in 1883 and Southern France, Italy and Spain in 1884; in this last country alone the disease caused 57,000 deaths out of 160,000 cases during 1885. In 1883 a French mission including Pasteur's assistants, Roux, Nocard, and Thuillier, and a German mission under the leadership of the great Robert Koch himself arrived in Egypt to investigate the cause of the disease. The epidemic, however, had almost run its course by the time the two commissions began their studies. One of the last sporadic cases was that of Thuillier, who by a cruel irony of fate died in Alexandria of the most violent form of cholera. The French mission returned home while the German group proceeded to Calcutta, where the epidemic was still raging. There, in December 1883, Koch isolated the cholera bacillus. The microbial cause of the disease was thus identified, but the mystery of its capricious course was not solved thereby. Cholera once more invaded Europe shortly after Koch's discovery, causing 8605 deaths out of 17,000 cases during an outbreak in Hamburg from August to November 1892. An epidemic of large proportions also occurred in Russia in 1907–1919.

The many theories that have been evolved to account for the explosive and erratic behavior of cholera epidemics are of interest only in illuminating the length to which imagination will go in order to satisfy the human urge for explaining natural events, even when the essential knowledge is still lacking. It is generally held at the present time that man constitutes the reservoir from which new infections are initiated, a view based upon the fact — first recognized by Koch — that certain individuals who do not show any symptoms of the disease can carry the cholera bacilli and transmit them to susceptible persons. The bacilli harbored by these apparently healthy "carriers" are assumed to be capable of initiating widespread epidemics when other conditions still shrouded in mystery are satisfied. To many hygienists and students of public health, it is precisely the understanding of this "epidemic climate" which constitutes the real problem, the riddle of contagion, and they regard this unsolved problem as more important than the mere discovery of some new bacillus in explaining the spread of infection. This attitude is symbolized by the picturesque career of the great German hygienist, Max von Pettenkofer, who opposed a "soil theory" of cholera to the purely bacteriological theory of his rival Robert Koch.

Like Pasteur, Pettenkofer had been trained as a chemist. At first very active in the field of analytical chemistry, then of medicinal chemistry, he had become more and more concerned with the applications of chemical knowledge to physiology and pathology, and more especially to hygiene and public health. Nevertheless, he retained for a long time a lively interest in all aspects of chemistry. For example in 1863, precisely during the period when Pasteur was lecturing to the students of the Paris School of Fine Arts on the chemical basis of oil painting, Pettenkofer undertook a study of the oils, pigments and varnish used by the old masters, in order to determine the cause of the alarming alterations which were then threatening the paintings in the galleries of Munich. However, it is as the high priest of hygiene that Pettenkofer won the admiration and love of the world, and particularly of his fellow citizens in Munich. He regarded hygiene as an all-embracing philosophy of life, concerned not only with an abundant supply of clean water and air, but also with trees and flowers because they contributed to the well-being of men by satisfying their aesthetic longings. Although microorganisms played only a small part in his philosophy of public health, he persuaded the Munich city fathers to have clean water brought in abundance from the mountains to all houses, and to have the city sewage diluted downstream in the Isar, all in the name of salubrity and aesthetic hygiene. With these steps, the great cleaning-up of Munich began. The mortality of typhoid fell from 72 per 1,000,000 in 1880 to 14 in 1898. Munich thus became one of the healthiest of European cities, thanks to the efforts of this energetic and public-minded citizen – who was entirely unimpressed by the germ theory of disease.

As years went on, Pettenkofer devoted more and more attention to the epidemiology of cholera, and taught that certain changes in the soil were of primary importance in establishing an "epidemic climate." While admitting that the disease had a certain specific cause, a *materies morbi,* he emphasized the importance of local, seasonal, and individual conditions which had to be satisfied before the infection could occur. After 1883, he admitted that the Koch bacillus was the specific cause of cholera but retained his conviction that the new discovery had not solved the problem, and that the bacillus alone could not produce the disease.

So convinced was he that he resolved to prove his thesis by ingesting cholera bacilli. He obtained a culture freshly isolated from a fatal case of the epidemic then raging in Hamburg and, on October 7, 1892, swallowed a large amount of it on an empty stomach, the acidity of which had been neutralized by drinking an adequate quantity of sodium carbonate; these were the very conditions stated by Koch as most favorable for the establishment of the disease. The number of bacilli ingested by Pettenkofer was immensely greater than that taken under normal conditions of exposure, and yet no symptoms resulted except a "light diarrhea," although an enormous proliferation of the bacilli could be detected in the stools. Shortly thereafter, several of Pettenkofer's followers, including Emmerich and Metchnikoff, both of whom were soon to become important investigators in bacteriological science, repeated the experiment on themselves with the same result. None of them doubted that the bacillus discovered by Koch was the cause of cholera. The experiment merely demonstrated that infectious diseases and epidemics are complex phenomena involving, in addition to the infective microorganisms, the physiological state of the patient, the climate and the environment, the social structure of the community, and countless other unsuspected factors. The implantation of bacilli, like the planting of seed, does not necessarily insure a growth.

Pasteur had come into contact with the complexities of the problem of infection during his studies on silkworms. He knew that the discovery of the microbial agent of a disease was only one link in the solution of the riddle and that, as a mere isolated fact, it was of little use to the physician, and of limited intellectual interest. These reasons probably played an important part in his decision to devote his energy to the elucidation of the mechanisms by which microorganisms can cause disease and by which they are carried from one individual to another. This type of preoccupation may account for the subjects which he selected for his studies. Anthrax, chicken cholera, swine erysipelas, were not the most important or most dramatic subjects from the viewpoint of man's immediate interests, but they lent themselves to experimentation better than human diseases.

Pasteur's first papers on animal pathology are replete with observations and theories which indicate that, intellectually, he was chiefly concerned with the mechanism of the reactions between parasite and infected host. As he was beginning to work on this problem, however, his experiments on chicken cholera unexpectedly revealed the possibility of vaccinating against infectious diseases. Immediately, the prospect of this development took precedence over his other scientific interests and from then on he directed all his experimental work to the problem of vaccination. Twenty years earlier, he had abandoned theoretical studies on the mechanisms of fermentation to deal with the practical problems involved in the manufacture of wine and vinegar; he was young then, and life permitted him to come back after 1871 to the speculations of his early years, and to formulate in more definite terms the nature of the fermentation process which he had envisioned in 1861. When he discovered vaccination, in 1882, he was sixty years old, and had only six years of active work left before disease struck him again. The labors and struggles of this last phase of his scientific life never gave him the opportunity of returning to the epidemiological problems and to the mechanisms of toxemia, which appear as sketchy visionary statements in the notes that he published between 1877 and 1882. Historians of bacteriology have neglected this aspect of Pasteur's work. It is very probable, however, that, had not circumstances channeled his efforts into the dazzling problems of vaccination, the study of the physiological and biochemical aspects of infection might have yielded results which now remain for coming generations to harvest.

The shepherds of the Beauce country had noticed that sheep allowed to graze on certain pastures were likely to contract anthrax, even after the fields had been abandoned for years, as if a curse had been placed on them. In the center of France certain "dangerous mountains" were also known to farmers as being unfit to pasture their animals for the same reason. The existence of these anthrax fields had been the source of many objections to the view that the rods occurring in the blood of sick animals were the cause of the disease. Why should the spread of the contagion from one animal to the other be limited to certain pastures, to these "accursed fields" or "dangerous mountains"? If the contagion were due to the transmission of the anthrax rods by direct contact between animals, or through the air, or by the intermediary of flies, as Davaine believed, why should these agencies of transmission be restricted by the hedges or stone walls enclosing particular fields? Davaine had no answer to these embarrassing questions.

The discovery by Koch that anthrax bacilli produce resting forms, the spores, which can survive for prolonged periods of time without losing their ability to produce disease, made it likely that these spores might often behave as agents of transmission. Pasteur was prepared for this theory by his earlier experience in the silkworm nurseries. There, he had seen the spores of the flacherie germs survive at least a year and germinate again in the nurseries the following spring. Turning his attention to anthrax, he first established in the laboratory that sheep fed on fodder artificially contaminated with anthrax spores developed the symptoms and lesions of the natural disease; then he set himself the task of elucidating how animals became infected under field conditions.

It was first necessary to determine whether anthrax spores really existed in the "accursed fields." In the Beauce country, the shepherds were in the habit of burying the animals right in the fields where they had died. Although one could reasonably assume that the anthrax rods or their spores were present for a time in the pit, it was not an easy task to demonstrate that they actually survived in the soil, where everything else undergoes decomposition. The demonstration was achieved by suspending suspected soil in water, letting it settle, collecting the fine particles and then heating them at 80° C. to kill any vegetative bacterial forms. When the heated material was injected into guinea pigs, several of them died of anthrax, thus demonstrating that the spores were capable of surviving in the soil. In fact, living spores could be found in soil near pits in which animals had been buried twelve years before.

How could sheep come into contact with the spores buried in

the ground with the dead animals? The spores have no motility and cannot by themselves reach either the surface of the soil or the plants ingested by the animals. Pasteur guessed that earthworms might bring spores up from the lower layers, and he proved his hypothesis by collecting worms from the soil over a burial pit containing infective remains. Moreover, he noticed an interesting correlation between the geology of a region and the prevalence there of anthrax in farm animals; for example, the disease was unknown where the topsoil was thin and sandy, or in the chalky soil of the Champagne country where the conditions are not favorable for the life of earthworms.

Another puzzling fact found its ready explanation in terms of the germ theory. It had long been known that mortality was the highest where animals grazed on the fields after the harvest of cereal crops. Pasteur recognized that the dried stubble and chaff left standing in the fields often inflicted upon the animals superficial wounds which, unimportant by themselves, gave the anthrax spores a chance to become established in the body and to initiate infection. He was very proud of this discovery and often referred to it. Although of small practical importance, it was probably for him a symbol of all the subtle factors, usually undetected, which control the manifestations of the germ theory of disease and conceal its operations to the unprepared mind.

These facts, so obvious once they had been recognized, led to the formulation of simple rules for the prophylaxis of anthrax. Pasteur never tired of advising the farmers not to abandon the dead animals in the pastures, but to destroy them by burning or by burying them in special grounds where sheep and cattle would not be allowed to graze.

Most of the early investigations on anthrax were carried out in the farms of the Beauce country for Pasteur, so often accused by his medical opponents of being merely a "laboratory scientist," was always ready to move into the field when the work demanded it. Roux has described the intimate contact between the laboratory and the farms during the anthrax campaign:

"For several years in succession, at the end of July, the laboratory of the Rue d'Ulm was abandoned for Chartres. Chamberland and I settled there with a young veterinarian, M. Vinsot. . . . Pasteur came every week to give the directives for the work. What pleasant memories we have kept of the campaign against anthrax in the Chartres country! Early in the morning, there were visits to the flocks of sheep scattered over all the vast Beauce plateau glimmering under the August sun; post-mortem examinations were performed at the slaughterhouse of Sours, or in the farmyards. The afternoon was devoted to bringing the notebooks up to date, writing to Pasteur, and getting ready for the new experiments. The days were rich in activity, and how interesting and healthy was this bacteriology in the open air!

"On the days when Pasteur came to Chartres, the lunch at the Hôtel de France did not last long, and we immediately proceeded by carriage to Saint-Germain, where M. Maunoury had placed his farm and his herds at our disposal. We discussed during the trip the experiments of the preceding week and the new ones to be undertaken. Upon arrival, Pasteur would hasten to the sheep parks. Motionless near the gates, he would observe the experimental animals with that sustained attention from which nothing could escape; for hours in succession, he would keep his gaze fastened on a sheep that he thought diseased. We had to remind him of the hour and show him that the spires of the Chartres cathedral were beginning to fade into the night before he could make up his mind to leave. He would question farmers and helpers, and listened in particular to the opinions of shepherds who, on account of their solitary life, devote all their attention to the herds and often become acute observers.

"No fact appeared insignificant to Pasteur; he knew how to draw the most unexpected leads from the smallest detail. The original idea of the role of earthworms in the dissemination of anthrax was thus born one day when we were walking through a field in the farm of Saint-Germain. The harvest was in and there remained only the stubble. Pasteur's attention was drawn to a part of the field where the earth was of different color; M. Maunoury explained that sheep dead of anthrax had been buried there the preceding year. Pasteur, who always observed things at close range, noticed at the surface a multitude of those small casts of soil such as are ejected by earthworms. He then conceived the idea that in their endless trips from the lower levels, the worms bring up the anthrax spores present in the earth, rich in humus, that surrounds the cadavers. Pasteur never stopped at ideas, but immediately proceeded to the experiment. . . . The earth extracted from the intestine of one of the worms, injected into guinea pigs, forthwith gave them anthrax."

It was during the same period that Darwin on his country estate also became interested in earthworms and came to look upon them as the silent but effective toilers of the soil, a view which he developed in his book *Formation of Vegetable Mould, through the Action of Worms, with Observations on Their Habits.* There is an atmosphere of idyllic and pastoral poetry in the picture of these two scientist-philosophers, the one combating disease and the other formulating the concept of evolution, both discovering natural laws while observing earthworms in the shadows of Gothic cathedrals.

Pasteur began in 1878 the study of chicken cholera, a disease that despite its name bears no relation to human cholera. The course of chicken cholera differs profoundly from that of anthrax. When an epidemic attacks a barnyard, it spreads through it with extreme rapidity, killing most of the birds within a few days. Normal chickens injected with pure cultures of the chicken cholera bacillus always die within forty-eight hours, often in less than twenty-four. The mere feeding of contaminated food or excrements is sufficient to establish a disease with a course almost as rapidly fatal. Rabbits are equally susceptible, and, like chickens, uniformly contract the infection when exposed to the chicken cholera bacillus.

In contrast with chickens and rabbits, adult guinea pigs exhibit a peculiar resistance to the infection. These animals develop an abcess which remains localized and which may persist for prolonged periods of time before it opens and heals spontaneously, without at any time disturbing the general health and appetite of the animal. This slow and retrogressive course of the infection in the guinea pig is not due to a change in the virulence of the bacillus, for chickens and rabbits die of the acute form of the disease when inoculated with a minute amount of the abcess material. Pasteur immediately saw the implications of these facts for the problem of epidemiology:

"Chickens or rabbits living in contact with a guinea pig suffering from such abcesses might suddenly become sick without any apparent change in the health of the guinea pig itself. It would be sufficient that the abcesses open and spread some of their contents onto the food of the chickens or rabbits. Anyone observing these facts and ignorant of the relationship that I have just described would be astounded to see the chickens and rabbits decimated without any apparent cause, and might conclude that the disease is spontaneous. . . . How many mysteries pertaining to contagion might some day be explained in such simple terms!"

Three years later, while studying swine erysipelas in the South of France, Pasteur made an observation which revealed that this epidemic disease was caused by a microorganism pathogenic not only for swine, but also for other animal life.

"Shortly after our arrival in the Vaucluse, in November 1882, we were struck by the fact that the raising of rabbits and pigeons was much neglected in this district because these two species were, at frequent intervals, subject to destructive epidemics. Although no one had thought of connecting this fact with swine erysipelas . . . experiments soon showed that rabbits and pigeons died of a disease caused by the erysipelas microorganism."

Thus, it became obvious that one animal species could serve as a reservoir of infection for another species, or even for man. The subsequent development of epidemiology was to provide many examples of the fact that wild and domesticated animals can act as natural reservoirs of certain infectious agents: the part played by rodents in the dissemination of plague and typhus, by rabbits in the infection of man with tularemia, by monkeys in the maintenance of yellow fever in the South American jungle, by domesticated birds in the spread of parrot fever, by the vampire bat in the transmission of rabies to man and to large animals, are examples which illustrate the importance of the animal reservoir problem in the transmission of disease.

The fact that certain ostensibly healthy individuals harbor infective microorganisms is also of great importance in maintaining a constant source of infection. After recovery from a disease, mild or severe, men or animals often continue to carry the causative agent and can transmit it to susceptible individuals. Chicken cholera revealed to Pasteur the existence of this "carrier" state. He observed that a few birds now and then resisted the epidemic and survived for prolonged periods, constantly releasing the virulent bacilli in their excreta. Moreover, certain chickens which appeared extremely resistant, and did not exhibit any general symptoms of disease, showed on the surface of their body a persistent abscess containing large numbers of virulent bacilli. Like the guinea pigs mentioned above, these birds were carriers of the infective agent and they constituted a constant reservoir of infection.

There is overwhelming evidence that the "carrier state" is of paramount importance in determining the initiation of new outbreaks. The notorious "typhoid Mary" was a cook who, throughout her life, remained a carrier of typhoid bacilli and unwittingly brought about outbreaks of the disease among those with whom she associated. Carriers of diphtheria bacilli, of virulent streptococci, and of many other infectious agents are a constant source of danger in exposed communities, and of concern for the public health officer. As mentioned earlier in this chapter, the carrier state doubtless plays an important part in the initiation of epidemics of Asiatic cholera, and Pasteur's prophetic observations on the animal reservoirs of swine erysipelas and of chicken cholera provide a pattern according to which many obscure facts of epidemiology find at least a partial explanation.

The germ theory of fermentation and of disease was based on a belief in the specificity and permanence of the biological and chemical characteristics of microbial species. Under the influence of Cohn and Koch, this concept of specificity became a rigid doctrine; each microorganism was claimed to be unchangeable in its form and properties, and to remain identical with its precursors under all circumstances. Pasteur first recognized that this concept had to be somewhat modified during the course of his studies on fermentation. As will be remembered, he had observed that the mold *Mucor mucedo,* which grew in a filamentous form in the presence of air, became yeastlike and behaved as an "alcoholic ferment" under anaerobic conditions. The mold returned immediately to its original morphology and physiological behavior as soon as adequate aeration was again provided, so this sort of change was a readily reversible process.

The study of the chicken cholera bacillus revealed another type of transformation, more profound because more lasting, which was so definitely in conflict with the dogma of the fixity of microbial species that it must have been at first very disconcerting and the source of great worry. Pasteur found that cultures of chicken cholera could lose their ability to produce disease and that, moreover, they retained this modified or "attenuated" character through subsequent generations. Thus, the chicken cholera bacillus could be virulent, or not, while the other characteristics by which it was ordinarily identified remained unchanged. Shortly thereafter, Pasteur also observed a similar transformation (loss of virulence) in the causative agents of anthrax, swine erysipelas, lobar pneumonia, and rabies. Since then, this phenomenon has been observed with practically all microbial agents of disease. Virulence is not a constant and permanent attribute of certain microbial species, but a variable property which can be lost, and then again recovered, sometimes at the will of the experimenter.

As soon as he became convinced of the validity of his observations, Pasteur dismissed from his mind the rigid views he had held concerning the fixity of biological behavior of microorganisms, and immediately turned his attention to the consequences that this change of virulence might imply for the problem of infection. We shall describe in a following chapter the use which he made of attenuated cultures to vaccinate against infectious diseases. Let us consider at the present time the significance of the alterations in virulence in the study of epidemiological problems.

The loss of virulence of the chicken cholera bacillus had been discovered by a chance observation. With great skill Pasteur worked out empirical techniques for deriving from several virulent microorganisms, modified forms which had more or less completely lost the ability to cause disease. Of equal interest was the discovery that attenuated cultures could be restored to maximum virulence by "passing" them through certain animals. For example, a culture of chicken cholera which had lost its virulence for chickens was found to be still capable of killing sparrows and other small birds and, when passed repeatedly from sparrow to sparrow, finally regained its virulence for adult chickens. He obtained fully attenuated anthrax bacilli innocuous for adult guinea pigs but still capable of killing the newly born. When these bacilli were passed from the newly born to two-day-old animals, then from those to three-day-old, and so on, the culture progressively regained its full virulence and soon became capable of killing adult guinea pigs and sheep. Strange as these results appear, they serve to illustrate how much effort and ingenuity Pasteur was willing to expend in establishing experimentally the phenomenon of the instability of virulence.

Even more remarkable was the discovery that, in certain cases, the virulence could be changed not only quantitatively, but also qualitatively. Thus the pneumococcus first isolated from human saliva was very virulent for the rabbit, and only slightly so for the adult guinea pig; and yet it could be rendered less virulent for the former animal and much more so for the latter, merely by passing it through newly born guinea pigs. The results obtained with the microorganism of swine erysipelas were also very striking. When the bacillus recovered from a hog was inoculated into the breast of a pigeon, the bird died in six to eight days; by inoculating the blood of this first pigeon into a second, then from the second to a third, and so on, the virulence increased progressively for the pigeon and at the same time for the hog. If, however, the bacillus was inoculated into a rabbit and then passed from rabbit to rabbit, its virulence increased for the rabbit but at the same time decreased for the hog, to such an extent that ultimately the microorganism became unable to cause disease in the very animal host from which it had been isolated originally.

Pasteur believed that these phenomena of variation were of great importance in the epidemiology of various infectious diseases. He suggested that epidemics might arise from the increase in virulence of a given microorganism, and also in certain cases from its ability to acquire virulence for a new animal species:

"Thus, virulence appears in a new light which may be disturbing for the future of humanity, unless nature, in its long evolution, has already experienced the occasions to produce all possible contagious diseases — a very unlikely assumption.

"What is a microorganism that is innocuous for man, or for this or that animal species? It is a living being which does not possess the capacity to multiply in our body or in the body of that animal. But nothing proves that if the same microorganism should chance to come into contact with some other of the thousands of animal species in the Creation it might not invade it, and render it sick. Its virulence might increase by repeated passages through that species, and might eventually adapt it to man or domesticated animals. Thus might be brought about a new virulence and new contagions. I am much inclined to believe that such mechanisms explain how smallpox, syphilis, plague, yellow fever, *et cetera* have come about in the course of the ages, and how certain great epidemics appear from time to time."

Symbiosis and parasitism are two apparently opposing manifestations of interrelationships between living beings. In symbiosis, two organisms establish a partnership which is of mutual benefit; in lichens, for example, two microscopic organisms — an alga and a fungus — live in association, the former synthesizing the chlorophyll, which absorbs from the sun the energy required for the assimilation of carbon dioxide in the air, the latter microorganism extracting water, minerals, and perhaps certain essential organic substances from the soil or from the plant which it uses for support. There are also many examples of symbiosis between microorganisms and higher plants or animals. Orchids require the presence of a fungus for the germination of their seeds; in legume plants, the nodules which occur on the roots are growths of bacteria which borrow water, minerals and carbon compounds from the plant and supply the latter in return with nitrogen derived from the air. In parasitism, by contrast with symbiosis, one of the members of the association exploits the other without contributing anything useful to its welfare.

The distinction between symbiosis and parasitism is not always well defined, nor perhaps constant. It appears possible that, in the general order of natural events, symbiotic relationships are now and then upset, with the result that one of the partners takes exclusive advantage of the association and becomes a true parasite; on the other hand, parasitism may be the first step of natural relationships, and may slowly evolve into that co-operative association which we call symbiosis or partnership. If the evolution from parasitism to symbiosis is the general trend in nature, optimism is then justified, and only patience is required to see man become man's helpful partner. If, on the contrary, parasites have evolved from once helpful partnerships, it demands much faith to believe that man will reverse the order of nature or that the ancient saying *Homo homini lupus*[1] will ever become obsolete.

Whatever its origin, parasitism implies that the parasite must find in its "host" conditions favorable for growth: adequate food, proper temperature and other essential living requirements. On the whole, bacteriologists have paid little heed to these physiological aspects of the problem of infection, despite the fact that infectious disease is clearly an example of parasitism. It is of special interest, therefore, to find that Pasteur attempted to analyze in biochemical terms the mechanistic basis of the parasitic behavior of microbial agents.

[1] "Man is man's worst enemy."

As will be remembered, he had grown yeast and certain bacteria in nutrient fluids of known chemical composition; under his inspiration Raulin had defined, with great detail, the nutritional requirements of the fungus *Aspergillus niger,* thus providing the pattern according to which the nutrition of other microorganisms could be studied. Somewhat later, Pasteur had recognized that many of the microbial agents of disease had more complex requirements and grew well only when supplied with certain types of organic substances. However useful, this information was not sufficient to throw light on the nutritional conditions required by pathogenic agents for multiplication in the animal body; and, in fact, this problem is still unsolved today despite much increased knowledge. Pasteur, nevertheless, bravely attempted to apply nutritional concepts to the phenomena of parasitism, and he considered the possibility that immunity might result from the exhaustion in the host of some component essential to the growth of the pathogen. He even imagined that cancers might consist of altered tissue cells competing successfully with normal cells for the nutritive elements brought by the blood, and suggested means of treatment based on this theory. Naïve as these views were, they deserve respect as the first statement of the problem of the nutritional relationship between parasite and invaded host.

Pasteur's preoccupation with the influence of body temperature on microbial multiplication came to light in his famous controversy with Colin concerning the susceptibility of chickens to anthrax. Colin, a professor at the Veterinary School of Alfort, had acquired a certain notoriety by constantly opposing Pasteur's views before the Academy of Medicine. In a slow, monotonous and sour voice, he would endlessly reiterate his doubts concerning the validity of the evidence against spontaneous generation, for the role of microorganisms in putrefaction, on the etiology of anthrax. Pasteur having stated that birds, and notably hens, could not contract anthrax, Colin had hastened to say that nothing was easier than to give this disease to hens. This was in July 1877. Pasteur, who had just sent Colin a culture of the anthrax bacillus, begged that he would bring him in exchange a hen suffering from that disease, very likely with the malicious hope of exposing some technical error on the part of his opponent. The story of this episode was told to the Academy of Medicine in March 1878.

"At the end of the week, I saw M. Colin coming to my laboratory, and even before I shook hands with him, I said, 'Why, you have not brought me that diseased hen!' . . . 'Trust me,' answered M. Colin, 'you shall have it next week.' . . . I left for vacation; on my return, and at the first meeting of the Academy which I attended, I went to M. Colin and said, 'Well, where is my dying hen?' 'I have only just begun experimenting again,' said M. Colin; 'in a few days I shall bring you a hen suffering from anthrax.' . . . Days and weeks went by, with fresh insistence on my part and new promises from M. Colin. One day, about two months ago, M. Colin acknowledged that he had been mistaken, and that it was impossible to give anthrax to a hen. 'Well, my dear colleague,' I told him, 'I will show you that it *is* possible to give anthrax to hens; I shall myself, one day, bring to you at Alfort a hen which shall die of the disease.'

"I have told the Academy this story of the hen which M. Colin had promised in order to show that our colleague's contradiction of our findings on anthrax had never been very serious."

In reply, Colin stated before the Academy: "I regret that I have not been able as yet to hand to M. Pasteur a hen dying or dead of anthrax. The two that I had bought for that purpose were inoculated several times with very active blood, but neither of them fell ill. Perhaps the experiment might have succeeded later, but, one fine day, a greedy dog prevented that by eating up the two birds, whose cage had probably been badly closed."

On the Tuesday following this incident, Pasteur emerged from the Ecole Normale, carrying a cage containing three hens, one of which was dead, and drove to the Academy of Medicine. After having deposited his unexpected load on the desk, he announced that the dead hen had been inoculated with anthrax two days before – at twelve o'clock on Sunday, with five drops of culture of the anthrax bacillus – and had died on Monday at five o'clock, twenty-nine hours after the inoculation. This result was the outcome of an original experiment. Puzzled by the fact that the hens were refractory to anthrax, he had wondered whether this resistance might not be due to the body temperature of the birds, known to be higher than that of animals susceptible to the disease. To test this idea, hens were inoculated with anthrax and then placed in a cold bath in order to lower their temperature. Animals so treated died the next day with their blood, spleen, lungs, and liver filled with bacilli. The white hen which lay dead on the floor of the cage was evidence to the success of the experiment. To show that it was not the prolonged bath which had killed it, a speckled hen had been placed in the same bath, at the same temperature and for the same time, but without infection; this bird was in the cage on the desk, extremely lively. The third hen, a black one, had been inoculated at the same time as the white hen, with the same culture, using ten drops of culture instead of five, to make the experiment more convincing; but it had not been subjected to the bath treatment and had remained in perfect health.

A fourth experiment was carried out later to establish whether a hen, infected with anthrax and allowed to contract the disease by being placed in a cold bath, would recover if allowed to reestablish its ordinary body temperature by being removed from the bath early enough. A hen was taken, inoculated and cooled in a bath, until it was obvious that the disease was in full progress. It was then taken out of the water, dried, wrapped in cotton wool and placed at a temperature sufficient to allow rapid restoration of normal body temperature. To Pasteur's great satisfaction, the hen made a complete recovery. Thus, the mere fall of temperature from 42° C. (the normal temperature of hens) to 38° C. was sufficient to render birds almost as receptive to infection as rabbits or guinea pigs.

Unconvinced by this experiment, or moved by his antagonism to Pasteur, Colin suggested on July 9, 1878, that the dead hen which had been laid on the desk of the Academy in the preceding March meeting might not, after all, have died of anthrax.

As had Liebig and Pouchet in earlier years, Colin thus opened himself to the riposte. Pasteur immediately extended to him the challenge of submitting their differences to a commission of the Academy, with the understanding that Colin himself would perform the post-mortem and microscopic examination of the dead bird. Pasteur's experiments were repeated on July 20, and naturally yielded the results that he had forecast. Colin ungraciously signed the commissioner's statement that hens inoculated with a culture of anthrax, then cooled in a water bath, died with a large number of anthrax bacilli in their blood and tissues.

Despite the apparent simplicity of the experiment, the effect of temperature on the susceptibility of chickens to anthrax is certainly a more complex phenomenon than Pasteur assumed it to be. True enough, the cooling of chickens by immersion in cold water brought their body temperature down to a level compatible with the growth of the anthrax bacillus, but at the same time it probably interfered with the performance of normal physiological mechanisms, thus increasing the susceptibility of the animals to infection. The results, nevertheless, were of interest as being the first experimental demonstration that environmental factors influence the course of infection, and that the presence in the body of a pathogenic agent is not necessarily synonymous with disease.

A few months later, Pasteur discussed before the Academy of Medicine another example of the influence of physiological factors on the behavior of microbial parasites. This new example, even more convincing to his audience because it had a more direct bearing on human infections, concerned the relation of oxygen to the role of the *vibrion septique* as an agent of disease. In contrast with anthrax bacilli, the vegetative cells of the *vibrion septique* cannot live in the presence of oxygen and are actually killed by it; only the spores survive aeration. The *vibrion septique* is widely distributed in nature, normally present in the intestinal tract of some animals, often present also in soil. In the intestinal canal it is protected from the toxic effect of air by the presence of immense numbers of other bacteria which are capable of utilizing the last trace of oxygen, but it has no chance to multiply in normal tissues, or on a clean wound open to air. How, then, can it become established in tissues and cause disease? According to Pasteur, this happens when the conditions in the wound or in the tissues are such that they limit the access of air, or when large numbers of other bacteria exhaust the oxygen from it. Then, the *vibrion septique* finds a favorable environment and produces its deadly toxin in contact with the susceptible tissues.

"Let a single clot of blood, or a single fragment of dead flesh, lodge in a corner of the wound sheltered from the oxygen of the air, where it remains surrounded by carbon dioxide . . . and immediately the septic germs will give rise, in less than twenty-four hours, to an infinite number of vibrios multiplying by fission and capable of causing, in a very short time, a mortal septicemia."

It is common experience that insects or worms can attach themselves to man, animals or plants, deriving thereby food and maintenance and causing at the same time annoyance and irritation, often injury and sometimes death. The first contagious disease shown to be caused by a minute parasite was probably itch (scabies), in which a barely visible anthropod insect (*Sarcoptes scabiei*) burrows a microscopic tunnel into the human epidermis. In this case, the meaning of the term "parasite" appears obvious and its application to disease justified. Physicians and experimenters found little difficulty in extending the concept of parasitism from the attack by insects to infections caused by fungi. In the ergot disease of rye, the *mal del segno* of silkworms, the favus and *herpes tonsurans* of man, one could imagine that the disease was due to some direct injury inflicted by the fungus parasite on the superficial tissues of the victim. In the case of bacterial diseases, however, it became much harder to form a concrete picture of the parasitic relationship. How could such microscopic beings, detectable in the body fluids and tissues only by the most exacting microscopic study, do damage to the powerful and well-organized body structures of man and animal? What weapons could they use to inflict injury profound enough to express itself in disease and death? Pasteur offered some tentative and preliminary answers to these questions in his early papers on contagion, but unfortunately he was prevented from developing them further by the pressure of his subsequent studies on vaccination.

He regarded disease as a physiological conflict between the microorganism and the invaded tissue. According to him, for example, the anthrax bacilli compete for oxygen with the red blood corpuscles and cause them to suffer a partial asphyxia; the dark color of the blood and of the tissues, which is one of the most characteristic signs of anthrax at the time of death, would thus be an expression of oxygen deficiency. In the case of chicken cholera, he assumed that "the microbe causes the severity of the disease and brings about death through its own nutritional requirements. . . . The animal dies as a result of the deep physiological disorders caused by the multiplication of the parasite in its body."

Pasteur established that disease-producing microorganisms can also cause symptoms and death by secreting soluble poisons. He passed the blood of an animal infected with anthrax through a plaster filter in order to remove the anthrax bacilli from it. When added to fresh normal blood this filtrate brought about an immediate agglutination of the red cells similar to that which occurs in the animal body during the course of the natural infection. This was the first indication that physiological disturbances can be caused by the products of bacterial growth, even in the absence of the living microorganisms themselves.

Even more convincing was the demonstration that the causative agent of chicken cholera produces a soluble toxin. One of the most striking symptoms of this disease is the appearance of somnolence in the birds before death.

"The diseased animal is strengthless, tottering, with drooping wings. The feathers of the body are raised and give it the form of a ball. An invincible somnolence overcomes the animal. If one compels it to open its eyes, it behaves as if coming out of a deep slumber and soon closes its eyelids again; usually death comes after a silent agony without the animal having moved at all. At most, it will beat its wings for a few seconds."

A culture grown in chicken broth, filtered so as to free it of living germs, is unable to cause true chicken cholera. However, injection under the skin of large amounts of this filtrate reproduces in the bird many of the symptoms of the natural disease. "The chicken . . . takes the shape of a ball, becomes motionless, refuses to eat and exhibits a profound tendency to sleep similar to what is observed in the disease produced by the injection of the living microbe itself. The only difference consists in the fact that sleep is lighter than in the real disease; the chicken wakes up at the slightest noise. This somnolence lasts approximately four hours; then the chicken again becomes alert, raises its head, and clucks as if nothing had happened. . . . Thus, I have acquired the conviction that during the life of the parasite there is produced a narcotic which is responsible for the symptom of sleepiness characteristic of chicken cholera."

Although Pasteur was inclined to believe that death was caused by the multiplication of the microorganisms in the body of the fowl, and not by the effect of the soluble toxin, he concluded his remarkable observation by the following words: "I shall attempt to isolate the narcotic, to determine whether it can produce death when injected in sufficient dose, and to see whether, in this eventuality, death would be accompanied by the pathological lesions characteristic of the natural disease."

This sentence could have heralded a new phase in Pasteur's scientific life. He had struggled hard to prove that contagious diseases were caused by living microorganisms. As soon as this fact had been established, he had asked himself the next question. Through what mechanism do these living agents cause disease? This query had brought him back to the analysis of disease in terms of chemical reactions. He had postulated that the life of the infectious agent interfered with the biochemical processes of the infected animal; he had demonstrated the production of soluble toxins, and had planned to "isolate" them as chemical substances. Prosecution of these two aspects of the problem would have led him into the most profound questions pertaining to the pathogenesis of infectious disease, questions which to a large extent remain unanswered today.

Unfortunately, time was running out and there were other pressing problems to be solved. The memoir in which Pasteur had described the soluble toxin of chicken cholera was devoted chiefly to a discussion of immunity against the disease, and to the possibility of vaccinating against it. This problem monopolized the energy of his remaining years, and he never came back to those visionary concepts which had thrown the first light on the mechanism of the physiological interrelationships between living things.

CHAPTER XI

Medicine, Public Health and the Germ Theory

> False facts are highly injurious to the progress of science, for they often endure long; but false views, if supported by some evidence, do little harm, for everyone takes a salutary pleasure in proving their falseness.
>
> — DARWIN

THE germ theory of disease constitutes one of the most important milestones in the evolution of medicine. It dispelled some of the mystery and much of the terror of contagion; it facilitated and rendered more precise the diagnosis of disease; it provided a rational basis for the development of prophylactic and therapeutic procedures. These great achievements should not lead one to assume that progress in the control of infectious disease dates from the bacteriological era. In reality, many of the most devastating scourges have been conquered without the benefit of laboratory research, and some have even disappeared without any conscious effort on the part of man.

In the course of recorded history, overwhelming epidemics have arrested invading armies on the march, decimated populations, disorganized the social fabric, changed the pattern of civilizations — but mankind has survived. Life has proved flexible enough to triumph over yellow fever, influenza, typhus, plague, cholera, syphilis, malaria, even when there were available no effective measures to combat disease. Less dramatic, but fully as astonishing as the spontaneous and often sudden termination of the great epidemics, is the continuous downward trend of certain diseases in the course of centuries.

Leprosy was once universally prevalent and remains today a widespread and destructive disease in many parts of the world. Witness to its importance in Biblical times are the precise laws in Leviticus regulating the behavior of the lepers, and of society toward them. The Hebrew belief in the contagiousness of leprosy survived in medieval times and led the Church to pronounce the leper dead to the world, leaving him only the consolation of immortality. By a symbolic ritual, the "unclean" was ordered to keep away from his fellow men, a measure which probably contributed to minimizing the spread of contagion. Homes of mercy were established all over Christendom to care for the lepers as well as to isolate them, these "lazarettos" having been precursors of our hospitals. Perhaps as a result of this segregation, probably also because of a general improvement in the standards of living, leprosy has been on the decline throughout Europe since the Renaissance, and is now practically nonexistent in our communities. There are still many "uncleans" among us, but it might be toward the control of syphilis and gonorrhea that the teaching of Leviticus would be directed today.

Not so long ago, tuberculosis was the Great White Plague, the "captain of all men of death" for the white race. In Boston, New York, Philadelphia and Charleston — in London, Paris and Berlin — all available statistics reveal tuberculosis mortality rates of 500 or higher per 100,000 inhabitants in the year 1850. Some time around 1860, the number of deaths from the disease began to decrease in Europe and North America, and it has continued to decrease ever since except for brief interruptions in the downward trend, interruptions associated with two world wars. In 1947, tuberculosis mortality rates were below 40 per 100,000 population in several countries and were still decreasing. Thus, in many places, the toll of deaths due to tuberculosis had decreased more than tenfold in less than a century, a spectacular event that has excited endless discussions among students of public health. The decrease began before the discovery of the tubercle bacillus, long before there were available any specific methods of prevention or cure. Western civilization is slowly, but perhaps not surely, gaining in its fight over tuberculosis, without being too certain of the circumstances to which it owes its success. For equally mysterious reasons scarlet fever is on the wane; fifty years ago a frequent cause of death, it is today a relatively mild disease. Syphilis spread through Europe like a prairie fire during the fifteenth and sixteenth centuries. Its course was then rapid, and often fatal—unlike that of the frequently mild and slowly progressive disease of our days. In this case, again, Western civilization took the disease in stride and learned to live with it. Indeed, the contagion may have made European culture burn with a brighter light, if it be true, as claimed by certain medical writers, that a correlation exists between syphilis and genius.

Only feeble hypotheses have been offered to account for the fact that society has gained the upper hand over certain diseases without knowing anything as to their cause or mode of transfer. It is usually assumed that better nutrition, housing and sanitation, as well as other improvements in the general standard of living, have played a large part in the conquest of leprosy and tuberculosis. This view certainly contains much truth, but there is also an element of human conceit in attributing the disappearance of certain diseases only to technological improvements. Many factors affect the course of epidemic cycles and some of them are beyond human control for the time being. As a population, the rats of Bombay have become resistant to the plague bacillus which has been present among them for many centuries, whereas the rats of New York, Paris and London are still susceptible to it; perhaps, like twentieth-century man, the Bombay rat prides himself on the achievements of his civilization in having overcome rat plague.

The conquest of malaria provides the most convincing evidence that material civilization can wipe out certain infectious diseases unaided by microbiological or other medical sciences. The Campagna Romana was free of malaria as long as Roman hearts and muscles were robust enough to drain its marshes. Two centuries ago, malaria was rampant in the Ohio valley, and pioneers suffered or died of it while clearing the green forests. It has virtually disappeared today, because malaria always recedes before a vigorous agrarian society. Its disappearance from our Middle West is only an accidental by-product of the clearing of the land, and not the result of a planned campaign. Extensive farming rendered the country unsuitable for the mosquitoes that transmit malaria, and the disease became extinct, just as did many of the forms of wild life native to the primitive forests.

All leaders of wandering men, of roaming tribes or conquering armies, have had to become sanitarians to prevent the spread of the diseases of filth. Moses enacted strict sanitary regulations for camp life before he could lead his people across the desert; the wandering Jew codified in the Old Testament many of the precepts which are still essential to the control of crowd diseases. As stated by the historian Garrison: "The ancient Hebrews were, in fact, the founders of prophylaxis, and the high priests were true medical police."

Although plague long constituted the major menace to European life, it had almost disappeared from Western Europe by the nineteenth century. Cholera and typhoid fever became the outstanding diseases associated with filth, while typhus also remained prevalent, especially in jails. Through the efforts of public-minded citizens, most of them not physicians, around 1850 society slowly began to take an active interest in a more salubrious life — clearing slums, eliminating filth, providing fresh air and abundant, clean water. Edwin Chadwick first sold to England the "sanitary idea," the concept that it is possible by controlling the social environment to suppress the forces of disease instead of accepting them as an inevitable fate. The extreme degree of filth with which the reformers had to cope can be imagined from the following account left by an observer who visited the tenements of Glasgow in 1855. "We entered a dirty low passage like a house door, which led from the street through the first house to a square court immediately behind, which court, with the exception of a narrow path around it leading to another long passage through a second house, was used entirely as a dung receptacle of the most disgusting kind. Beyond this court, the second passage led to a second square court, occupied in the same way by its dunghill; and from this court there was yet a third passage leading to a third court and third dung heap. There were no privies or drains there, and the dung heaps received all filth which the swarm of wretched inhabitants could give; and we learned that a considerable part of the rent of the houses was paid by the produce of the dung heaps." A similar situation was found at Inverness. "There are very few houses in town which can boast of either water closet or privy, and only two or three public privies in the better part of the place exist for the bulk of the inhabitants." At Gateshead, "The want of convenient offices in the neighborhood is attended with many very unpleasant circumstances, as it induces the lazy inmates to make use of chamber utensils, which are suffered to remain in the most offensive state for several days, and are then emptied out of the windows." These conditions had their counterpart in every country and were described for New York City in the survey prepared by Stephen Smith in 1865. It told of streets littered with garbage and paper. Youngsters armed with brooms made a small income by sweeping a path through the muck for those who wanted to cross Broadway near City Hall.

Even in the midst of prevailing filth, the individual can to some extent protect himself against cholera, typhoid and dysentery by a never-ending attention to the water that he drinks, the food that he eats, and the objects that he touches. It is told for example that in the Philippines the orthodox Chinese who had retained the ancestral habit of drinking nothing but tea made from boiled water remained free of cholera during the epidemics which killed the Filipinos surrounding them. But this eternal vigilance is hardly compatible with a normal life, and the control of filth diseases obviously had to come from a general improvement of hygiene. This was the point of view emphasized by Chadwick in a celebrated report on the *Sanitary Condition of the Labouring Population of Great Britain,* published in 1842. He concluded:

"That the various forms of epidemic, endemic, and other disease caused, or aggravated, or propagated chiefly amongst the labouring classes by atmospheric impurities, produced by decomposing animal and vegetable substances, by damp and filth and close overcrowded dwellings, prevail amongst the population in every part of the kingdom, whether dwelling in separate houses, in rural villages, in small towns, in the larger towns – as they have been found to prevail in the lowest districts of the metropolis.

"That such disease, wherever its attacks are frequent, is always found in connexion with the physical circumstances above specified, and that where those circumstances are removed by drainage, proper cleansing, better ventilation and other means of diminishing atmospheric impurity; the frequency and intensity of such disease is abated; and where the removal of the noxious agencies appears to be complete, such disease almost entirely disappears.

"The primary and most important measures, and at the same time the most practicable, and within the recognized province of administration, are drainage, the removal of all refuse of habitations, streets, and roads, and the improvement of the supplies of water."

Although unaware of the role of microorganisms as agents of disease, the men who brought about the "great sanitary awakening" often succeeded in introducing practices of community life which greatly limited the spread of contagion. Suffice it to mention again the German hygienist Max von Pettenkofer, who did so much to rid Munich of typhoid fever without the benefit of chlorine or of vaccination, simply by cleaning up the city and providing pure water. It was at that time, also, that Florence Nightingale effected her reforms of hospital sanitation during the Crimean War, and laid the foundation for her campaign against unhygienic conditions in the British Army in India. Unbeliever in disease germs that she was, she nevertheless knew how to control most of them.

Eradication of certain diseases has been achieved by a conscious attack not on the causative microbial agent, but on its insect vector. It is well known that the opening of the Panama Canal became a possible enterprise only after General Gorgas had rid the surrounding tropical country of every breeding place for the mosquitoes which transmit yellow fever and malaria. Similarly, typhus, one of the most devastating infections of all previous armed conflicts, was rendered insignificant during the last World War by the systematic delousing of exposed individuals and by the widespread use of the insecticide DDT. In these cases, the control techniques did not directly affect the microbial agent of the disease but only the insect that transfers it to man.

Societies have attempted to protect themselves against the spread of infection by the enactment of quarantine policies. In the time of great epidemics, men were forbidden to move from the stricken areas into unaffected districts; ships were not allowed to unload their passengers until the threat of contagion had disappeared. Today the protection of ropes, to prevent the passage of rats ship-to-shore, and the treatment of airplanes with insecticides after landing, in an attempt to destroy mosquitoes, are examples of quarantine measures based on factual knowledge of the modes of spread of infections.

It is questionable, however, whether quarantine as formerly practiced ever played a significant part in minimizing the spread of great epidemics of plague and cholera. Convinced as he was that microorganisms by themselves could not cause the disease unless many other environmental factors were also present, Pettenkofer naturally minimized the value of these measures and cited numerous examples of their failure. There are today many students of public health who share his skepticism. The existence of reservoirs of contagion, of the apparently healthy "carriers" mentioned in the preceding chapter, imposes severe technical limitations to any attempt at preventing the entrance of the infective microorganisms in a community. Rabies constitutes perhaps the only case for which there is convincing evidence that certain countries have been successful in protecting themselves against the introduction of a human disease. Except in Central

America, where the vampire bat can transfer the virus of rabies to man and animals, the disease is contracted chiefly through the bite of rabid dogs and wolves. By practicing a severe control over the introduction of dogs from the outside, as well as over the stray animals within their borders, England, Australia and Germany have managed to protect themselves more or less completely against the disease. Here again, this achievement was independent of the rise of the microbiological sciences.

Thus, many techniques for the partial control of infectious disease had been developed before the bacteriological era. The possibility of approaching the problem of control from several different angles stems from the fact that the establishment of disease is dependent upon many unrelated factors, involving the infective microorganism, its physical and biological carriers and vectors, the physiological and psychological conditions of the individuals exposed to it, as well as the physical and social characteristics of the environment. This very multiplicity of factors often makes it possible to attack infectious disease at several independent levels. The germ theory led to a more accurate understanding of the circumstances under which host and parasite come into contact, and thus permitted the formulation of rational control policies of greater effectiveness than those devised empirically in the past. Knowledge of the properties and behavior of the infective microorganism often suggested means to attack it, either before or after it had reached the human body. Such is the practice of immunization – which consists in establishing a specific resistance against a given contagious disease by exposing the body, under very special conditions, to the infective agent, to an attenuated form of it, or to one of its products. The science of immunity is one of the most direct outcomes of the germ theory; and it is the more surprising, therefore, to realize that immunization had been practiced during antiquity, long before anything was known of the role of microorganisms as agents of disease; vaccination against "Oriental sore" and against smallpox are among the most successful and ancient achievements of preventive medicine.

There exists, in many Eastern countries, an infection of the skin known under the name of "Oriental sore" or "Aleppo boil," caused by the protozoan *Leishmania tropica.* It develops following the bite of insects, often leaving unsightly sores. As an attack of the disease confers lasting immunity, it became the practice in endemic areas to infect young children on concealed parts of the body, in order to prevent disfiguring lesions of the face. In a similar manner, inoculation with smallpox was widely practiced in ancient Oriental civilizations. Jenner introduced vaccination with cowpox in 1796, and this procedure is, even today, one of the most effective examples of preventive immunization. So important was Jenner's achievement in stimulating Pasteur's work that it is best to reserve the detailed discussion of it for a separate chapter.

For the reasons outlined in the preceding pages, it is far less simple than commonly believed to assess the effect of the germ theory on the control of infectious diseases. The number of deaths due to typhus, cholera, typhoid, tuberculosis, had begun to decrease at a very appreciable and, in certain cases, at a startling rate before the causative agents of these diseases had been discovered. This statement is not intended to minimize the importance of the revolution which microbiological sciences brought about in medical thinking, but rather to provide a historical basis on which to describe the nature of this revolution, and to evaluate its consequences for human health.

The problems of surgical infections, childbirth fever and intestinal diseases offer striking illustrations of the influence of the germ theory on the growth of medicine and hygiene.

In the past, infections had always been the chief cause of the mortality following operations of any sort. Of the 13,000 amputations performed in the French Army during the Franco-Prussian War in 1870–1871, no less than 10,000 proved fatal. Here and there, individual surgeons attempted to lessen mortality by cleanliness and by the employment of special washes for wounds, but all these attempts were empirical and in general did not avail. It was Pasteur's demonstration that bacteria were responsible for fermentations and putrefactions which gave the clue that Lister followed to reform surgical practice.

Lister's attention was called to Pasteur's work on the role of microorganisms in putrefaction sometime around 1864, by the chemist Anderson. He was well prepared to understand the significance of Pasteur's observations because, as mentioned earlier, his father had early made him familiar with the microbial world. If, as Lister postulated, microorganisms cause wound suppuration, just as they cause fermentation and putrefaction, they must be excluded at all costs from the hands of the surgeon, from his instruments, and from the very air surrounding the operating field. To achieve this Lister used a spray of phenol throughout the operations, taking his lead from the fact that this substance was then employed for the disinfection of sewage and excreta. Within a short time, he acquired the conviction that his antiseptic technique prevented suppuration and permitted healing "by first intention" in the majority of cases.

This antiseptic method was based on the hypothesis, derived from Pasteur's writings in the 1860's, that wound contamination originated chiefly from microorganisms present in the air. As he began to frequent hospital wards, however, Pasteur became more and more convinced that the importance of the air-borne microorganisms had been exaggerated and that the most important conveyors of infection were the persons who took care of the sick. He emphasized this point of view in a famous lecture delivered before the Academy of Medicine.

"This water, this sponge, this lint with which you wash or cover a wound, deposit germs which have the power of multiplying rapidly within the tissues and which would invariably cause the death of the patient in a very short time, if the vital processes of the body did not counteract them. But alas, the vital resistance is too often impotent; too often the constitution of the wounded, his weakness, his morale, and the inadequate dressing of the wound, oppose an insufficient barrier to the invasion of these infinitely small organisms that, unwittingly, you have introduced into the injured part. If I had the honor of being a surgeon, impressed as I am with the dangers to which the patient is exposed by the microbes present over the surface of all objects, particularly in hospitals, not only would I use none but perfectly clean instruments, but after having cleansed my hands with the greatest care, and subjected them to a rapid flaming, which would expose them to no more inconvenience than that felt by a smoker who passes a glowing coal from one hand to the other, I would use only lint, bandages and sponges previously exposed to a temperature of 130° to 150° C."

This memorable statement has become the basis of aseptic surgery, which aims at preventing access of pathogenic agents to the operative field rather than trying to kill them with antiseptics applied to the tissues.

One might think that, by 1878, the germ theory would be sufficiently well-established to make Pasteur's warnings needless. In reality, the sense of aseptic technique was still at that time completely foreign to many enlightened physicians, as is revealed by the following account by Loir: "One day, at the Hôtel Dieu, Professor Richet was asked by Pasteur to collect pus from one of the surgical cases. He was doing his ward rounds with a soiled white apron over his black dress suit. Interrupting himself, he said, 'We are going to open this abscess; bring me the small alcohol lamp which M. Pasteur used yesterday to flame the tube in which he collected some pus for his experiment. We shall now sacrifice to the new fashion and flame the scalpel,' and with a wide gesture, which was characteristic of him, he wiped the scalpel on the soiled apron twice, and then attacked the abscess."

In contrast to the carelessness of his medical colleagues, Pasteur carried his concern for aseptic precautions to the most extreme degree. The odd advice to the surgeons that they flame their hands before operating on their patients reflected a procedure which was part of routine technique in his laboratory until 1886. Pasteur's habit of cleaning glasses, plates and silverware with his napkin before every meal is easier to understand when placed in the atmosphere created by the recent discovery of disease germs. He had shown that the *vibrion septique,* commonly present in the intestinal content of animals and in soil, could also be the cause of violent death if it reached susceptible tissues. He had seen in the blood and organs of women dying of childbirth fever a streptococcus which was similar in appearance to that found in many fermenting fluids. In this bewildering new world which unfolded before him, there were at first no criteria to judge where danger might be lurking. True enough, he was aware of the fact that in addition to the dangerous microorganisms, there are many which are completely innocuous, but as no techniques were then available to differentiate the black sheep from the white, he deemed it advisable to exert the utmost caution in everyday life.

Before the advent of the germ theory, the problem presented by childbirth fever was in many respects similar to that of wound infections. Out of 9886 pregnant women who came for confinement at the Maternité Hospital in Paris between 1861 and 1864, 1226 died of the disease, and the situation was as tragic in all lying-in hospitals of Europe. In Boston, Oliver Wendell Holmes had taught in 1843 that childbirth fever was an infectious disease, and that the infection was carried by the hands of the physician or midwife from one patient to another. There was much opposition to his theory from Meigs of Philadelphia, who resented what he considered Holmes's imputation that the physician's hands were not clean, and who quoted a number of cases of infection that had occurred in the practice of the great Dr. Simpson of Edinburgh, an "eminent gentleman." To this, Holmes replied: "Dr. Simpson attended the dissection of two of Dr. Sidney's cases (puerperal fever), and freely handled the diseased parts. His next four childbed patients were affected with puerperal fever, and it was the first time he had seen it in his practice. As Dr. Simpson is a gentleman, and as a gentleman's hands are clean, it follows that a gentleman with clean hands may carry the disease."

Holmes's warnings were unheeded, as were those of his contemporary, Semmelweis, who preached the same gospel in Budapest. Semmelweis had become convinced that childbirth fever was a wound infection caused by the contamination of the raw surface left in the uterus after the birth of the child, and that the infection was transmitted by the unclean hands of the physicians and students who examined the women during labor. Merely by requiring students to wash their hands in a solution of chloride of lime before making an examination, Semmelweis succeeded in decreasing the death rate in his service by 90 per cent. Nevertheless he also was opposed by his colleagues; tormented by hostility and injustice everywhere, he lost his mind and died without having convinced the medical world of his discovery.

It appears incredible, today, that physicians should have remained blind for so long to the contagiousness of childbirth fever, and that one of the authoritative speakers in the Paris Academy of Medicine could speak with scorn of contagion as late as 1879. It was in the course of this discussion that Pasteur dared to interrupt the speaker and sketched on the blackboard the germs — streptococci — which are the most common cause of the disease. Acceptance of the germ theory made of childbirth fever a preventable disease. Cleanliness became the supreme virtue of the lying-in hospital when finally physicians recognized that the infection could be carried to the patient by her attendants.

The public-minded citizens who had championed the great sanitary awakening of the nineteenth century had attributed to filth the crowd diseases — particularly the intestinal disorders. Pure water, pure food, pure air and pure soil appeared to them as an adequate formula to prevent disease and promote health. It was obvious to many physicians, however, that the problem was not so simple. All had observed that many rural areas remained free of infectious fevers such as tuberculosis, typhoid or cholera despite the overwhelming prevalence of filth. It was also familiar knowledge that disease often reigned in places where the advocates of "pure" and salubrious living conditions had apparently every reason to be satisfied. The view that filth is not synonymous with disease was defended in England by William Budd, the greatest epidemiologist of the nineteenth century, who had been the first to establish beyond doubt that intestinal fever was caused by a "virulent poison cast off by the diseased intestine" and capable of propagating itself. Budd took the opportunity of "the Great Stench" of London – which occurred during the hot summer months of 1858 – to illustrate in a forceful manner the fact that organic putrefaction, alone, cannot cause disease.

"The Occasion was no common one. An extreme case, a gigantic scale in the phenomena, and perfect accuracy in the registration of the results – three of the best of all the guarantees against fallacy – were all combined to make the induction sure. For the first time in the history of man, the sewage of nearly three million people had been brought to seethe and ferment under a burning sun, in one vast open cloaca lying in their midst. The result we all know: Stench so foul, we may well believe, had never before ascended to pollute this lower air. Never before, at least, had a stink risen to the height of an historic event. Even ancient fable failed to furnish figures adequate to convey a conception of its thrice Augean foulness. For many weeks, the atmosphere of Parliamentary Committee rooms was only rendered barely tolerable by the suspension before every window, of blinds saturated with chloride of lime, and by the lavish use of this and other disinfectants. More than once, in spite of similar precautions, the law courts were suddenly broken up by an insupportable invasion of the noxious vapour. The river steamers lost their accustomed traffic, and travellers, pressed for time, often made a circuit of many miles rather than cross one of the city bridges.

"For months together, the topic almost monopolized the public prints. Day after day, week after week, *The Times* teemed with letters, filled with complaint, prophetic of calamity, or suggesting remedies. Here and there, a more than commonly passionate appeal showed how intensely the evil was felt by those who were condemned to dwell on the Stygian banks. At home and abroad, the state of the chief river was felt to be a national reproach. 'India is in revolt, and the Thames stinks,' were the two great facts coupled together by a distinguished foreign writer, to mark the climax of a national humiliation.

"Members of Parliament and noble lords, dabblers in sanitary science, vied with professional sanitarians in predicting pestilence. But, alas for the pythogenic theory, when the returns were made up, the result showed not only a death rate below the average, but, *as the leading peculiarity of the season,* a remarkable diminution in the prevalence of fever, diarrhoea and the other forms of disease commonly attributed to putrid emanations."

After Koch had discovered the cholera vibrio in 1883 and Gaffky identified the typhoid bacillus in 1884, it became obvious that even the dirtiest water or most polluted atmosphere would not cause cholera or typhoid if it did not contain the specific causative microorganism, and obvious also that the worst agents of disease could lurk in the "cleanest" and most transparent water. This knowledge made it possible to plan for the supply of safe water on a more rational basis, the criterion of safety being no longer the absence of foul smells, but the freedom from living agents of disease. To achieve this end, sources of uncontaminated water were secured wherever possible, arrangements were made for adequate filtration, and chlorine was added to water in concentrations sufficient to kill the vegetative forms of bacteria. The understanding of the nature of contamination also permitted the method of water purification to be adapted to changing circumstances. Thus, because the cysts of the amoeba which causes dysentery are more resistant to chlorine than are bacteria, the sterilization of water in certain tropical regions demands steps more drastic than those which suffice where these cysts are unlikely to occur. Microbiological sciences also provided convenient techniques for the control of the safety of water. Even where the amount of organic matter is too small to permit ready detection by chemical means, bacteriological analysis is often capable of revealing the presence of living organisms and thus provides a guide in tracing sources of contamination.

The causative microorganisms of typhoid and cholera have not changed, and men are still susceptible to them, yet the great epidemics of the past are not likely to occur again under normal conditions in our cities. The sanitarians of the mid-nineteenth century had made typhoid and cholera less frequent in the Western world; armed with bacteriological knowledge, the modern public health officer is now in a position to complete the victory and to gain absolute control over these diseases if the community is willing to support him. Thanks to the germ theory, the blind campaign against filth has been replaced by an attack on the sources of infection, based on knowledge of the nature and modes of transmission of the agents of disease. In the words of Charles V. Chapin, who was head of the Department of Public Health in Providence, Rhode Island, at the beginning of the present century:

"It will make no demonstrable difference in a city's mortality whether its streets are clean or not, whether the garbage is removed promptly or allowed to accumulate, or whether it has a plumbing law. . . . We can rest assured that however spick and span may be the streets, and however the policeman's badge may be polished, as long as there is found the boor careless with his expectoration, and the doctor who cannot tell a case of polio from one of diphtheria, the latter disease, and tuberculosis as well, will continue to claim their victims. . . . Instead of an indiscriminate attack on dirt, we must learn the nature and mode of transmission of each infection, and must discover its most vulnerable point of attack."

Pasteur never took an active part in the formulation of public health regulations; he left to others the duty to administer the land which he had conquered. Chemotherapy — that is, the treatment of established disease by the use of drugs — is another field of medical microbiology which he did not till. He had not ignored it, but he did not believe that it was the most useful approach to the control of infection. "When meditating over a disease, I never think of finding a remedy for it, but, instead, a means of preventing it." This is a policy which enlightened societies are slowly learning to adopt, one which the wise men of China have understood—if it be true that they advise paying doctors to prevent sickness, rather than to treat it. It is also possible that Pasteur was kept from working on chemotherapy by another reason that appears as a casual sentence in one of his reports on the silkworm diseases:

"My experiments (on silkworms) have brought the knowledge of the prevailing diseases to a point where one could approach scientifically the search for a remedy. . . . However, discoveries of this nature are more the result of chance than of reasoned orderly studies.

"The discovery of the use of sulfur for treating the oidium of the grapevine was so little scientific that the very name of its author has remained unknown."

Pasteur was right in his opinion that useful drugs are usually discovered by accident, or at least by purely empirical methods, but he was wrong in believing that the discovery of the use of sulfur had remained unknown for this reason. If the names of those who first worked out methods for the treatment of plant diseases are so rarely mentioned, it is not because their work was unscientific, but because men consider of great importance only that which directly affects their own persons. Historians work hard to identify the individuals who introduced quinine in human medicine, but pay little attention to those who developed techniques to save our crops.

One could quote many examples to illustrate Pasteur's statement that the discovery of drugs has often been the result of chance. The beneficial effects of salicylic acid in rheumatic fever, and of digitalis in dropsy, were first recognized and utilized on the basis of empirical observations. The uses of quinine and of ipecacuanha (emetine) were discovered by American Indians long before anything was known of the cause of malaria and amoebic dysentery, diseases for which these drugs are so effective. The discovery of the usefulness of sulfonamides came out of the empirical testing of countless dyes in countless experimental animals infected with a variety of infectious agents; today, after fifteen years of intensive research, there is still doubt as to the mechanism by which these drugs control infection.

It is one of Pasteur's own accidental observations which ushered in the most spectacular phase of discoveries in the field of therapy of infectious disease. He had observed that cultures of anthrax bacilli contaminated with common bacteria often lose their ability to establish disease in experimental animals, and he rightly concluded that these common bacteria produced some substance inimical to the disease agent. Was it sheer luck, or the desire to comment on this interesting phenomenon, or real vision, that inspired him to predict a great future for his chance observation? "Neutral or slightly alkaline urine is an excellent medium for the bacilli [of anthrax]. . . . But if . . . one of the common aerobic microorganisms is sown at the same time, the anthrax bacilli grow only poorly and die out sooner or later. It is a remarkable thing that the same phenomenon is seen in the body even of those animals most susceptible to anthrax, leading to the astonishing result that anthrax bacilli can be introduced in profusion into an animal, which yet does not develop the disease. . . . These facts perhaps justify the highest hopes for therapeutics."

The hint was not lost. Immediately after him, and ever since, many bacteriologists have attempted to find in nature microorganisms capable of producing substances effective in the treatment of infectious disease. The story of this search does not belong here. The title of its most important chapter, "Penicillin," is sufficient to call to mind the accidental detection of a mold which inhibited the growth of staphylococcus, and then the organized effort of pathologists, bacteriologists, chemists and technologists to make the miraculous drug available to the world. Initially, it was a chance observation which revealed the existence of penicillin; but again it was true that "chance favors only the prepared mind." In this case, the mind favored by chance had been prepared by years of familiarity with bacteriological lore. Not only did the germ theory permit the discovery of penicillin; it also guided at every step those who worked to define the immense possibilities of the drug in the treatment of disease. Today, it still guides the search for other substances capable of interfering with the pathogenic behavior of the microbial agents of infection.

In addition to suggesting that certain common microorganisms might be used in the therapy of infection, Pasteur had also the extraordinary idea of advocating the utilization of microbial life for the control of animal and plant parasites. The first suggestion of this nature concerns phylloxera, a plant louse that was then infesting and ruining the vineyards of France and of the rest of Europe. It appears as a casual laboratory note dictated to Loir by Pasteur in 1882:

"To find a substance which could destroy phylloxera either at the egg, worm, or insect stage appears to me extremely difficult if not impossible to achieve. One should look in the following direction.

"The insect which causes phylloxera must have some contagious disease of its own and it should not be impossible to isolate the causative microorganism of this disease. One should next study the techniques of cultivation of this microorganism, to produce artificial foci of infection in countries affected by phylloxera."

Pasteur never tried to establish the practical usefulness of this suggestion in the case of phylloxera, but he came back to the idea five years later under the following circumstances. During the latter part of the nineteenth century, the settlers in Australia and New Zealand introduced rabbits and hares from Europe into their countries. The land and climate proved so favorable to the rabbits that these animals multiplied at an extraordinary rate, reaching immense numbers and destroying crops and pastures. Hunting, trapping and poisoning proved without avail against the new plague. So great was the destruction of crops that the Government of New South Wales offered in August 1887 a prize of £25,000 to anyone demonstrating an effective method for the extermination of rabbits.

In November, Pasteur wrote a long letter to the editor of the Paris newspaper *Le Temps,* where he had read the announcement of the prize, and outlined his views on the subject:

"So far, one has employed mineral poisons to control this plague. . . . Is not this the wrong approach? How could mineral poisons deal with animals that multiply at such an appalling rate? The poisons kill only at the place where one deposits them. Is it not preferable to use, in order to destroy living beings, a poison also endowed with life, and also capable of multiplying at a great speed?"

"I should like to see the agent of death carried into the burrows of New South Wales and of New Zealand by attempting to communicate to rabbits a disease that might become epidemic."

He pointed out that chicken cholera is extremely fatal to rabbits and could be given to the animals by feeding infected foodstuffs, and he suggested practical techniques by which the method could be applied on a large scale in the field.

In January 1888 he reported in the *Annales de l'Institut Pasteur* several laboratory experiments proving the susceptibility of rabbits to infection by feeding and by contact, and he suggested the following procedure: "Cut the grass around the rabbit burrows and gather it with rakes in a place readily accessible to the rabbits, before they come out in the evening. This grass, properly contaminated with culture of the chicken cholera bacillus, would be eaten by the animals as soon as they came across it."

He received at that time from Madame Pommery, owner of the champagne firm, a letter advising him that rabbits had become a great nuisance in the wine cellars, and that none of the means used against them had succeeded in checking their multiplication. Pasteur immediately sent Loir to the Pommery estate to carry out the antirabbit campaign that he had outlined. On Friday December 23, Loir spread the culture of chicken cholera on alfalfa around the burrows. Madame Pommery wrote on December 26: "Saturday morning (the day following the contaminated meal), nineteen dead rabbits were found outside the burrows. . . . On Monday morning sixteen more cadavers were found, and no living rabbit could be seen. Some snow had fallen during the night and yet no rabbit tracks were to be found near the cellars."

Further correspondence from Madame Pommery, on January 5, revealed that the cadavers of many rabbits could be found wherever one looked in the burrows, and confirmed the full success of the test.

On the strength of these results, Pasteur sent Loir to Australia to organize a campaign of destruction of rabbits by creating an epidemic of chicken cholera among them, but the test was never carried out, as the Department of Agriculture of Australia refused to give the necessary authorization. Loir nevertheless stayed in Sydney for several years, organizing a microbiological institute for the Australian government and conducting a program of vaccination of farm animals against anthrax.

A few attempts, patterned after Pasteur's experiments on the effect of chicken cholera on rabbits, have since been made to control animal and plant plagues by the use of other microbial parasites. Best known are those utilizing bacteria pathogenic for rats and mice, and also for certain plant pests. Although encouraging results have been obtained, they have not lived up to the early expectations. It is relatively easy to cause the death of animals with infected food, but it is extremely difficult to establish an epidemic with a progressive course. A few years ago, an attempt was made in Australia to introduce the virus of infectious myxomatosis on an island infested with rabbits. In this case again, the disease did not become established in the rabbit population although myxomatosis is known to be a highly fatal disease for these animals. As pointed out repeatedly in preceding pages, the spread of any infection is conditioned by a multitude of factors, many of them unknown; epidemics often break out in a mysterious manner, but they also subside spontaneously for equally obscure reasons. If this were not the case, leprosy, tuberculosis, plague, cholera, typhus, influenza, poliomyelitis and countless other scourges would have long ago annihilated the human race. The factors which limit the spread of epidemics have been responsible so far for the failure of the microbiological warfare devised by Pasteur to control animal plagues. Fortunately they limit also the destructive potentialities of microbiological warfare between men, at least until more knowledge is available.

There is a tragic irony in the fact that one of the last of Pasteur's experimental studies should have been devoted to the utilization of a technique by which contagious disease can be used to destroy life. Today, further progress in the control of infection depends to a large extent upon a more thorough understanding of the factors which govern the spread of epidemics and it is this very knowledge which is also needed to make of biological warfare the self-reproducing weapon of future wars. This prospect, however, should not be held as an argument to minimize the beneficial results of microbiological sciences. For, as Francis Bacon said, "If the debasement of arts and sciences to purposes of wickedness, luxury, and the like, be made a ground of objection, let no one be moved thereby. For the same may be said of all earthly goods; of wit, courage, strength, beauty, wealth, light itself, and the rest."

The centers of medical enlightenment in classical Greece were the rival schools located on the islands of Cos and Cnidus. Cnidian medicine was based on the diagnosis of the different diseases, which it attempted to describe and classify as if they constituted well-defined entities. In some respects, the bacteriological era is the fruition of this ancient biology: the different fevers, by themselves, may not appear to be absolutely independent and separate entities, but the specific microorganisms which cause them certainly are.

Medicine in Cos was concerned with the patient rather than with the disease, and considered the environment as of decisive importance in conditioning the behavior and performance of the body. This doctrine is symbolized in the person, legendary or historical, of Hippocrates, and was codified in his treatise on "Air, Water and Places," where there occurs no clear reference to contagion. Medicine remained Hippocratic in inspiration until late in the nineteenth century, and its progress has long been hindered by the lack of awareness that microbial parasites can become established in the fluids and tissues of the body, and cause profound alterations of structure and functions. Hippocratic medicine had failed to take into account the fact that microorganisms constitute one of the most important factors of man's environment.

After 1877, the pendulum swung widely in the opposite direction, and physicians, as well as the lay public, became obsessed with the thought of disease germs. Today, many medical scholars lament the fact that, under the influence of the germ theory, too much emphasis has been placed on the microorganisms which cause disease, and too little on the effects exerted by hereditary constitution, climate, season and nutritional state on susceptibility to infection. In this justified criticism there is often implied the belief that Pasteur, who was not a physician, was responsible for a distortion of medical thinking. In reality, the complete sacrifice of the physiological to the bacteriological point of view is not Pasteur's guilt but that of the medical bacteriologists who followed him during the so-called "Golden era of bacteriology." True enough, Pasteur had to limit his own experimental work to the study of microorganisms and of their activities, but this limitation was the consequence of the shortness of days and of life, and not of the narrowness of his concepts. On many occasions, he referred to the importance of constitution and environment for the occurrence of disease, and to his desire to investigate them. Unfortunately, he was prevented from doing it by the fierceness of the controversies concerning the participation of microorganisms in disease, and by the enormous amount of experimental work required to bring unassailable proof of his views. This effort monopolized all his energies, even though it did not satisfy his genius.

For the sake of effective experimentation, Pasteur designed his studies on infection and vaccination in such a way that the virulence of the microorganism or the state of acquired immunity was the dominant factor in his tests. But many a time — although this was seldom recalled by later workers — he referred in passing to the effect of environmental factors, and to the significance of constitutional susceptibility and resistance, on the course and outcome of contagious disease. He had shown that the susceptibility of chickens and of rabbits to anthrax could be respectively increased or decreased by lowering or raising the body temperatures of these animals. He had postulated that resistance to infection could depend on the absence of certain chemical elements in the tissues. He had repeatedly emphasized that different races, or even different individuals within one given species, exhibit varying degrees of "vital resistance," and that this resistance can be still further modified by changing the conditions of life.

"If you place this child [born of tuberculous parents] under good conditions of nutrition and of climate, you have a good chance to save him from tuberculosis. . . . There exists, I repeat it, a fundamental difference between the disease in itself and its predisposing causes, the occasions which can bring it about. . . .

"How often the constitution of the wounded, his weakened condition, his mental state . . . are such that his vital resistance is not sufficient to oppose an adequate barrier to invasion by the infinitely small."

All through his studies on silkworms, as we have pointed out, he devoted much attention to the influence of general hygienic conditions in the nurseries, and he came to believe that this was the most important approach to the control of pébrine and flacherie.

"If I were . . . to undertake new studies on silkworms, I should like to concern myself with the conditions which increase their general vigor. . . . I am convinced that it would be possible to discover means to give the worms a higher level of physiological robustness and increase thereby their resistance to accidental maladies. . . .

"To increase the vigor of the silkworms by exposing the eggs to the cold of winter or to an artificial cold would be an achievement of very great importance."

Rudimentary as these thoughts are, they reveal clearly how much importance he attributed to the physiological state of well-being as a factor in resistance.

His concern for this problem led him to devote his first lecture on physics and chemistry at the Ecole des Beaux Arts in Paris "to the important question of the sanitation and ventilation . . . of hospitals, theaters, schools, private dwellings and meeting rooms." Far from being hypnotized with the idea that microorganisms are the only factors of importance in medicine, Pasteur knew that men as well as animals, in health or in disease, must always be considered as a whole and in relation to their environment. Medicine can best help the patient by co-operating with the *vis medicatrix naturae.*

Many needful discoveries remain to be made before the role of microorganisms in disease is completely known and controlled; the Pasteurian chapter is not closed, and will never be forgotten. But acceptance of the germ theory of disease was only one step in the evolution of medicine. Knowledge of the existence and properties of the microbial parasites is making it easier to study the fundamental processes of the living body, its intrinsic strength and weaknesses, its reaction to the environment. Medicine can again become Hippocratic now that contagion is no longer a mysterious and unpredictable threat to the life of man. Thanks to the germ theory, it has become possible to analyze with greater profit the part played by nature and nurture in health and in disease, as well as the pervasive influence of "Air, Water and Places."

CHAPTER XII

Immunity and Vaccination

> Arts and sciences are not cast in a mold, but are formed and perfected by degrees, by often handling and polishing, as bears leisurely lick their cubs into shape.
>
> — MONTAIGNE

SMALLPOX was probably introduced into Europe from the Orient by the Crusaders in the sixth century and by the Saracens when they invaded Spain. It increased in prevalence from the sixteenth century onward and became the most important infectious disease of that time. According to reports, only five out of every thousand persons escaped infection, and one out of four died of it in seventeenth-century England. More than half the population had obvious pockmarks, and blindness due to smallpox was of common occurrence; Macaulay has depicted in graphic terms the atmosphere of terror engendered by "the Scourge" in those days. "The smallpox was always present, filling the churchyards with corpses, tormenting with constant fears all whom it had not yet stricken, leaving on those whose lives it spared the hideous traces of its power, turning the babe into a changeling at which the mother shuddered, and making the eyes and cheeks of a betrothed maiden objects of horror to the lover."

Although the frequency of pocking and blindness during the seventeenth and eighteenth centuries may have been exaggerated, there is no doubt that smallpox was then a deadly and loathsome disease all over Europe. The enthusiasm displayed by Thomas Jefferson in a letter that he wrote to Jenner congratulating him on the discovery of vaccination gives a measure of the importance of smallpox in the society of his time:

> Medicine has never before produced any single improvement of such utility. . . . You have erased from the calendar of human afflictions one of its greatest. Yours is the comfortable reflection that Mankind can never forget that you have lived; future generations will know by history only that the loathsome smallpox has existed, and by you had been extirpated.

Smallpox had invaded the American continent with the Conquistadores early in the sixteenth century, a Negro slave in Hernando Cortez's army being credited with transmitting the infection to the Inca populations in Mexico. The disease spread among them without restraint and was perhaps more effective than Spanish arms and valor as an instrument of conquest:

> For sixty days it raged with such death-bringing virulence that the period of the raging of *hueyzahuatl*, or Great Pest, fixed itself as a central point in the chronology of the natives. In most districts half the population died, towns became deserted, and those who recovered presented an appearance which horrified their neighbors. . . .[1]

The story repeated itself when the North American Indians in their turn came into contact with the European invaders. Like tuberculosis and alcoholism half a century later, smallpox played havoc with the red man and contributed to his defeat by the whites, who were more resistant to the forces of destruction they had brought with them. The Europeans soon realized that they had an unsuspected ally in smallpox and did not hesitate to use it willfully to further their aims. Seeing Indian villages and tribes decimated by the new scourge, the invaders tried to accelerate the spread of the infection by introducing contaminated objects into the settlements of their enemies. The following quotations from official colonial documents leave no doubt that the European

[1] Quoted in Stearn, W. E. and Stearn, A. E., *The Effect of Smallpox on the Destiny of the Amerindian* (Boston: Bruce Humphries, Inc., 1945.)

soldiers and the colonists were aware of the contagiousness of smallpox and of the susceptibility of the Indians to the disease.

> You will do well to try to inoculate the Indians by means of blankets as well as to try every other method that can serve to extirpate this exorable race.
>
> Out of our regard for them [i.e., two Indian chiefs] we gave them two blankets and a handkerchief out of the smallpox hospital. I hope it will have the desired effect.
>
> I will try to inoculate . . . with some blankets that may fall into their hands, and take care not to get the disease myself.[2]

Moralists consider it a sign of the degeneracy of our times that scientists dare discuss the possibility of biological warfare for the next armed conflict. They lack historical memory, for conquerors have never concerned themselves with moralists, nor waited for scientists to use the forces of nature for the prosecution of their plans. So it was that long before the atomic bomb and the spreading of bacteria through the air were ever thought of, the soldiers of the European kings and the New England Puritans used smallpox to destroy the Indians; and before them the medicine men and soldiers of early civilizations had learned to poison or contaminate wells and food supplies. There is nothing new under the sun.

It has been known from all antiquity that second attacks of smallpox are rare and that persons who have once had the disease can nurse patients with safety. This knowledge led to the idea that, since it was almost impossible to avoid the infection, it might be desirable to have it at one's own convenience. Thus grew the practice of "inoculation" or "variolation," which consisted in inoculating individuals in a good state of health with pustule material from mild cases of smallpox and in placing them under conditions believed to allow the disease to run its course with a minimum of risk. Inoculation against smallpox is said to have been practiced in China, India and Persia since remotest times — but it was not until 1717 that Lady Montagu, wife of the British am-

[2] Stearn, W. E., and Stearn, A. E., *op. cit.*

bassador to Turkey, introduced it from Constantinople into Europe. Variolation was first practiced in America by a Dr. Boylston of Boston in 1721. He had been encouraged by his friend Cotton Mather, who had learned from slaves that the practice was common in Africa. The disease induced by inoculating smallpox virus under the skin of a well person was usually mild and had a low mortality; it was as contagious as normally contracted smallpox, however, and required the isolation of the patient for a number of weeks. For this reason, friends arranged to be inoculated at the same time and to spend the period of seclusion in each other's company. Despite all care, the practice of variolation remained dangerous and never gained wide acceptance.

It is claimed that Sanskrit texts mention that an attack of cowpox exerts a protective effect against subsequent exposure to smallpox, but this fact was lost sight of until the observation made by Jenner at the end of the eighteenth century. Edward Jenner was a country practitioner of lively and inquiring mind, as witnessed by his notebooks in which records of weather, facetious epigrammatic verses, and entertaining drawings compete for space with accounts of the habits of birds and of the doings of his patients. His observations on natural history had given him scientific distinction and he had been elected a Fellow of the Royal Society. In particular, he was known for having seen a young cuckoo bird pitch a newly hatched sparrow out of the nest, and for having recognized on the back of the young cuckoo a peculiar depression "formed by nature for the design of giving a more secure lodgment to the egg of the hedge sparrow or its young one when the young cuckoo is employed in removing either of them from its nest."

In eighteenth-century England there was some belief that persons having cowpox, an infection which presented some similarity to smallpox, were thereby rendered incapable of contracting the latter disease. It is reported that Jenner was led to study the matter by the statement of a Gloucestershire dairymaid who had come to him as a patient. When he suggested that she was suffering from smallpox, she immediately replied: "I cannot take the smallpox because I have had the cowpox." Jenner attempted to give scientific foundation to the popular belief by observing the reaction of cowpoxed persons to inoculation with smallpox. There were many chances to make such observations, for cowpox was then a fairly common disease and it was an accepted medical practice to inject smallpox into human beings for prophylactic purposes. Jenner reported that the local reaction was transitory, that no fluid-containing vesicle was produced on the skin, and that there was no constitutional disturbance. He thus satisfied himself, if not all others, "that the cowpox protects the human constitution from the infection of the smallpox."

In May 1796, he gave cowpox to James Phipps, a boy eight years old, and later inoculated him with virulent smallpox virus. The boy failed to contract smallpox and Jenner hastened to report this epoch-making observation:

"The first experiment was made upon a lad of the name of Phipps in whose arm a little vaccine virus was inserted, taken from the hand of a young woman who had been accidentally infected by a cow. Notwithstanding the resemblance which the pustule, thus excited on the boy's arm, bore to variolous inoculation yet as the indisposition attending it was barely perceptible, I could scarcely persuade myself the patient was secure from the Small Pox. However, on his being inoculated some months afterwards, it proved that he was secure."

Thus was introduced into the Western world the practice of immunization against smallpox by the injection of virus material originating in the cow; the word "vaccination," under which the method came to be known, is derived through "vaccine" from *vacca*, a cow.

After having shown that inoculation with cowpox could "vaccinate" against smallpox, Jenner had experienced anxiety for fear it might always remain necessary to return to the cow to obtain the vaccine. He believed that cowpox originated in the cow from the hands of a milker infected with smallpox, and that the human disease became transformed into cowpox by passage through the animal. So confident was he of having discovered a technique for eliminating smallpox that he wondered whether cowpox material might not become unavailable for human vaccination. He therefore attempted to vaccinate from arm to arm, in the hope that cowpox virus would not lose its vaccinating potency, nor acquire excessive virulence by being passed through the human body. He inoculated a child with material from the teat of a cow; from the sore on this child's arm another child was inoculated and the process was repeated from one child to another, up to the fifth removed from the cow. Three of these children were later inoculated with smallpox and all three proved resistant to the disease, demonstrating the possibility of carrying the vaccination from arm to arm.

Were Jenner's own observations really adequate to justify such an important conclusion? Some epidemiologists have questioned it and one of them, Greenwood, has referred to Jenner's writings as "just the sort of rambling, discursive essay, containing acute observations mixed up with mere conjectures, which an unsystematic field naturalist might be expected to produce." In fact, Jenner's first paper on his discovery was rejected by the Royal Society in 1797 with a friendly admonition that such incomplete studies would injure his established reputation.

Jenner extended his paper, and published it as a pamphlet in 1798 under the title *An Inquiry into the Causes and Effects of the Variolae, a Disease Discovered in Some of the Western Counties of England, Particularly Gloucestershire, and known by the Name of Cow Pox.*

It is not without interest that, like Jenner's first paper on vaccination, John Snow's first report, demonstrating that cholera is water-borne, had to be published at the expense of its author. Official academies are more likely to exhibit enthusiasm over the improvements of the commonplace than to recognize the unexpected when it is first brought to them. But academic indifference did not keep Jenner from becoming convinced of the absolute effectiveness of vaccination and he stated his faith in no uncertain terms. "At present, I have not the most distant doubt that any person who has once felt the influence of perfect cowpox matter would ever be susceptible to that of the smallpox."

However, many accidents soon occurred that could have shaken his faith. As the method began to be widely used, many patients developed bad ulcers at the site of vaccination, and some contracted smallpox despite the treatment. In May 1811 the Honorable Robert Grosvenor, whom Jenner himself had vaccinated ten years before, fell ill with an extremely severe case of smallpox. The young man recovered but his case created an enormous sensation in medical and lay circles and led to bitter controversies. Nevertheless, the efficacy of vaccination soon became widely accepted and rewards in addition to fame came to Jenner for his discovery. In 1802, Parliament voted him £10,000 and again in 1806, £20,000 for his achievement, while learned societies joined with sovereigns in paying him honor.

Jenner soon had many followers in England but it was perhaps in America that the method received the most vigorous support. Benjamin Waterhouse in Boston took up the cudgels for vaccination, as Cotton Mather had done for inoculation almost a century earlier. Having received vaccine virus from England, he vaccinated his own family in July 1800 and dared expose his children to infection in the smallpox hospital in order to demonstrate that they were immune. In 1801, he sent some of Jenner's vaccine to President Thomas Jefferson, who had his own family vaccinated, as well as some of their neighbors and a few Indians.

Opposition to Jenner continued in some medical circles even after cowpox vaccination had become established throughout the world as a standard practice. As late as 1889, Creighton, eminent English historian of infectious diseases, dismissed Jenner as a "vain, imaginative, loose-thinking person" and his claims as mere roguery. In his book *The Wonderful Century,* Wallace described vaccination as one of the dark spots of the age, not only denying its efficacy, but also expressing great doubts as to its innocuousness. Although no trained epidemiologist would take this view today, all recognize that it is not a simple matter to evaluate statistically the effectiveness of vaccination. Like other infectious diseases, smallpox probably undergoes fluctuation unpredictable both in prevalence and in severity. As vaccination is now widely practiced throughout the civilized world, it is difficult to determine how the disease would behave in an unvaccinated population; hence it is impossible to determine exactly what role vaccination plays in the control of the disease.

For a balanced judgment of the case, we may turn to the conclusions reached by Greenwood in his study of vaccination.

"I have, indeed, as an individual, almost as strong a conviction that recent vaccination is a thoroughly adequate defence against the risk of taking smallpox as anti-vaccinists have that it is worthless as a defence and otherwise pernicious, but I know of no data by means of which I could estimate the measure of such advantage. . . . I conclude by inferring from the statistical evidence which I have discussed that Jenner was, directly or indirectly, the means of saving many hundreds of thousands of lives. That is a less grandiose conclusion than some others have reached, yet, I submit, quite enough to entitle Englishmen to take pride in the recollection that Jenner was their countryman."

The success of vaccination encouraged many efforts during the middle of the nineteenth century to use similar methods of preventive inoculation against other diseases such as measles, plague, syphilis and pleuropneumonia of cattle. Indeed, a proposal to inoculate all the youth of France with syphilis actually reached the Paris Academy of Medicine, where it gave rise to a lively discussion. One of the chief proponents of this measure was Auzias-Turenne who, as we shall presently see, played some indirect part in the formulation of Pasteur's views on the problems of immunity.

Like many others who preached the germ theory of disease before Pasteur and Koch, Auzias-Turenne is now forgotten. And yet his views, which he presented indefatigably in many articles and lectures, so impressed some of his contemporaries that they republished them after his death in the form of a large book entitled *La Syphilisation.*

Auzias-Turenne advocated immunization of the youth of the

world against syphilis by inoculating material from soft chancre, which he regarded as an attenuated form of the disease. According to him, soft chancre bore to syphilis the same relation that cowpox does to smallpox. It was with this message that he expected to gain immortality, but he also had much to say concerning other infectious diseases such as anthrax, cholera, smallpox, rabies, and pleuropneumonia of cattle. His statements illustrate the point of view held by some medical men before Pasteur and Koch had definitely established that the "viruses of disease" were living microorganisms. Auzias-Turenne accepted as established the variability in virulence of infective agents, the possibility of immunizing against certain contagious diseases, and the view that immunity was due to the exhaustion in the body of some substance required by the causative agent of the disease. All these concepts were to be recast in experimental terms by Pasteur two decades later.

From a lecture delivered in October 1865 before the Academy of Medicine and reprinted in *La Syphilisation,* one can glean the following statements which summarize the philosophy of infection reached by Auzias-Turenne:

> The virus is always identical to itself, variable in intensity, transmissible, *i.e.,* capable of reproducing itself after a given time of incubation in the proper organism. . . .
>
> Viruses derive what they need from the infected organism and often end by exhausting the latter . . . they either destroy it or abandon it for lack of food. . . .
>
> Viruses can undergo variations in intensity. . . .
>
> A virus can be regenerated in a good terrain in which it multiplies, whereas it can be weakened by an unfavorable terrain. . . . But a good terrain becomes exhausted when it carries the virus for too long a time. . . .
>
> Viruses are transmissible. They pass from one individual to the other like parasites. . . .
>
> Contagion presupposes a direct contact of the virus with the organism. Infection does not involve a direct contact; the virus may be carried through a medium which is usually the atmosphere. The virus survives in it without being de-

composed and thus passes from one place or one individual to another.

We have established . . . that inoculation can be used for the prevention of contagious pleuropneumonia of cattle. We shall now show that it can also be used as a therapeutic measure.

Let us walk into a stable where pleuropneumonia is prevailing. . . . The animals can be divided into three groups.

1. Those in which the symptoms of the infection are already evident. . . .
2. Those in which the disease exists in a state of latency or incubation. . . .
3. Those which have not yet been affected by the virus. . . .

Let us inoculate into these animals some virus taken from the lung. . . . In animals of the first group . . . the inoculation will be without detectable effect. It will have a curative effect in the animals of the second group . . . and the effect will be preventive in the animals of the third group.

La Syphilisation was published in 1878 and, through family relationships, a copy fell into the hands of the young Adrien Loir, nephew of Pasteur, who immediately gave the book to his uncle. According to Loir, Pasteur was much interested in Auzias-Turenne's writings. He kept the book at home in a special drawer of his desk, and often read it, even copying whole sentences from it. It is possible that Pasteur was encouraged to attempt immunization of dogs and human beings after infection with the rabies virus by Auzias-Turenne's claims on therapeutic immunization in cattle pleuropneumonia. In 1878, however, he was well along in his studies on immunity and had already begun to work on rabies; Auzias-Turenne's book, therefore, probably served him as a concrete basis on which to anchor his meditations at home rather than as the source of new ideas.

Speaking of the plague that destroyed one fourth of the Greek population during the Peloponnesian Wars, Thucydides reported that "the sick and dying were tended by the pitying care of those who had recovered, because they knew the course of the disease and were themselves free from apprehensions. For no one was ever attacked a second time, or not with a fatal result."

Pasteur knew that one attack of certain diseases conferred a definite immunity against another attack of the same disease and he had been much impressed by Jenner's discovery of vaccination. The problem of immunity was constantly in his mind and he continually wondered why Jennerian vaccination had remained an isolated fact in medicine. There was, at this time, much debate in the Paris Academy of Medicine concerning the relation of smallpox to cowpox. Were the two diseases completely independent one from the other, or was the latter, as believed by Jenner, a form of smallpox which had become modified by passage through the cow? Pasteur followed the debate with intense interest and used to tell his collaborators: "We must immunize against the infectious diseases of which we can cultivate the causative microorganism."

It was an accident which gave the clue to the solution of the problem. Pasteur had begun experiments on chicken cholera in the spring of 1879, but a trivial difficulty came to interrupt the work after the summer vacation; the cultures of chicken cholera bacillus that had been kept in the laboratory during the summer failed to produce disease when inoculated into chickens. A new, virulent culture obtained from a natural outbreak was inoculated into new animals, and also into the chickens which had resisted the old cultures. The new animals, just brought from the market, succumbed to the infection in the customary length of time, thus showing that the culture was very active. But to the surprise of all, and of Pasteur himself, almost all the other chickens survived the infection. Be it the result of his reading and incessant pondering on the facts of immunity, or a product of the creative imagination which so often permitted him to guess the solution of a problem without adequate evidence, Pasteur immediately recognized in this accidental occurrence an analogy with cowpox vaccination.

The simple observation that chickens inoculated with an avirulent culture of the chicken cholera bacillus were thereby rendered resistant to a fully virulent culture was made seventy years ago and its consequences have continued to grow in importance ever since. To make more emphatic the analogy between his and Jenner's discoveries, Pasteur chose to describe the new phenomenon under the name of "vaccination." Thus, as is the wont of many words, the meaning of "vaccination" had evolved from the description of a concrete procedure into the expression of an abstract concept. By transferring to man pox material obtained from the cow, Jenner had so modified the human constitution as to render it no longer receptive to smallpox. Pasteur recognized that this effect was the manifestation of a general law, and that the old cultures of chicken cholera which had become "attenuated" during the summer had brought about a transformation of the animal economy which made it less receptive to the virulent form of the microorganism. Jenner's discovery was only a special case of a general immunization procedure; vaccination was the art of specifically increasing the resistance of the body to an inimical agent.

The discoveries of Jenner and Pasteur have implications which transcend immunological science. They reveal in what subtle manner and how profoundly the nature of living things can be affected by influences that reach them from the external world. Man or fowl, once having received a minute amount of material from cowpox or from the culture of a bacterium, are indelibly marked by this apparently trivial experience; they thereby become somewhat different living beings. There exists a biochemical memory which is no less real than the intellectual and emotional memory, and perhaps not essentially different. Immunological science has provided techniques to detect, and in a certain measure to control, a few of those permanent alterations. At the other end of the spectrum of human reactions, experimental psychology is beginning to investigate the permanent alterations of the psyche which often result from apparently trivial external events, alterations which experience and literature have long recognized. Who can doubt that the gap which today separates the immunologist from Pavlov and Marcel Proust will someday be breached, that a time will come when one can recapture and reconstruct from a distant biological *Temps perdu* the complex of biochemical happenings which make of each living thing a unique event in nature?

"It is characteristic of experimental science," Pasteur wrote, "that it opens ever-widening horizons to our vision." While perceiving in the distance the promised land toward which he travels, the scientist must accept with humility his slow, limited and tedious task; he may dream of the green pastures of natural philosophy, but he must till patiently the small patch in space and time where fate has placed him. This, Pasteur accepted, and he set himself with diligence to clear the new land which had just been revealed to him.

Pasteur's discovery had merely suggested that one could obtain a vaccine capable of insuring protection against chicken cholera; it only extended to a bacterial disease a phenomenon already known in the case of the virus disease smallpox. There was at that time no factual information concerning the origin of cowpox; and the speculative view that it was a form of smallpox modified by passage through the cow had not led to any technique by which other agents of disease could be attenuated for the purpose of immunization. Pasteur realized immediately that his observations on chicken cholera brought the phenomena of immunity within the range of study by microbiological techniques. As he could cultivate the causative bacillus of chicken cholera in vitro, and as attenuation of the bacillus had occurred spontaneously in some of his cultures, Pasteur became convinced that it should be possible to produce vaccines at will in the laboratory. Instead of depending upon the chance finding of naturally occurring immunizing agents, as cowpox for smallpox, vaccination could then become a general technique, applicable to all infectious diseases. Within the incredibly short period of four years, Pasteur succeeded in demonstrating the practical possibilities of this visionary concept in the cases of chicken cholera, anthrax, swine erysipelas and rabies.

He had attributed the attenuation of the chicken cholera bacillus to the deleterious effect of air and particularly of oxygen during aging of the culture. Indeed cultures maintained in glass tubes sealed off in the flame retained their virulence for several months whereas they lost their activity more rapidly in tubes closed only with cotton plugs. By taking advantage of these observations, Pasteur obtained a series of cultures of intermediate virulence that he could grow in his bouillons, maintaining unchanged their characteristic degree of attenuation and their vaccinating properties. As mentioned in a preceding chapter, these findings appeared in conflict with the doctrine of the fixity of microbial species. Nevertheless, as soon as he became convinced of its validity, Pasteur turned the fact of variability of virulence into one of the central tenets of his subsequent investigations and applied it systematically in order to obtain attenuated cultures for the purpose of vaccination.

He had once made an observation which suggested that vaccination against anthrax might be possible. A group of eight sheep had been maintained for a prolonged period of time in a pasture where an animal dead of anthrax had been buried. When inoculated with a virulent anthrax culture, several of these animals had survived, whereas normal sheep had all died of the same inoculation. As Pasteur knew that chickens fed upon food contaminated with the chicken cholera bacillus do not invariably die, and that those which survive are often found resistant to subsequent inoculation with a virulent culture, he postulated that ingestion of the microorganism determined in certain cases a mild disease which induced a state of resistance against severe infection. Later studies suggested that cows which had survived an attack of anthrax could withstand inoculation with large amounts of virulent anthrax material. These facts incited Pasteur to attempt to prepare a vaccine capable of producing immunity against anthrax without inducing severe disease. Immediately, an unexpected difficulty arose when attempts were made to attenuate the culture, for the anthrax bacillus produces spores which cannot readily undergo any modification. It was therefore necessary to prevent the formation of spores and at the same time to keep the bacilli alive. This was first accomplished by adding certain antiseptics to the culture and later by keeping it in a shallow layer at 42°–43° C. After eight days under these conditions the bacilli became harmless for guinea pigs, rabbits and sheep. Before completely losing their virulence, however, they passed through all degrees of attenuation, and each of these could be preserved by cultivation in ordinary media as had been done in the case of the chicken cholera organism.

Pasteur found it advisable to conduct anthrax vaccinations in two steps. First a preparatory inoculation of a culture of very low virulence was given, followed twelve days later by a more virulent "second vaccine" which conferred a higher level of immunity. This technique invariably made possible the protection of guinea pigs, rabbits and sheep against infection with the most virulent form of anthrax bacilli. Within a few months, Pasteur undertook to demonstrate the effectiveness of prophylactic vaccination by a large-scale public test on farm animals. This was the famous experiment of Pouilly le Fort, and the details of this dramatic episode will be described later. A few weeks after his triumph at Pouilly le Fort, Pasteur was the star of the International Medical Congress in London, and there he propounded, in the course of his address, the use of the words "vaccine" and "vaccination" to render homage to "the merit and immense services rendered by one of the greatest men of England, your Jenner."

In association with Thuillier, he had also undertaken the study of swine erysipelas. The isolation of the bacterial organism was easy, but its attenuation to a level adequate for practical vaccination presented new problems owing to the different susceptibilities of the various races of pigs. The remarkable fact was established, however, that the bacterium became attenuated for the pig by passage from rabbit to rabbit and it was the culture adapted to this animal species that became the source of the vaccine used for immunization of pigs on a large scale.

Thus, three different methods of attenuation had to be worked out for the first three bacterial vaccines developed in Pasteur's laboratory: aging of the culture for chicken cholera, cultivation at high temperature for anthrax, and passage through rabbits for swine erysipelas. This achievement will appear little short of miraculous to anyone familiar with the technical problems involved. It is difficult to comprehend how Pasteur and his collaborators found it possible, in the course of three years, to work out the practical techniques of vaccination while still struggling to formulate the very concept of immunization. This is even the more startling in view of the fact that they continued at the same time to investigate the etiological problems of infection and were already engaged in the study of rabies.

There is something odd in the selection by Pasteur of rabies as the next subject for his immunological studies. The disease was even then of relatively minor importance, claiming in France at most a few hundred deaths a year; the example of Germany and Australia had clearly shown that simple police and quarantine regulations for the control of dogs were sufficient to minimize its incidence still further. Moreover, there were no clues concerning its etiology; experimentation with it was laborious and expensive, and seemingly ill-adapted to the solution of theoretical and practical problems.

It has been claimed that Pasteur was attracted to the study of rabies through the vividness of childhood memories. He had never forgotten the impression of terror produced on him when a rabid wolf charged through the Jura, biting men and beasts on his way, and he had seen one of the victims cauterized with a red-hot iron at the blacksmith's shop near his father's house. The persons who had been bitten on the hands and head had succumbed to hydrophobia, some of them with horrible suffering; there were eight victims in the immediate neighborhood of Arbois and for years the whole region remained in dread of the mad wolf.

It is possible that this experience of his youth may have influenced Pasteur's decision, but alone it could not have determined it. Rabies had long had a firm hold on public imagination and was the epitome of terror and mystery. It was therefore well suited to satisfy Pasteur's longing for romantic problems, as Renan hinted in his usual subtle manner during the speech welcoming Pasteur at the French Academy of Letters. "It is rabies which preoccupies you today. You will conquer it and humanity will owe to you deliverance from a horrible disease and also from a sad anomaly: I mean the distrust which we cannot help mingling with the caresses of the animal in whom we see most of nature's smiling benevolence." Thus, rabies combined a supreme challenge to the experimenter and his method, an occasion to deal with one of the apparently inscrutable problems of nature, and the chance to capture the interest of the medical and lay public by a spectacular achievement. In fact, Pasteur was right in the selection of this seemingly hopeless problem. The Pouilly le Fort experiment had rendered the public conversant with the doctrine of immunization, but it was the prophylaxis of rabies which made of microbiological science an established religion, and surrounded its creator with the halo of sainthood.

We have pointed out in a preceding chapter that experimentation with rabies demanded the development of entirely new methods. The incubation period of the disease was greatly shortened and rendered more predictable by inoculating the infective matter directly into the nervous tissue. When the virus was passed through the brain of rabbits its virulence increased for these animals and the incubation period became progressively shortened to six days. Pasteur referred to the virus so stabilized as "fixed virus." On the other hand, passage through a series of monkeys attenuated the virus for dogs, rabbits and guinea pigs. Thus it became almost as easy to experiment with rabies as with bacterial infections, even though nothing was known of the causative agent of the disease. Armed with these techniques, Pasteur was now ready to apply himself to the development of a vaccine.

It is certain that many different schemes were imagined by Pasteur for the attenuation of the virus, but only the one upon which he settled is known to us. Fortunately, some of the circumstances under which he arrived at the practical solution of his problem have recently been made public by Loir. Unknown to

Pasteur, Roux was studying at the time the length of survival of the rabies virus in the spinal cord. For this purpose, he had placed infected cord in a flask with two openings, the cord hanging inside and attached to the stopper which closed one of the openings. Pasteur once walked into the incubator where Roux's flasks had been placed, accompanied by Loir.

"At the sight of this flask, Pasteur became so absorbed in his thoughts that I did not dare disturb him, and closed the door of the incubator behind us. After remaining silent and motionless a long time, Pasteur took the flask outside, looked at it, then returned it to its place without saying a word.

"Once back in the main laboratory, he ordered me to obtain a number of similar flasks from the glass blower. The sight of Roux's flasks had given him the idea of keeping the spinal cord in a container with caustic potash to prevent putrefaction, and allowing penetration of oxygen to attenuate the virus. The famous portrait painted by Edelfeldt shows Pasteur absorbed in the contemplation of one of these flasks."

Thus was born the first technique of attenuation of the rabies virus. The method consisted in suspending in dry, sterile air the spinal cords of rabbits which had died from the injection of fixed virus. In the course of about two weeks, the cord became almost nonvirulent. By inoculating into dogs emulsions of progressively less attenuated cord, it was possible to protect the animal against inoculation with the most virulent form of virus. A dog receiving infected spinal cord dried for fourteen days, then the following day material thirteen days old, and so on until fresh cord was used, did not contract rabies, and was found resistant when inoculated in the brain with the strongest virus. In other words, it was possible to establish immunity against rabies in fifteen days.

Under normal conditions of exposure, rabies develops slowly in man as well as in animals. For example, a man bitten by a mad dog ordinarily does not display symptoms of the disease until a month or more after the bite. This period of incubation therefore appeared long enough to suggest the possibility of establishing resistance by vaccinating even after the bite had been inflicted. Experiments made on dogs bitten by rabid animals, and then treated with the vaccine, gave promising results. Would the same method be applicable to human beings bitten by rabid animals and still in the incubation period of the disease?

The story of the mental anguish which Pasteur experienced before daring to proceed from animal experiments to the treatment of the human disease has been often told. The thought of injecting into man rabid virus, even though attenuated, was terrifying. Furthermore, the procedure went counter to one of the medical concepts of the time, namely, that one could not deal with virus once it had become established within the animal body. It was bound, therefore, to stir up great and justified opposition from conservative physicians. In fact, the opposition to the application of the rabies treatment to human beings did not come only from the medical world at large, but even from Pasteur's own laboratory. Roux, in particular, felt that the method had not been sufficiently tested in animals to justify the risk of human trial and refused to sign with Pasteur the first report of treatment. He ceased to participate in the rabies study and resumed his association with the laboratory only when Pasteur became the object of bitter attacks in the Academy of Medicine.

The decision to apply rabies vaccination to man was forced upon Pasteur when a young boy, Joseph Meister, was brought to him for treatment. The physiologist Vulpian and the physician Grancher assured Pasteur that the nature of the bites made it likely that the boy would contract fatal rabies, and that the evidence derived from experimentation in dogs was sufficient to justify attempting the treatment. Grancher took the medical responsibility of the case and from then on managed the program for the treatment of rabies in human beings under Pasteur's close supervision.

Joseph Meister, aged nine, was brought from Alsace to Pasteur July 6, 1885, suffering from bites inflicted by a rabid dog on the hands, legs, and thighs. On July 7, sixty hours after the accident, the boy was injected with rabbit spinal cord attenuated by fourteen days' drying. In twelve successive inoculations he received stronger and stronger virus until on July 16 he received an inoculation of virulent cord removed the day before from the body of a rabbit that had died following inoculation with fixed virus. Joseph Meister exhibited no symptoms and safely returned to Alsace. He later became gatekeeper of the Pasteur Institute. In 1940, fifty-five years after the accident that gave him a lasting place in medical history, he committed suicide to escape being compelled to open for the German invaders the crypt where Pasteur is buried.

The second case treated by Pasteur was that of Jean Baptiste Jupille, aged fifteen, a shepherd of Villers-Farlay in the Jura. Seeing a dog about to attack some children, Jupille seized his whip and attempted to drive it away, but was severely bitten; he finally managed to wind his whip around the muzzle of the animal and to crush its skull with his wooden shoe. The dog was subsequently declared rabid, and Jupille was brought to Paris for treatment six days after being bitten. He survived, and his deed was commemorated in a statue which stands today in front of the Pasteur Institute in Paris.

These two dramatic successes encouraged numerous patients to come to Pasteur for treatment after being bitten by animals known or presumed to be rabid. By October 1886, fifteen months after Joseph Meister had first been treated, no fewer than 2490 persons had received the vaccine. Thus, like Jenner, Pasteur saw his method become an established practice within a short time of its inception, but as had been the case with smallpox vaccination, the rabies treatment was immediately attacked as valueless, and capable of causing the very disease which it was designed to control. There are few physicians who now believe that either smallpox or rabies vaccination can be a likely source of danger to the patient when properly administered, but much question has been raised concerning the effectiveness of rabies treatment. Before discussing this problem, however, it is necessary to retrace our steps somewhat and consider certain practical aspects of the different methods of vaccination discovered in Pasteur's laboratory.

As soon as he had worked out the technique of anthrax vaccination, Pasteur expressed the desire for an opportunity to apply it to farm animals under field conditions. Anthrax was then a disease of great economic importance and the possibility of protecting against it constituted a lively subject of discussion in veterinary circles. The germ theory was still in its infancy and few were the physicians and veterinarians who had any concept of the scientific meaning of immunization. Among those who discussed the discovery, there were a handful who were dazzled by Pasteur's achievements, and many more who had only scorn for the odd claims of one whom they regarded as a conceited chemist, unversed in true medical thinking. It is a fact not without interest that Rossignol, the man who took the initiative in organizing the first practical field test of immunization, in 1881, was one of the critics of the germ theory. One may well suspect that his avowed desire to serve the cause of truth was not unmixed with the hope that he would gain from the experiment the prestige of having been the champion of classical medicine at a time when it was threatened by the invasion of microbiological doctrines. In January of the same year Rossignol had written in a sarcastic vein: "Microbiolatry is the fashion and reigns undisputed; it is a doctrine which must not even be discussed, particularly when its pontiff, the learned M. Pasteur, has pronounced the sacramental words, *I have spoken.* The microbe alone is and shall be the characteristic of a disease; that much is understood and settled; henceforth the germ theory must take precedence over the clinical art; the microbe alone is true, and Pasteur is its prophet."

During the spring of 1881, Rossignol succeeded in enlisting the support of many farmers of the Brie district to finance a large-scale test of anthrax immunization. Pasteur was well aware of the fact that many veterinarians and physicians saw in the test a welcome occasion to cover the germ theory with ridicule; nothing could set in bolder relief, therefore, his confidence and gameness of spirit than his acceptance of the incredibly drastic terms of the protocol submitted to him. Rossignol publicized the program of the test widely and the experiment thus became an event of international importance. It took place in the presence of a great assembly of people of all kinds, including the Paris correspondent of *The Times* of London, Mr. De Blowitz, who for a few days focused the eyes of his readers throughout the world on the small village of Pouilly le Fort. The following account is quoted from Roux, who participated actively in the preparation and execution of the experiment.

"The Society of Agriculture of Melun had proposed to Pasteur a public trial of the new method. The program was arranged for April 28, 1881. Chamberland and I were away on vacation. Pasteur wrote us to return immediately, and when we met him in the laboratory, he told us what had been agreed upon. Twenty-five sheep were to be vaccinated, and then inoculated with anthrax; at the same time twenty-five unvaccinated sheep would be inoculated as controls; the first group would resist; the second would die of anthrax. The terms were exacting, no allowance was made for contingencies. When we remarked that the program was severe, but that there was nothing to do except carry it out since he had agreed to it, Pasteur replied: 'What succeeded with fourteen sheep in the laboratory will succeed with fifty in Melun.'

"The animals were assembled at Pouilly le Fort, near Melun, on the property of M. Rossignol, a veterinarian who had originated the idea of the experiment and who was to supervise it. 'Be sure not to make a mistake in the bottles,' said Pasteur gaily, when, on the fifth of May, we were leaving the laboratory in order to make the first inoculations with the vaccine.

"A second vaccination was made on May 17, and every day Chamberland and I would go to visit the animals. On these repeated journeys from Melun to Pouilly le Fort, many comments were overheard, which showed that belief in our success was not universal. Farmers, veterinarians, doctors, followed the experiment with active interest, some even with passion. In 1881, the science of microbes had scarcely any partisans; many thought that the new doctrines were pernicious, and rejoiced at seeing Pasteur and his followers drawn out of the laboratory to be confounded in the broad daylight of a public experiment. They expected to put an end with one blow to these innovations, so compromising to medicine, and again to find security in the sane traditions and ancient practices that had been threatened for a moment!

"In spite of all the excitement aroused by it, the experiment followed its course; the trial inoculations were made May 31, and an appointment was arranged for June 2 to determine the result. Twenty-five hours before the time decided upon, Pasteur, who had rushed into the public experiment with such perfect confidence, began to regret his audacity. His faith was shaken, as though he feared that the experimental method might betray him. His long mental tension had brought about this reaction which, however, did not last long.[3] The next day, more assured than ever, Pasteur went to verify the brilliant success which he had predicted. In the multitude at Pouilly le Fort, that day, there were no longer any skeptics but only admirers."

The experiment had consisted in the inoculation of twenty-four sheep, one goat and six cows, with five drops of a living attenuated culture of anthrax bacillus on May 5. On May 17, all these animals had been reinoculated with a culture less attenuated. On

[3] It has been related that, on the day after the inoculation of the animals with the virulent challenge dose, a message was brought to Pasteur advising him that some of the vaccinated sheep appeared sick. Despite his confidence in the results of the laboratory experiments, he was under such great nervous tension that he immediately lost heart. Blinded by emotion, and refusing to consider the possibility that he had been mistaken, he turned to Roux who was present and accused him in violent words of having spoiled the field test by carelessness. As his wife was trying to quiet him down, pointing out that they had to start early the next morning for Pouilly le Fort, he replied that he could not expose himself to the sarcasm of the public and that Roux should go alone and suffer the humiliation since he was responsible for the failure.

A telegram received during the night brought reassurance as to the progress of the test; the next morning as Pasteur's group arrived at the railroad station, the enthusiastic welcome of the public gave them a forewarning of the complete success. Standing in his carriage, Pasteur turned to the crowd and exclaimed in a triumphant voice "Well, then! Men of little faith!"—*Biologie de l'Invention,* by Charles Nicolle. (Paris, Felix Alcan, 1932, p. 64.)

May 31 all the immunized animals had been infected with a highly virulent anthrax culture, and the same culture had been injected as well into twenty-nine normal animals: twenty-four sheep, one goat, and four cows. When Pasteur arrived on the field on the second day of June with his assistants Chamberland, Roux and Thuillier, he was greeted with loud acclamation. All the vaccinated sheep were well. Twenty-one of the control sheep and the single goat were dead of anthrax, two other control sheep died in front of the spectators, and the last unprotected sheep died at the end of the day. The six vaccinated cows were well and showed no symptoms, whereas the four control cows had extensive swellings at the site of inoculation and febrile reactions. On June 3, one of the vaccinated sheep died. It was pregnant and an autopsy suggested that it had succumbed on account of the death of the foetus but without showing any symptom of anthrax. Pasteur's triumph was thus complete.

It has often been stated that the success of the Pouilly le Fort experiment was merely the result of tremendous luck, and that the chances of ever reproducing it are small. This is an error of fact, for similar experiments were repeated on several occasions and with equal success when carried out by Pasteur himself or done exactly according to his instructions.

In July 1881 an experiment patterned after that of Pouilly le Fort took place at the Lambert farm near Chartres, with the only modification that, to render the test even more drastic and more convincing, the sheep were inoculated not with a broth culture of the anthrax bacillus, but with the blood of an animal dead of anthrax. The result was the same as that at Pouilly le Fort: absolute resistance of the vaccinated animals, and death of the controls.

In January 1882 and again in June 1882 Pasteur described the results of experiments in which the animals had been subjected to infection by contact and by feeding, under natural conditions of exposure; in these cases again, the protection of the vaccinated animals was absolute. Identical results were also obtained by workers outside of France, showing that wherever Pasteur's method of vaccination was faithfully applied and the challenging infection test carried out within the time which he prescribed, the animals were protected.

There were, of course, some failures. Some, like the one reported from the veterinary school of Turin, were due merely to the fact that inexperienced workers had inoculated anthrax blood contaminated with the *vibrion septique*. Despite the fact that such accidents were the object of long and bitter controversies, they taught nothing new and need not be considered further. More significant were the sarcastic criticisms of Robert Koch, who maintained that, on account of the imperfection of Pasteur's techniques and because of the shortness of the immunity produced, anthrax vaccination was not a practical proposition. It is somewhat disheartening to see the great German bacteriologist attack Pasteur's discovery at the level of technical imperfections, without recognizing the immense theoretical importance and practical implications of the new procedure. Nevertheless, some of Koch's criticisms were justified and deserve attention.

Because of the pressure of work in Pasteur's laboratory, the preparation of the anthrax vaccine for wholesale distribution had been transferred to a small annex under Chamberland's supervision. Unknown to Pasteur, Chamberland had taken the initiative of adding to each bottle of vaccine a small amount of culture of the hay bacillus (*Bacillus subtilis*). When, by accident, Pasteur became aware of this modification in his technique, he guessed that Chamberland's purpose had been to minimize any further attenuation of the vaccine by adding to it a microorganism capable of absorbing all available oxygen from the bottle, for Pasteur held steadfastly to the view that contact with oxygen was one of the most effective methods to bring about the attenuation of virulence. It is almost sure that the presence of the hay bacillus in the bottles of anthrax vaccine had been spotted in a German laboratory and that it accounted for Koch's scathing remarks concerning the purity of the vaccine. It is also possible that the culture of hay bacillus used by Chamberland may have exerted a toxic effect on the anthrax bacillus, causing its total or partial inactivation. This would have resulted in a weakening of the vaccinating potency and in some of the failures reported from the field. Of deeper significance were Koch's objections concerning the shortness of the immunity produced by the treatment. Pasteur soon became aware of this limitation, and he emphasized the necessity of repeating the vaccination every year, preferably in March before the natural disease became established.

The complete success of Pasteur's own vaccination experiments was dependent upon an absolute respect for a number of minute technical details. The vaccines used had to be of the correct degree of attenuation; if too virulent, they could cause disease and even death in a number of animals; if too attenuated, they failed to establish an adequate degree of resistance. Moreover, the high level of resistance was usually short-lived, and in order to duplicate the Pouilly le Fort results, it was necessary to challenge the animals within a limited period of time after vaccination. Most experimenters failed to appreciate the importance of some of these details and attributed Pasteur's absolute success to luck. Like the experienced cook, the seasoned investigator often makes use in his work of a vast body of ill-defined but none the less real knowledge which never finds its way into the published description of experimental procedures. Pasteur had this know-how of the experimental method to an extreme degree; he owed it to a complete mastery of the smallest details of his experimental world, and to an immense persistence in repeating endlessly the experiment which he was intent on perfecting. "Allow me," he once said to a group of students, "to give you the advice which I have attempted always to follow in my own work, namely remain as long as possible with the same subject. In everything, I believe, the secret of success is in prolonged efforts. Through perseverance in one field of investigation, one succeeds in acquiring what I am inclined to call the instinct of truth."

As soon as he became convinced of the prophylactic efficacy of anthrax vaccination, Pasteur undertook to make himself the promoter of the new method. In order to convince those who wished to touch and to see before believing, he arranged for immunization experiments to be repeated in different places in France and abroad. To the secluded life in the laboratory where the studies on rabies had already begun, he now added a public life not less active, involving detailed analysis of the results of field experiments, replies to the demands for information, answers to the complaints, and defense in the face of criticism and sly attacks, as well as of open warfare.

As early as 1882, less than two years after the discovery of the attenuation technique, Pasteur was in a position to report on the results obtained with 85,000 vaccinated animals. In 79,392 sheep, the mortality from anthrax had fallen from 9.01 per cent among uninoculated to 0.65 per cent among inoculated. Thanks to prodigious efforts, anthrax vaccination soon became an established practice. By 1894, 3,400,000 sheep and 438,000 cattle had been vaccinated with respective mortalities of 1 and 0.3 per cent under natural conditions of field exposure. Just as the demonstration of the pathogenic role of the anthrax bacillus had been the touchstone of the germ theory of disease, it was the vaccination against anthrax that revealed to the medical and lay mind the practical possibilities of the new science of immunity.

Pasteur's vaccination method involved two inoculations at intervals of twelve days with vaccines of very critical potency, the second being more virulent than the first; moreover vaccination had to be repeated every year in the spring to remain effective. This method is costly and consequently its use is restricted by economic factors, but these limitations do not in any way minimize the importance of Pasteur's achievement. He had demonstrated, once and for all, that immunization against infectious diseases was a possibility. Each microorganism, each type of infection, would present new problems to be solved within the framework of the factors conditioning the course and spread of the particular disease under consideration, but the faith that immunity could be established against any infectious agent by artificial means has never faltered since the days of Pouilly le Fort.

* * *

Pasteur's next success was the immunization against swine erysipelas with a culture attenuated by passage through rabbits. Between 1886 and 1892, over 100,000 pigs were immunized in France, while the number exceeded 1,000,000 in Hungary from 1889 to 1894. It is, however, the antirabies treatment which is usually quoted as Pasteur's greatest triumph and claim to immortality, and which established the hold of microbiological sciences on the practice of medicine.

As early as October 1886, one year after the first application of the rabies treatment to Joseph Meister, Pasteur could report that there had been only ten failures out of 1726 bitten persons of French nationality who had been subjected to treatment by inoculation. Up to 1935, 51,107 patients had been inoculated in the Pasteur Institute of Paris, with 151 deaths, a mortality of only 0.29 per cent. These excellent results have been confirmed in all parts of the world. Yet, the application of the rabies treatment immediately became and remained for several years the subject of violent objections on the grounds that it was ineffective and moreover dangerous; Pasteur was accused of having infected patients with fatal rabies. These accusations are almost certainly unjustified, although it is now known that the repeated injection of nerve tissue can, under certain circumstances, give rise to paralytic symptoms which Pasteur's critics would have regarded as the effect of the rabies virus. On the other hand, opinion is still divided as to the effectiveness of Pasteur's antirabies immunization. Most epidemiologists believe that the treatment has saved far fewer lives than it was credited with at the time of its discovery; and some even doubt that it has any value at all because few of the human beings bitten by mad dogs ever develop rabies. The very existence of these startling views, similar to those reported earlier concerning smallpox vaccination, emphasizes the inadequacy of our knowledge concerning the natural history of infectious diseases. The technical reasons which account for this state of affairs cannot be discussed here, beyond restating that the existence of the many unrelated and uncontrollable factors which condition the spread and course of contagious diseases often makes it extremely difficult to evaluate convincingly the effect of prophylactic or therapeutic measures.

Some tragic failures were recorded following the successful treatments of Meister and Jupille. On November 9, 1885, there was brought to Pasteur the little girl Louise Pelletier who had been bitten on the head by a mountain dog thirty-seven days before. The nature of the wounds and the time elapsed since the bite convinced Pasteur that the treatment would almost certainly fail, and he knew that any failure would provide ammunition to the enemies of his method. Nevertheless, he could not resist the prayers of the child's parents, and against his better judgment he consented to treat her. The first symptoms of hydrophobia became apparent on November 27, eleven days after the end of the treatment, and Louise Pelletier died. She was the first casualty of the antirabies treatment, one which was often and unfairly played up by Pasteur's opponents. A few years later M. Pelletier made the following statement in a letter concerning the circumstances of his daughter's death:

"Among great men whose life I am acquainted with . . . I do not see any other capable of sacrificing, as in the case of our dear little girl, long years of work, of endangering a great fame, and of accepting willingly a painful failure, simply for humanity's sake."

As the number of patients applying for treatment increased, so naturally did the number of failures and the frequency of opposition from physicians. Most, even among Pasteur's followers, held the view that the method was not adequately worked out for human application. Some even accused Pasteur and his collaborators of homicide by imprudence. Grancher, who performed the rabies inoculations and therefore had to bear much of the brunt of the fight, has described the atmosphere of hostility that surrounded the Pasteur camp:

> These same men, fervent disciples of Pasteur, hesitated to follow him on this new ground of antirabies treatment. I can still hear Tarnier speaking, as we walked out of those memorable meetings at the Academy of Medicine where

Pasteur's adversaries accused him and his disciples of homicide.

"My dear friend," Tarnier told me, "it would be necessary to demonstrate, by repeated experiments, that you can cure a dog, even after intracranial inoculation; that done, you would be left in peace."

I replied that these experiments had been made; but Tarnier did not find them numerous enough, and still he was one of Pasteur's friends.

I felt disaffection and embarrassment grow all around us, not to speak of the anger which was brooding under cover. One day, I was at the Medical School for an examination. . . . I heard a furious voice shouting, "Yes, Pasteur is an assassin." I walked in, and saw a group of my colleagues, who dispersed in silence.

And this was not Professor Peter, who had at least the courage of his opinion. And his opinion was as follows. "During the first ten months of the Pasteur treatment, the method was ineffective. Now that it has been modified, it has become outright dangerous. Pasteur confers on the persons whom he inoculates the hydrophobia of rabbits – 'laboratory rabies.' "

These assertions were based upon the type of clinical symptoms exhibited by a few patients who had succumbed despite the treatment. In vain did we point out that rabies was not yet known in all the variety of its symptoms. In vain Vulpian would point to numerous facts of paralytic rabies reported before the advent of the treatment. Pasteur's adversaries replied that paralytic rabies was transmitted to man by the injection of the spinal cord of animals dead of paralytic rabies. . . .

Certain political and medical journals as well as a number of politicians and the Antivivisectionist League conducted a violent campaign against Pasteur. Even in the colleges of Paris, students would split into Pasteurians and anti-Pasteurians and engage in fights.

In the meantime, the laboratory was bending under the weight of the demands placed upon it. I was in charge of the inoculations and prepared the statistics with the help of Chantemesse and Charrin; Roux carried out the many tests required to establish the presence of rabies in biting dogs, and his activity, multiplied by the hard work of Viala, could hardly cope with this huge task. It was therefore quite impossible to satisfy Tarnier's request and to take up again the experiments of vaccination of dogs after inoculation by the intracranial route.

Furthermore, at this date of January 1887, there was no laboratory in Europe equipped to repeat Pasteur's experiments.

Pasteur's health had suffered from continuous exertion, from anxiety over the results of the antirabies treatment, and from these endless and bitter struggles. As he began to exhibit symptoms of cardiac deficiency, his doctors Villemin and Grancher persuaded him to leave Paris for the South at the end of November 1886, but the attacks against him and his method did not subside during his absence. A suit was threatened against the laboratory by the father of a young patient who had died after receiving the treatment. Medical testimony had been obtained that the child had died of the type of hydrophobia characteristic of the rabbit paralytic disease, and that Pasteur and Grancher were therefore responsible for it. It was at that critical time that Roux, who had been estranged from the laboratory and even avoided seeing Pasteur, had returned to share in the common danger and help weather the storm.

This incident brought Roux into contact with a young doctor, Georges Clemenceau, who was to become a center of turmoil in French political life during the early part of the twentieth century, and to gain international fame as the Tiger of France during the First World War. Clemenceau was a physician whose life was torn between his medical interests, his free thinking and radical philosophy, and his passionate love of freedom. He had taken sides against the antirabies treatment and was exploiting the case mentioned above in the political press. His opposition to rabies inoculation probably originated from the fact that he was hostile to Pasteur's conservative views in politics and that, like most other leftist thinkers, he had favored the theory of spontaneous generation. Half a century later, in January 1930, it was Roux who was selected to deliver before the Academy of

Medicine the obituary speech after Clemenceau's death. He related that in March 1924 the aged Tiger had asked him to call at his apartment. "He questioned me at length on the nature of fermentation and on the role of microbes in the transformation of matter. . . . The next day, I received from him a special delivery letter asking for further information on the subject of fermentation. Thus Clemenceau appeared to me as a philosopher, revising with full serenity his scientific view of the world in the twilight of his life."

As mentioned above, it is probably true that the antirabies treatment may bring about paralytic symptoms in a few cases, although these are not necessarily due to the active virus present in the vaccine. Fortunately, these accidents are extremely rare and it is almost certain therefore that the accusations directed against Pasteur on this score were unjustified. There was perhaps more ground for the attacks aimed at the efficacy of the treatment. Pasteur's statistics, which have been repeatedly confirmed, indicated that more than 99.5 per cent of the individuals bitten by rabid dogs fail to die of the disease if treated by his method. Judged in such terms, the therapeutic result appears remarkable, but in reality it is difficult to evaluate the significance of this figure because the chance of an exposed person's contracting rabies if left unvaccinated is unknown. It appears probable that man possesses a high resistance to the rabies virus and that the chance of fatal infection is exceedingly small, so small indeed that proof of the utility of the treatment is difficult to obtain. The discussions held on this subject before the Paris Academy of Medicine in the 1880's reveal that some French clinicians of the time held similar views. In an impassionate and able plea against the Pasteur treatment, the Parisian clinician Michel Peter stated his case in the following terms:

> Rabies in man is a *rare disease, exceedingly rare;* I have seen only two cases in the course of thirty-five years of hospital and private practice, and all my colleagues in hospitals in the city, as well as in the country, count in units and not in tens (let alone hundreds) the cases of human rabies which they have observed. In order to amplify the beneficial

> effects of his method and to mask its failures, it is M. Pasteur's interest to believe that the annual mortality of rabies in France is higher than it really is. But these are not the interests of truth.
>
> Do you wish to know for example how many individuals have died of rabies in Dunkirk in a period of twenty-five years? Only *one*. And do you wish to know how many have died of it in this same city in one year, since the application of the Pasteur method? *One* died of rabies.

It would be difficult to determine whether belief in the rarity of human rabies was then prevalent among physicians, or whether Peter was misrepresenting the situation to bolster his thesis. In any case, there cannot be any doubt that Pasteur had some justification in not sharing his colleague's view of the rarity of the disease. He remembered well the rabid wolf in Arbois and the eight victims who had succumbed to hydrophobia following bites on the head and hands. Later the reading of official reports had confirmed the impression left on him by this childhood experience. An official inquiry had concluded that, of 320 cases studied, 40 per cent had died after bites from rabid dogs. In another report from the Sanitary Department of the City of Paris, the mortality rate had been estimated at 16 per cent. Finally, at the very time that Joseph Meister was under Pasteur's care, five persons were bitten by a rabid dog near Paris, and every one of them had died of hydrophobia. It is of little surprise, therefore, that Pasteur and most of his contemporaries should have been overwhelmed by the low mortality of 0.5 per cent among humans receiving the antirabies treatment.

The difficulty of evaluating quantitatively the prevalence and severity of rabies was well expressed by the English commission charged with the duty of investigating the validity of Pasteur's claims in 1888:

> (1) It is often difficult, and sometimes impossible, to ascertain whether the animals by which people were bitten, and which were believed to be rabid, were really so. They may have escaped, or may have been killed at once, or may have been observed by none but persons quite incompetent to judge of their condition.

(2) The probability of hydrophobia occurring in persons bitten by dogs that were certainly rabid depends very much on the number and character of the bites; whether they are on the face or hands or other naked parts; or, if they have been inflicted on parts covered with clothes, their effects may depend on the texture of the clothes, and the extent to which they are torn; and, in all cases, the amount of bleeding from the wounds may affect the probability of absorption of virus.

(3) In all cases, the probability of infection from bites may be affected by speedy cauterizing or excision of the wounded parts, or by various washings, or other methods of treatment.

(4) The bites of different species of animals, and even of different dogs, are, probably, for various reasons, unequally dangerous. Last year, at Deptford, five children were bitten by one dog and all died; in other cases, a dog is said to have bitten twenty persons of whom only one died. And it is certain that the bites of rabid wolves, and probable that those of rabid cats, are far more dangerous than those of rabid dogs.

The amount of uncertainty due to these and other causes may be expressed by the fact that the percentage of deaths among persons who have been bitten by dogs believed to have been rabid, and who have not been inoculated or otherwise treated, has been, in some groups of cases, estimated at the rate of only 5 per cent, in others at 60 per cent, and in others at various intermediate rates. The mortality from the bites of rabid wolves, also, has been, in different instances, estimated at from 30 to 95 per cent.

All students of rabies appear in agreement on a few points. The chance of an individual contracting the disease depends to a large extent upon the depth and location of the bite inflicted by the rabid animal, and the bite of a mad wolf is very likely to cause rabies. In our communities, most wolves are behind the gates of the zoos, and a young Louis Pasteur of today would have no chance of seeing that wild animal roaming about the countryside of Arbois. Uncontrolled and stray dogs also have become scarce; like modern man under normal circumstances, they lead a lazy and comfortable life. Fed on ground-up diets and on well-cooked bones, most of them have lost the habit, if not the profound instinct, of biting deeply into living flesh; ugly dogs, once common on lonely farms, are almost nonexistent today. The rabies virus, and the susceptibility of man to it, have probably not changed, but the social circumstances under which man encounters the virus may be sufficiently different to have altered somewhat the expected course of the disease since Pasteur studied it.

Fortunately for Pasteur's peace of mind, his work on rabies was immediately investigated by the official English commission mentioned above, which repeated the animal experiments in England and analyzed the results of the human treatment in Paris. Its report, issued in July 1888, confirmed Pasteur's experimental findings by stating:

> 1. That the virus of rabies may certainly be obtained from the spinal cords of rabbits and other animals that have died of that disease.
>
> 2. That, thus obtained, the virus may be transmitted by inoculation through a succession of animals, without any essential alteration in the nature, though there may be some modifications of the form of the disease produced by it.
>
> 3. That, in transmission through rabbits, the disease is rendered more intense; both the period of incubation, and the duration of life after the appearance of symptoms of infection being shortened.
>
> 4. That, in different cases, the disease may be manifested either in the form called dumb or paralytic rabies which is usual in rabbits; or in the furious form usual in dogs; or in forms intermediate between, or combining both of these, but that in all it is true rabies.
>
> 5. That the period of incubation and the intensity of the symptoms may vary according to the method in which the virus is introduced, the age and strength of the animal, and some other circumstances; but, however variable in its intensity, the essential characters of the disease are still maintained.
>
> 6. That animals may be protected from rabies by inoculations with material derived from spinal cords prepared after M. Pasteur's method. . . .

The Commission also investigated Pasteur's clinical records in Paris and carried out a detailed inquiry in the homes of ninety patients who had received the antirabies treatment; while emphasizing the difficulty of evaluating the normal fatality following bites by rabid animals, it expressed confidence in the value of Pasteur's results.

> Thus, the personal investigation of M. Pasteur's cases by members of the Committee was, so far as it went, entirely satisfactory, and convinced them of the perfect accuracy of his records. . . .
>
> From the evidence of all these facts, we think it certain that the inoculations practiced by M. Pasteur on persons bitten by rabid animals have prevented the occurrence of hydrophobia in a large proportion of those who, if they had not been so inoculated, would have died of that disease. And we believe that the value of his discovery will be found much greater than can be estimated by its present utility, for it shows that it may become possible to avert, by inoculation, even after infection, other diseases besides hydrophobia. . . .

Peter remained unimpressed by the report, remarking facetiously, "The most curious point in this story is that the Report of the English Commission does not conclude, as one might have expected, by recommending the establishment of a Pasteur Institute in London, but instead recommends, as a means of rabies prevention, a more rigorous enactment of police regulations on dogs."

And, indeed, the English Commission was correct in its practical conclusions as well as in its evaluation of the importance of Pasteur's work. Even granted that the antirabies treatment had saved the lives of a few human beings, this would have been only meager return for so much effort, and for so many animals sacrificed on the altar of man's welfare. The same result could have been obtained, at much lower cost, by the muzzling of dogs and by the training of their owners to keep them under control. It is on much broader issues that Pasteur's achievements must be judged. He had demonstrated the possibility of investigating by rigorous techniques the infectious diseases caused by invisible, noncultivable viruses; he had shown that their pathogenic potentialities could be modified by various laboratory artifices; he had established beyond doubt that a solid immunity could be brought about without endangering the life or health of the vaccinated animals. Thanks to the rabies epic, men were to be immunized against yellow fever and several other widespread virus diseases; even more important, immunization had become recognized as a general law of nature. Its importance for the welfare of man and animals is today commonplace, but only the future will reveal its full significance in the realm of human economy.

The acquisition of immunity to an invading parasite is, in many ways, one of the most extraordinary phenomena of life. Man and animals can become resistant to what would otherwise be fatal infective doses of the causative agents of many infectious diseases, as a result of prior exposure to these agents; the immunity is specific and it is lasting, sometimes for a few months, often for many years. What is the nature of this change that transforms selectively the behavior of a living being toward a small fragment of the Universe?

Pasteur had a ready answer to this question, one that for a time he considered convincing because it presented analogies with some of his previous experiences. He had observed that each microorganism has exacting nutritional requirements, the anthrax bacillus growing well in neutralized urine, the chicken cholera organism in chicken broth. By analogy, he imagined that the sheep is susceptible to anthrax, and the dog not sensitive to the same disease, because the former animal provides an adequate growth medium for the specific bacillus, and the latter does not. Pasteur had also recognized that the chicken cholera bacillus refused to multiply in a medium in which a culture of it had already been grown. For similar reasons, he felt, microbial agents of disease refuse to grow in a body which they have previously invaded: this body, like the medium, has been depleted by the first invasion of some factor essential for growth.

"One could imagine that cesium or rubidium are elements necessary for the life of the microbe under consideration, that there exists only a small amount of these elements in the tissues of the animal, and that this amount has been exhausted by a first growth of this microbe; this animal, then, will remain refractory until its tissues have recuperated these elements. As they are scarce, a long time may elapse before recuperation is adequate." This "exhaustion theory" was by no means new; it had been suggested by Tyndall and in particular by Auzias-Turenne almost two decades earlier. Pasteur himself soon recognized, however, that the theory was incompatible with some of the facts of immunity and he quickly discarded it, as he was always ready to abandon concepts that were not fruitful of new discoveries.

He then turned the argument around: "Many microbes appear to produce in the media where they grow substances that have the property of inhibiting their further development. Thus, one can consider that the life of the microbe, instead of removing or destroying certain essential components present in the body of animals, on the contrary adds new substances which could prevent or retard its later growth." This was not an idle speculation. He had tried in 1880 to separate such an inhibitory principle from cultures of the chicken cholera bacillus. Although this attempt had failed, he was eager to pursue the hypothesis further. "I believe, today, that the attempt should be repeated in the presence of pure carbon dioxide, and I shall not fail to try it." This was late in 1885. The controversy on rabies vaccination was increasing in violence, and time was getting short. In 1888, Pasteur suffered a new attack of paralysis and had to abandon experimental work. Had he been able to work a few years longer, he would certainly have recognized that his new hypothesis did not yet fit the facts, although it was getting closer to the truth. There are indeed produced in the body, as a result of infection or of immunization, substances which may interfere with the development of the infective agent in the tissues; these substances, however, are not produced by the microorganism but by the invaded body itself, as a response to the infectious process. Pasteur never engaged in the experimental analysis of this "immune response" of the host, but he lived to see one of its greatest triumphs in the development of diphtheria antitoxin, an achievement to which his assistant Roux contributed much important work in the newly created Pasteur Institute.

Pasteur believed that the protective effect of vaccination resulted from the multiplication of the attenuated cultures in the body, a view which has been amply confirmed. He had also suspected, however, that the immune reaction was not necessarily dependent upon the living processes of the parasites but might be directed against certain of their constituents or products. Should this prove to be the case, he felt, one might use for vaccination these constituents or products of the microbial cell, instead of living attenuated cultures. Thus, after having discovered that culture filtrates of the chicken cholera bacillus contained a nonliving soluble toxin, he injected the culture filtrate into birds hoping to immunize them against the disease, but failed.

Early in the course of the rabies work, he suspected that attenuation of the infected spinal cord did not consist in a change in the intrinsic virulence of the rabies virus, but was the result of progressive decrease in the number of living particles. "The progressive increase in the length of incubation of the disease induced . . . by our spinal cords desiccated in contact with air, is due to the decrease in quantity of rabies virus in these cords and not to a decrease in their virulence." This conclusion led him to a further hypothesis. He postulated that immunization might be due not to the living virus itself but to a nonliving substance which retained its immunizing power even after the virus had been killed by prolonged desiccation. In other words, he believed in the existence of "a vaccinating substance, associated with the rabies virus." Interestingly enough the first record of this extraordinary thought is dated from a meeting of the Academy of Letters as early as January 29, 1885. In the course of a discussion on the Dictionary of the Academy, Pasteur wrote the following note: "I am inclined to believe that the causative virus of rabies may be accompanied by a substance which can impregnate the nervous system and render it thereby unsuitable for the growth of the virus. Hence rabies immunity. If this is the case, the theory might be a general one; it would be a stupendous discovery." On August 20, 1888, at the end of his active scientific life, he reported preliminary experiments suggesting that antirabies immunity could be induced in dogs by injecting infected spinal cord rendered noninfectious by heating for forty-eight hours at 35° C. "The heated cord which had become noninfectious was still effective as a chemical vaccine." Indeed, he went so far as to state, "It will not be long before the chemical vaccine . . . of rabies is known and utilized." This aspect of Pasteur's work is never discussed in textbooks, and it appears worth while therefore to quote at length the views that he presented in 1888 in the first issue of the newly created *Annales de l'Institut Pasteur.*

> How could one explain without assuming the existence of the rabies vaccinating substance the fact that . . . two dogs each inoculated under the skin with the content of ten syringes of a very virulent virus . . . became at once resistant to rabies? How is it possible that the large amount of rabies virus introduced under the skin does not start multiplying here and there in the nervous system if, at the same time, there were not introduced a substance reaching this system even faster, and placing it under conditions such that it is no longer capable of allowing the growth of the virus . . . ?
>
> Some will ask why inoculation by trephination always induces rabies, and never the refractory state. . . . The true difference between the two routes of inoculation appears to be that the inoculation under the dura mater permits the introduction of only very small amounts of virus and consequently of its vaccinating substance, amounts insufficient to induce the refractory state, whereas much larger amounts can be introduced under the skin.

Only experienced immunologists can appreciate the visionary character of these statements that acquired their full significance only after fifty years of research in the virus field. It is possible that the mechanism of resistance to rabies perceived by Pasteur in the dim light of his time will become more obvious when his findings are interpreted with the help of modern knowledge. Effective immunization with killed filtrable viruses, and demonstration of the phenomenon of interference, are technical achievements only of the past decade and there is an exciting atmosphere of archeological discovery in detecting their first expression in these hesitating statements of the founder of immunology.

A few months later, in the last presentation of experimental work from his own laboratory before the Paris Academy of Sciences, Pasteur reported a few sketchy observations concerning the possible existence in anthrax blood of a vaccinating substance. He had injected repeatedly into rabbits the blood withdrawn from infected animals and heated at 45° C. for several days; although the heated blood was presumably free of living bacteria, it seemed to confer upon the animals a certain degree of immunity. The experiments had had to be interrupted in the fall of 1887 because of Pasteur's ill-health. When he returned to Paris the following spring, he was a broken man, unable to pick up the tools. After having accepted every scientific challenge, having fought with facts, men and infirmity, he finally had to give up.

Despite the inconclusiveness of these last observations on the vaccinating substances of rabies and of anthrax, there is a great human beauty in the spectacle of Pasteur getting ready at the end of his life to start on a new intellectual adventure. Most of his popular scientific triumphs had been gained by demonstrating the participation of a vital principle in chemical and pathological processes; he had shown that fermentation, putrefaction and disease were caused by living microbial agents; that immunity could be established with attenuated living germs of disease. Thanks to him, a new land had been discovered and was being conquered; busy men were at work to settle and exploit it. But the old explorer was on his way again, blazing new trails. Living microorganisms were the cause of disease as well as of immunity, but how and through what agencies did they perform these prodigious feats? The Sibylline thoughts on the vaccinating substance of rabies, the crude observations on the immunizing power of heated anthrax blood, were gropings towards the new continent where the chemical controls of disease and immunity were hidden. There, Pasteur would have joined hands with his old opponents, Liebig and Claude Bernard. He had not, as they thought, forfeited the luminous and vigorous doctrine of modern physiology for some dusty and degenerate vitalistic philosophy. He had searched with curiosity and eagerness for the primary causes of natural phenomena and found them in living processes. But instead of submitting to Life, he had first learned to control and domesticate her, and was now ready to extract from the living entrails the secret of her power. It is only because human days are so short that he left his work unfinished.

CHAPTER XIII

Mechanisms of Discoveries

> We want the creative faculty to imagine that which we know.
>
> — SHELLEY

FOR THE SAKE of convenience, we have presented Pasteur's scientific work as a series of separate problems. In reality, these problems were never separated in his mind, their prosecution often overlapped in time, and he considered them as part of a whole, evolving one from the other.

Within this fundamental unity, one can recognize two definite chronological sequences in the questions which Pasteur chose for study. On the one hand, as we have shown, his emphasis shifted toward the solution of practical problems, away from large theoretical issues. True enough, he exhibited to the end the same acuity in relating experimental findings to questions of broad significance, but he found less and less time to develop those aspects of his discoveries which did not bear on practical matters of technology or medicine. On the other hand, his work shows an evolution from the physicochemical, through the chemical and biochemical, to the purely biological point of view. This is evident from the topics which he selected for investigation — first molecular structure, then the physiological mechanisms of fermentation, and finally the pathogenesis of infectious diseases.

Pasteur attempted to rationalize this evolution by attributing it to a compelling inner logic which had led him inevitably from one subject to the next — "Carried on, enchained should I say, by the almost inflexible logic of my studies, I have gone from investigations on crystallography and molecular chemistry to the study of microorganisms."

The chronological sequence of Pasteur's studies gives credence to the view that they are linked in an orderly manner within a progressively developing conceptual scheme, and the theory that it was a compelling logic which led him from crystallography to disease has been widely accepted. In reality, Pasteur's own writings provide evidence that the different aspects of his work did not stem one from the other, in a progressive and orderly manner, as would appear from the order of appearance of his major publications. His greatest discoveries were the fruits of intuitive visions and they were published in the form of short preliminary notes long before experimental evidence was available to substantiate them.

The dates of first publication of Pasteur's most important achievements reveal that the essential steps in discovery occurred at the very beginning of each of the periods which he devoted to the different fields of research. It was in 1848 – he was then twenty-six, and had just graduated from the Ecole Normale – that he published his findings and views concerning the relations between the crystal morphology of organic substances and their ability to rotate the plane of polarized light. All his subsequent publications until 1857 are essentially elaborations of these views; the discovery made by the student at the Ecole Normale was the propelling force for ten years of research activity by the young chemist.

In August 1857, shortly after having begun to work on fermentation, Pasteur presented in his preliminary paper on lactic acid a precise statement of the laws and methodology of a new science devoted to microorganisms and to the role they play in the economy of matter. Experimental evidence to substantiate these theoretical views kept him at work until 1875.

A special phase of this problem, namely the existence of anaerobic life and its relation to the intimate mechanism of fermentation, first appears in his publications of 1860. But although the statement that "fermentation is life without oxygen" dates from February 1861, it was not until 1872 that Pasteur presented extensive discussion of its biochemical significance.

The studies on spontaneous generation, on the manufacture of vinegar, wines, and beers, on the technique of pasteurization, which extend from 1860 to 1875, do not involve any new fundamental concept and are merely developments of his earliest theoretical views on the germ theory of fermentation.

The studies on silkworms illustrate in extreme degree Pasteur's successful use of "hunches" or intuition in the solution of scientific problems. Within two weeks after his arrival in Alais, he recommended the egg-selection method that was to lead to the practical control of pébrine. The four following years were devoted to working out the practical details of the method, demonstrating its effectiveness, and elucidating the nature of the disease. In this case, the different phases of the work followed in an order opposite to what might have been expected from a logical development.

It was in 1877 that Pasteur published his first studies on animal pathology. Two years later, he recognized the possibility of immunizing against chicken cholera, and, generalizing from this accidental observation, perceived its analogy with the procedure of vaccination against smallpox. From then on, he turned all his energies to the preparation of "vaccines" against various bacterial diseases, a pursuit occupying the balance of his scientific life.

Pasteur achieved his most startling results through bold guesses which permitted him to reach the solution of a problem before undertaking its systematic experimental study. Because he was well trained in the philosophy of the experimental method, he recognized that these guesses were nothing more than working hypotheses, the validity of which had to be verified and demonstrated by critical scrutiny, and which became useful only to the extent that operational techniques could be evolved to develop and exploit their logical consequences. Interestingly enough, the urge to overcome objections and contradictions, to triumph over his opponents, became in many cases a powerful incentive to the systematic accumulation of the proofs necessary to support theories that had first been affirmed without convincing evidence. In the work of Pasteur, logic is evident in the demonstration and exploitation of his discoveries, rather than in their genesis. It is the phase of his work devoted to the development of his ideas which makes the bulk of his long papers, and which gives the impression of orderly logical progression.

Pasteur's associates and contemporaries have emphasized his dreamy and intuitive nature; and Tyndall described his genius as a happy blending of intuition and demonstration. The use of intuition as a guide to discovery is perhaps a more common procedure than some exponents of the scientific method are inclined to believe. An extreme interpretation of Francis Bacon's writings has led to the view that the accumulation of well-established facts is sufficient to the elaboration of scientific truth, that facts speak for themselves and become automatically translated into general laws. It is indeed certain that the experimental method has a self-propelling force, and that many discoveries have been made by the routine and faithful application of its rules without an obvious use of hypotheses or intuition, but it is also true that scientific creation often involves the selection, from the wealth of amorphous data, of those facts which are relevant to a problem formulated in advance from abstract concepts.

In this respect the progress of science depends to a large extent upon anticipatory ideas. These give rise to the working hypothesis that constitutes the imaginative component and one of the mainsprings of scientific discovery. Before addressing himself to nature for a definite answer from results of experiments, every investigator formulates tentative answers to his problem. The experiment serves two purposes, often independent one from the other: it allows the observation of new facts, hitherto either unsuspected, or not yet well defined; and it determines whether a working hypothesis fits the world of observable facts. The precision and the frequency with which hypotheses hit the target of reality constitute a measure of the intuitive endowment of their author. Needless to say, successful guesses are not sufficient for the instrumentation of discovery. The scientist must also be able to demonstrate the validity and to exploit the consequences of his intuitions if they are not to be stillborn.

Only few of the great experimenters have described the mental processes by which they discovered new facts or formulated new generalizations. Some, it must be admitted, assure us that their method consists merely in the use of their eyes, their ears, and other physical senses to perceive and describe reality as it presented itself to them. This view is illustrated in the picturesque words of the physiologist François Magendie: "I am a mere street scavenger of science. With hook in hand and basket on my back, I go about the streets of science collecting whatever I find." Others have told a very different story. They report how a period of intense preoccupation with a given problem was followed by a flash of inspiration often occurring under odd circumstances, away from the bench or the desk, in the course of which the solution presented itself, ready-made, as emerging from some subconscious labor. Examples of inspired creations are common from the world of arts and letters, and many scientists, several of them Pasteur's contemporaries, have acknowledged a similar origin to their discoveries.

In the course of an address on his seventieth birthday, Helmholtz thus described how his most important thoughts had come to him. "After previous investigation of the problem in all directions . . . happy ideas come unexpectedly without effort, like an inspiration. . . . They have never come to me when . . . I was at my working table. . . . They come . . . readily during the slow ascent of hills on a sunny day."

According to William Thompson (Lord Kelvin), the idea of the mirror galvanometer occurred to him at a moment when he happened to notice a reflection of light from his monocle. The theories of the structure of the atom and of the benzene ring were formulated by Kekule under the following circumstances. He had been visiting a friend in London and was riding home on the last bus. Falling into a revery, he saw atoms flitting before his eyes, two coupled together, with larger atoms seizing the smaller ones, then still larger atoms seizing three and even four smaller atoms, all whirling around in a bewildering dance, the larger atoms forming a row and dragging still smaller atoms at the end of the chain. Arriving home, he spent the night sketching pictures of the "structural theory."

At the time of the discovery of the benzene ring theory, Kekule was working on a textbook. Turning from his desk toward the fireplace, he fell into a hypnotic state of mind, seeing the same atoms flitting again before his eyes, long rows of them assuming serpentine forms. All at once, one of the serpents seized his own tail "and whirled mockingly before his eyes." Flashing awake at once, Kekule began writing the benzene ring theory.

To the uninitiated, it appears even more remarkable that many mathematicians like Gauss, Poincaré and Einstein have traced some of their greatest discoveries to a sudden illumination. Einstein said, in *Physics and Reality:*

> There is no inductive method which could lead to the fundamental concepts of physics. Failure to understand this fact constituted the basic philosophical error of so many investigators of the nineteenth century. . . . We now realize with special clarity, how much in error are those theorists who believe that theory comes inductively from experience. . . .

Even Clerk Maxwell, probably the most rigorous and logical scientific mind of the nineteenth century, has emphasized that purely imaginative mechanical models and analogies are often the precursors of mathematical abstractions. As is well known, Faraday evolved many of his discoveries from the mechanical concept of lines of forces; for twenty-five years, he used and elaborated this model until the lines of forces became to him as real as matter, and he mentally constructed a model of the universe in such terms. Maxwell at first borrowed from Faraday a similar model of the electromagnetic field. True enough, he discarded its use after he had reached an adequate mathematical formulation of electromagnetism with its help, but he acknowledged his indebtedness to Faraday's mechanical concept and added: "For the sake of persons of different types of mind, scientific truth should be presented in different forms and should be regarded as equally scientific whether it appears in the robust form and vivid coloring of a physical illustration or in the tenuity and paleness of a symbolical expression."

Elsewhere, Maxwell attempted to analyze Faraday's method of discovery and admitted the possibility of apprehending truth by approaches vastly different from those usually understood under the name of scientific method. He considered that reality might be perceived not only through clear intellectual steps leading to well-understood relationships, but also through the apprehension of phenomena and events as a whole, before any analytical process has revealed the nature and relations of their component parts.

> Faraday's methods resembled those in which we begin with the whole and arrive at the parts by analysis, while the ordinary mathematical methods were founded on the principle of beginning with the parts and working up to the wholes by synthesis. . . .
>
> We are accustomed to consider the universe as made up of parts, and mathematicians usually begin by considering a single particle, and so on. This has generally been supposed the most natural method. To conceive of a particle, however, requires a process of abstraction, since all our perceptions are related to extended bodies, so that the idea that the *all* is in our consciousness at a given instant is perhaps as primitive an idea as that of any individual thing. Hence, there may be a mathematical method in which we move from the whole to the parts instead of from the parts to the whole.

Is not this apprehension of the whole responsible in part for some of the mysterious processes of intuition that have so often been claimed by men of science? Was it not such a process which made Gauss reply, when asked how soon he expected to reach certain mathematical conclusions, "that he had them long ago, all he was worrying about was how to reach them"?

In certain respects, Darwin used this unanalytical intuitive approach in formulating the theory of evolution based on natural selection. He became convinced of the fact of organic evolution – the variability of species – during his short stay in the Galapagos Islands, and the hypothesis of natural selection came to him in a flash while reading Malthus's essay on population. Twenty years elapsed before he would publish his theory, a period devoted to the accumulation of the detailed body of facts required to bolster his preconceived views.

"The imagination," he said, "is one of the highest prerogatives of man. By this faculty he unites former images and ideas independently of the will, and thus creates brilliant and novel results. . . ." The value of the products of imagination depends, of course, upon the number, accuracy, and clearness of the impressions on which they are based; it is also conditioned by the power of voluntarily combining these impressions, and by the judgment and taste used in selecting or rejecting involuntary associations.

The opposition to the *Origin of Species* came not only from churchmen and from those scientific quarters in which the fixity of species was then an unattackable dogma, but also from many who questioned the validity of Darwin's discovery because it had not been achieved by the Baconian method, and owed too much to imagination instead of depending solely upon objectivity and induction. This aspect of the opposition to Darwin strengthened Huxley's belief that the Baconian method was fruitless as an instrument of discovery, and that imagination and hypothesis were the most powerful factors in the development of science.

"Those who refuse to go beyond fact rarely get as far as fact; and anyone who has studied the history of science knows that almost every great step therein has been made by the anticipation of nature; that is, by the invention of a hypothesis which, though verifiable, often had little foundation to start with; and not infrequently, in spite of a long career of usefulness, turned out to be wholly erroneous in the long run." Huxley came to feel that

Bacon's "majestic eloquence and fervid vaticination" were yet, for all practical results concerning discovery, "a magnificent failure."

Yet the great Chancellor probably never meant that the unimaginative accumulation of facts is synonymous with science, but wanted only to affirm that imagination cannot function usefully without the help of accurate facts. The mirror galvanometer, the formula of the benzene ring, the theory of evolution, had not been generated from nothing, as a product of "pure" imagination; they were the fruits of an enormous growth of physical, chemical and biological knowledge that had been available to Thompson, Kekule or Darwin at the proper time for the formulation of a scientific synthesis. The few who reach the intuitive perception of truth must be preceded by the host of workers, most of them forgotten, whose role it has been to accumulate the facts that constitute the raw material of successful working hypotheses, of the intuitions of discovery. The immense wastefulness of organic life, which demands that thousands of germs perish so that one may live, has its counterpart in the processes of intellectual life; many must run, so that one or a few may reach the goal.

Because every discovery, even that which appears at first sight the most original and intuitive, can always be shown to have roots deep in the past, certain students of the history of science believe that the role of the individual in the advancement of knowledge is in reality very small. To support their views, they point out that many discoveries have been made simultaneously in different places, by different individuals working independently and unknown to each other.

Thus, the phenomenon of electromagnetic induction was discovered independently and almost simultaneously by Joseph Henry in America and by Michael Faraday in England. Similarly, the law of conservation of energy was suggested in 1844 by Grove in his essay "The Correlation of Forces"; it was implied in Faraday's equivalence of different forms of force in 1847; it was analyzed in clear terms by Helmholtz in Germany and by Joule in England. And before any of these, the French physicist Carnot, the Russian chemist Lomonosov and the German physician Mayer had arrived at essentially the same conclusion. Obviously, this second law of thermodynamics was as much a product of the preoccupation of the age as it was an expression of the genius of the men who formulated it. The periodic table of the chemical elements provides another example in which the accumulation of chemical knowledge became sufficient at a certain time to elicit in two independent workers, Lothar Meyer and Dmitrij Mendelejeff, the vision of an orderly relationship between the properties of atoms. In a similar manner, Darwin and Wallace reached simultaneously the conclusion that species of living organisms have evolved one from the other.

Further evidence that the progress of science depends less than is usually believed on the efforts and performance of the individual genius, is found in the fact that many important discoveries have been made by men of very ordinary talents, simply because chance had made them, at the proper time and in the proper place and circumstances, recipients of a body of doctrines, facts and techniques that rendered almost inevitable the recognition of an important phenomenon. It is surprising that some historian has not taken malicious pleasure in writing an anthology of "one discovery" scientists. Many exciting facts have been discovered as a result of loose thinking and unimaginative experimentation, and described in wrappings of empty words. One great discovery does not betoken a great scientist; science now and then selects insignificant standard bearers to display its banners.

For all these reasons, one cannot doubt that discovery is always an expression of the intellectual, social and economic pressure of the environment in which it is born. Nevertheless, each generation produces a few individuals who direct this pressure into meaningful channels and who discipline and harness the chaotic forces of their scientific age to create out of them the temples of knowledge. Science has her *nouveaux riches,* opportunists who exploit new fields of research opened by others, or those who merely profit from a discovery made by accident. But there are also in the kingdom of science visionary explorers, builders, statesmen and lawgivers. History demands that both groups be considered, for both play a part in the evolution of knowledge. However, it is only by studying the mental processes of the creators and men of vision that we can hope to decipher the mechanisms of discovery, and to understand the relation of our perception to the world of facts. Of all this, unfortunately, nothing is known; no one can predict who will formulate a new law or recognize a new fact, and there is as yet no recipe by which a scientific discovery can be made. Progress in the understanding of the intellectual factors involved will certainly be slow, for, like its literary and artistic counterparts, the process of scientific creation is a completely personal experience for which no technique of observation has yet been devised. Moreover, out of false modesty, pride, lack of inclination or psychological insight, very few of the great discoverers have revealed their own mental processes; at the most, they have described methods of work – but rarely their dreams, urges, struggles and visions.

Pasteur made a few remarks concerning those of his qualities which played a part in the unfolding of his astonishing scientific performance. Besides his reference to the logic which "enchained" him from one field of endeavor to another, he often mentioned his use of preconceived ideas, from which he derived the stimulus for many experiments – preconceived ideas which he was always willing to abandon when they did not fit the observed facts. He also emphasized his painstaking efforts in the laboratory, his patience, his persistence, his willingness to submit to the teachings of experiment even when they went counter to his theoretical views. But in reality, these statements do not reach the core of the problem. They tell nothing of what made Pasteur – and his peers in the Kingdom of Science – different from their contemporaries who also formulated hypotheses, checked them by the experimental method, labored diligently and faithfully, and yet failed to leave their footprints on the sands of history. We can see the mechanics of Pasteur's workings, but their inward urge remains hidden to us.

All those who saw Pasteur at work and in everyday life have emphasized how completely and exclusively he became engrossed in the problem at hand, and how great was his power of concentration. Visitors were unwelcome, laboratory associates must be few and silent, not even the family dinner or home atmosphere could interrupt the preoccupations of the day. So as to possess more completely all the details of the work done in his laboratory, he would insist upon writing down himself all experimental procedures and findings in the famous notebooks. When returning from some Academy meeting, he would go down to the animal quarters, tear from the cages the labels prepared by his collaborators, and make new ones in his own handwriting as if to identify his life more completely with the experiments. He had the ability, and the discipline, to focus all his physical and mental energies on a given target, and perhaps as a consequence he could recognize immediately all manner of small details pertaining to it. One gets the impression that the intense "field" of interest which he created attracted within his range all the facts—large and small—pertinent to the solution of the problem which was preoccupying him.

With his nearsighted eye, he was capable of seeing much that escaped others, a quality which he had probably possessed from his early youth. One can recognize it in the portraits which he painted in his early teens, and years of disciplined effort had merely served to intensify the priceless attribute.

His first scientific venture illustrates well his method of investigation. It was, as reported earlier, by intense pondering over the relation of optical activity to crystal morphology that he had imagined—seen with the mind's eye—that the crystals of optically active tartaric acid might display morphological evidence of asymmetry. And it was because of his gift of observation that he actually saw on the crystals the small asymmetric facets which his predecessors had failed to notice.

Pasteur remained throughout his life an immensely effective observer. He succeeded in differentiating the flacherie from the pébrine of silkworms because he had noticed and remembered that, during the 1865 season, certain broods of worms had ascended the heather in a peculiarly sluggish manner. He was led to guess the role played by earthworms in the epidemiology of anthrax by noticing their castings over the pits where animals had been buried. He gave a classical description of the symptoms of chicken cholera, and of the effects of its toxin. With his primitive microscopes and without staining techniques, he learned to distinguish the different microbial forms and to recognize the bacterial impurities correlated with faulty fermentations; he pointed out that the morphology of yeast varies somewhat with its state of nutrition; he noticed that the microorganisms present in the wine deposits looked larger because they fixed some of the wine pigments. Within two weeks after his arrival in Alais, he had learned to recognize the microscopic corpuscles of pébrine; he saw and described bacterial spores in the intestine of silkworms affected by flacherie before anything was known of the nature and physiological importance of these bodies. He described in precise terms the capsule of pneumococci and became skillful in detecting infection of brain tissue with the rabies virus, even though he had no knowledge of pathological anatomy.

The hours which he spent in silence looking at the object of his studies were not only periods of meditation. They were like long exposure times, during which every small detail of the segment of the world which he was contemplating became printed in his mind. Even more, they served to isolate, as it were, a section of the universe; and every component of it became organized with reference to his preoccupations. It was during these hours that were born between him and his experimental material those subtle but sure relations which blossomed into the intuitive preception of the "whole," characteristic of most of his discoveries.

But power of observation does not suffice to explain Pasteur's scientific performance. For he knew how to integrate any relevant observation into his conceptual schemes. So important was this peculiarity in the genesis of his discoveries that it appears worth while to recall one specific example, illustrating how concrete facts found their place in the worlds which he was constantly imagining.

In 1859, while observing under the microscope a drop of sugar solution undergoing butyric fermentation, Pasteur noticed that the microorganisms present in it became motionless at the edge of the drop, while those in its center remained actively motile. This accidental observation acted as a spark which fired the deep layers of his mind loaded with incessant questionings and ponderings concerning the nature of the fermentation mechanism. He was convinced that alcoholic fermentation, as usually observed, was dependent on the life of yeast but he also knew that the production of alcohol out of sugar did not involve the participation of oxygen. This indicated that, under certain conditions, life could proceed without oxygen, a conclusion in conflict with the doctrine then universally accepted that oxygen was the very breath of life. When Pasteur saw the butyric organisms become motionless as they approached the edge of the droplet he immediately imagined that they were inactivated by contact with the air. Indeed, experiment soon proved that they failed to multiply in aerated media, whereas they grew abundantly when oxygen was removed from the environment. In this case an apparently trivial fact found its place in Pasteur's meditations and led him to conclude that (*a*) life can exist without oxygen, (*b*) fermentations in general are metabolic reactions by which any cell can derive its energy from certain organic substances in the absence of oxygen, (*c*) the production of alcohol is only a particular case of the fermentation process and is the reaction by which yeast obtains energy under anaerobic conditions. All these extraordinary views, formulated as early as 1861, did not receive adequate experimental confirmation until 1872, when the studies on beer gave to Pasteur the occasion to establish their factual validity. Meditation on a general problem had been fertilized by an accidental observation and had given birth to a discovery; the systematic experimentation which followed served only to nurture and guide into adult development this child born of the mysterious union.

In many instances, discovery appears to have evolved from the fact that Pasteur had been made alert to the recognition of a phenomenon, because he was convinced a priori of its existence. Such was the genesis, as we have discussed in detail earlier, of the work on molecular asymmetry. Even more illustrative, perhaps, are the studies on the theory and practice of immunization. Much impressed by the facts that smallpox rarely occurs twice in the same individual and that one can protect against it by vaccination, Pasteur had formed very early the conviction that one should be able to immunize against other contagious diseases as well. It was this conviction which allowed him to grasp at once the significance of his accidental finding that birds inoculated with avirulent chicken cholera bacilli became resistant to inoculation with the fully virulent cultures. Because he had been anticipating such a fact, he postulated its analogy with Jenner's use of cowpox for vaccinating against smallpox, and extended the meaning of the word vaccination to include the new phenomenon. Discovery in this case was essentially the recognition of a natural law in the separate occurrence of two isolated facts which could be connected by the process of analogy.

Pasteur made frequent use of analogy as a source of ideas for his investigations. The dimorphic right- and left-handed quartz crystals served him as a model for the study of the optical activity of tartaric acids. In this case the analogy was only formal, since the optical activity of quartz resides in the crystal structure, whereas that of organic substances is a fundamental property of the molecule itself. Nevertheless it was sufficient to drive the first wedge into the analysis of the relation between asymmetry of molecular structure and optical activity. As he progressed in this study, Pasteur ceased thinking in terms of the quartz model, which was no longer useful in devising new experiments. Instead, he imagined that the organic molecule itself was an asymmetric body, and he tried to illustrate the concept of molecular asymmetry – which could not yet be described in chemical terms – by pointing out that the opposite members of asymmetric molecules bear to each other the same relation that the right hand bears to the left, each resembling the mirror image of the other.

The case of rabies offers a striking example of analogy, which, although based on false premises, led to successful experiments. Studies on chicken cholera, anthrax, and swine erysipelas had revealed the existence of attenuated cultures of bacteria which were unable to cause progressive disease, but still capable of vaccinating against the fully virulent organisms. Nothing was then known of the nature of the rabies virus and there was no evidence that it bore any relation to bacteria. Nevertheless, Pasteur attempted to attenuate it, as he had bacteria, and he found that the spinal cord infected with the virus lost most of its infective power during desiccation in the presence of air, while retaining its ability to immunize against the virulent disease. He soon realized, however, that despite its successful outcome his work had been built upon a false assumption. Desiccation had not caused a true diminution of the virulence of the rabies virus, but only a progressive decrease in the number of active virus particles. Whereas attenuation of the chicken cholera and anthrax cultures was truly due to a change in properties of the bacteria, the decrease in virulence of the spinal cord infected with rabies was due to the fact that there was less active virus left, and not to a change in the properties of the surviving virus. Although the hypothesis was erroneous, it had led him to the recognition of new and important phenomena.

In general, great investigators have left in writing only the ultimate form of their thoughts, polished by prolonged contact with the world of facts, and often with the world of men. It is because they reach us in this purified state that scientific concepts possess an awe-inspiring air of finality, which gives the illusion of a pontifical statement concerning the nature of things. Anyone interested in the performance of the human mind – out of mere curiosity, or for scholarly pursuit – welcomes the publication of the tentative and often crude sketches through which artists and writers evolve the final expressions of their ideals. However, scientific workers now consider it unbecoming and compromising to reveal their gropings towards truth, the blundering way in which most of them, if not all, reach the tentative goal of their efforts. This modesty, or conceit, robs scientific operations of much human interest, prevents an adequate appreciation by the public of the relative character of scientific truth, and renders more difficult the elucidation of the mechanisms of discovery, by placing exclusive emphasis on the use of logic at the expense of creative imagination. The raw materials out of which science is made are not only the observations, experiments and calculations of scientists, but also their urges, dreams and follies.

Loir has spoken of the fanciful mental constructions in which Pasteur indulged before undertaking a new problem. Of these scientific novels, only fragments remain. We have mentioned earlier the designing of equipment for submitting plant growth and chemical synthesis to the action of strong magnetic and electric fields, or to light rays inverted by mirrors, in the hope of creating living chemical molecules or of altering the properties of living beings. These attempts had the quality of an alchemist's quest and Pasteur acknowledged it – "One has to be senseless to undertake the projects in which I am now engaged." The beginning of the work on rabies seems to have been particularly fruitful in unwarranted hypothesis; bacteriologists will perhaps find entertainment in the account of one of Pasteur's early theories concerning the etiology of the disease. As will be remembered, he had isolated from the saliva of the first rabid child that he studied a virulent encapsulated microorganism – now known as the pneumococcus. Before recognizing that this organism did not bear any relation to rabies, he constructed the theory that the period of incubation of the disease was the time required for the dissolution or destruction in the tissues of the capsule surrounding the microorganism. Had this working hypothesis fitted the facts of rabies, it would have been, indeed, an exciting theory; but because it did not, scientific etiquette rules it bad taste to mention it in print. And yet, the elucidation of the mental processes involved in scientific discovery requires a knowledge of the hypotheses that miscarry, as well as of those which bear fruit.

Pasteur was well aware of the enormous role played by imagination in his scientific performance, and he repeatedly acknowledged it. He was, however, always eager to try to dissociate his dreams from reality, regarding the experimental method as a tool – almost infallible in skilled and honest hands – to weed out facts from fancy. Throughout his life he retained the ability to eliminate from his mind, once and for all, hypotheses which had proved incompatible with factual observations or experimental results.

"Preconceived ideas are like searchlights which illumine the path of the experimenter and serve him as a guide to interrogate nature. They become a danger only if he transforms them into fixed ideas – this is why I should like to see these profound words inscribed on the threshold of all the temples of science: 'The greatest derangement of the mind is to believe in something because one wishes it to be so.' . . .

"The great art consists in devising decisive experiments, leaving no place to the imagination of the observer. Imagination is needed to give wings to thought at the beginning of experimental investigations on any given subject. When, however, the time has come to conclude, and to interpret the facts derived from observations, imagination must submit to the factual results of the experiments."

Indeed Pasteur's imagination was always rich, often undisciplined, but the verification of his scientific concepts was so exacting and severe that he remains unsurpassed in the validity of his claims within the range of his experimental world.

In addition to his long practice of the art of experimentation and to his knowledge of its illusions and pitfalls, Pasteur had an imperious need for some form of solid conviction that only clean-cut and incontrovertible facts could give him. He liked to admire and to believe in institutions, men and facts. He despised vague philosophical and political doctrines, because he was uncomfortable in uncertainties. He loved the experimental method not so much because it revealed new philosophical outlooks on the Universe as because it could answer an unambiguous Yes or No to well-defined questions asked in unambiguous terms. Facing the skeptical Ernest Renan, who was receiving him at the Académie Française, he spoke of "this marvelous experimental method, of which one can say, in truth, not that it is sufficient for every purpose, but that it rarely leads astray, and then only those who do not use it well. It eliminates certain facts, brings forth others, interrogates nature, compels it to reply and stops only when the mind is fully satisfied. The charm of our studies, the enchantment of science, is that, everywhere and always, we can give the justification of our principles and the proof of our discoveries."

At the end of his life, Pasteur expressed regret at having abandoned his early studies on molecular structure. In this regret, there was perhaps the longing for the youthful days when the intoxication of discovery had first been revealed to him. There was also the faith, which he never gave up, that molecular asymmetry was in some way connected with the property of life; and to deal with the origin of life had certainly remained one of his haunting dreams. Beyond all that, however, was the fact that his early work was the symbol of all the problems which he had abandoned in his restless march forward. He had left many studies unfinished, although it was within his power to bring them to a more advanced stage of development and perfection. As if apologizing to his contemporaries and to posterity, he claimed that he had been "enchained" to an inescapable, forward-moving logic. And indeed, there was a definite logic in the sequence of his works; but this logic was not inescapable. His career might have followed many other courses, each one of them as logical, and as compatible with the science of his time and with the potentialities of his genius.

It is impossible, and indeed preposterous, to attempt to recast a life in terms of what it would have been had circumstances encouraged the expression of its potentialities in other channels of endeavor. Nevertheless, it may be of interest to consider some of the directions in which Pasteur could have directed his genius, if the pressure of other callings had not imposed upon him the tasks which assured his immortality. This is not *lèse-majesté,* for we shall do little more than elaborate on tentative projects of research which he himself is known to have suggested, in the form of casual remarks, and which only the shortness of days prevented him from developing further.

The fact that the rotation of the plane of polarized light by the solution of optically active organic substances is the greater, the larger the number of molecules encountered by the beam of light, convinced Pasteur that these organic molecules possess some sort of asymmetry. He found numerous analogies for this hypothesis, but the science of structural organic chemistry was not yet sufficiently developed to permit a definite explanation of the phenomenon. However, the interpretation of molecular asymmetry of organic compounds was provided – within a few years after he abandoned experimental work on the problem – by Couer's and Kekule's theory that the carbon atom has a normal valence of four. In 1874, van't Hoff in Holland and Le Bel in France proposed, simultaneously, the theory of arrangement of atoms in space – now known as stereoisomerism – which provided a simple chemical interpretation for Pasteur's findings. Thus, all the components required for the formulation of this new phase of chemical science, one of the greatest in its theoretical implications and practical consequences, became available at a time when Pasteur's physical energy, scientific imagination and knowledge of physical chemistry were at their highest level. Stereoisomerism could have evolved from the logical unfolding of his own scientific efforts.

Nor did he need to limit himself to the purely chemical consequences of his initial discovery. He had recognized that the left and right tartaric acids exhibited very different behavior toward living agents, and pointed out that the difference in taste of left and right asparagine was only one of several biological differences between these two substances. Observation of the specific behavior of certain enzymes with reference to the optical activity of organic compounds was well within the range of chemical knowledge of the second part of the nineteenth century, and would have brought Pasteur to experimental grips with the problem on which he spent so much romantic imagination: the relation of asymmetry to living processes.

Many different "logical sequences" could have stemmed also from the germ theory of fermentation. Because of his thorough chemical training, Pasteur could have separated and studied, both bacteriologically and chemically, a large variety of microbial processes which remained obscure long after him. The subsequent studies by others on the cycles of carbon, nitrogen, phosphorus, sulfur, and so on, in nature and particularly in the soil, required no technical procedures or theoretical knowledge beyond that which he possessed or could master. The relation of microbes to soil fertility, and to the general economy of matter, could have furnished ample material for a logical development of his life.

He abandoned early his spectacular studies on microbial nutrition, and it is only during the past two decades that microorganisms have again become tools of choice for the study of many nutritional problems. Yet, the discovery that carbohydrates, proteins, fats and minerals do not constitute the whole subject of the science of nutrition, and that vitamins are necessary components of a complete diet, could have come earlier, and more easily, had the lower forms of life, instead of animals, been used as test objects. It was within the logic of the germ theory to develop many theoretical aspects of the science of nutrition and many of its applications. Although Pasteur emphasized that the same fundamental metabolic reactions are common to all living things, he could have done much more to substantiate the doctrine of the biochemical unity of life, perhaps one of the most far-reaching concepts of modern times.

Understanding of microbial metabolism led him to formulate rational directives for the manufacture of vinegar, wine and beer, and his work inspired the leaders of the Carlsberg brewery in Copenhagen to establish a laboratory devoted to the improvement of brewing technology. But there are many other industries in which microorganisms do or could play a role. As Pasteur repeatedly emphasized, one can find in nature microorganisms adapted to the performance of almost any organic reaction and it is possible to domesticate microbial life just as plant and animal life have been domesticated in the course of civilization. In fact, Pasteur himself stated: "A day will come, I am convinced, when microorganisms will be utilized in certain industrial operations on account of their ability to attack organic matter." This prophecy has been fulfilled, and today organic acids, various solvents, vitamins, drugs and enzymes are produced on an enormous scale by microbial processes — all this a logical development of Pasteur's work.

It is a remarkable fact that many of the advances in the understanding of infectious diseases, which have occurred since Pasteur and Koch, have been made with techniques so simple, and often so empirical and so crude, that they could have come just as well out of the early bacteriological laboratories. For this reason, to list the possible lines of work which Pasteur could have elected to follow instead of devoting himself to the practical problems of vaccination, would be to review a large part of medical microbiology. A few examples will suffice.

Very early, he recognized that certain microorganisms commonly present in soil can affect the anthrax bacillus in such a manner as to render it unable to establish disease in animals. He suggested immediately that this phenomenon might lend itself to therapeutic applications, that saprophytic organisms might someday be used to combat infectious agents. Despite his prophetic vision, however, he neglected to exploit the therapeutic possibility which he had recognized. Had he chosen to follow this line of investigation, the techniques then available would have permitted the isolation from soil of the strains of Bacillus, Penicillium or Streptomyces which have since been shown to produce substances capable of inhibiting the anthrax bacillus both in the test tube and in the animal body. Chemical knowledge was then sufficiently developed to allow the purification of these therapeutic substances of microbial origin up to a point where they could be of some use in practice. Thus, bacteriotherapy might have been born in 1880. Better than any subsequent phase of Pasteur's work, it could have been, indeed, the logical synthesis of his training as a chemist and of his familiarity with saprophytic and pathogenic microorganisms. Instead, the exploitation of microorganisms as producers of therapeutic antimicrobial agents had to wait almost three fourths of a century before it became a practical reality. The fruit of Pasteur's logic ripened only when the land which he had explored was tilled by the bacteriologists and chemists of the twentieth century.

It is to immunity that Pasteur addressed himself as a method of control of infectious diseases. Although most of his work in this line dealt with the use of attenuated living vaccines for immunization, he might have approached the problem from a different angle. He had early recognized the importance of nonliving soluble bacterial toxins in the production of disease, and he could have selected antitoxic immunity as a goal for his efforts. Immunity to the toxins of diphtheria and tetanus was achieved within his lifetime, by methods no different from those available to him. The treatment of diphtheria by antitoxic serum was a logical development of his discovery of the toxin of chicken cholera. He could have attempted also to immunize with killed bacilli, instead of using the living attenuated microorganisms, a step which was taken in 1889 by Salmon and Smith while he was still alive. This approach, it appears, would have been very congenial to him – for he must often have longed to escape from the uncertainties attendant upon the use of living biological material. The modern trend of utilizing, for immunization, substances separated from killed bacilli, and amenable to purification and standardization by chemical procedures, would have satisfied his eagerness for well-defined methods. And in reality, he anticipated this development when he discussed the "chemical vaccines" for anthrax and for rabies, and saw in them the method of the future.

Pasteur often emphasized the great importance of the environment, of nutrition, and of the physiological and even psychological state of the patient, in deciding the outcome of the infectious process. Had the opportunity come for him to undertake again the study of silkworm diseases, he once said, he would have liked to investigate the factors which favor the general robustness of the worms, and thereby increase their resistance to infectious disease. This statement reveals the potential existence of yet another Pasteur, who would have focused the study of contagious disease toward the understanding of those physiological and biochemical factors which condition the course and outcome of infection. If circumstances had favored the manifestation of this aspect of his potential personality, instead of those traits which determined the dedication of his life to microbiology, the science of infectious disease might today have a complexion far different from that under which we know it. A logic of Pasteur's life centered on physiological problems is just as plausible as that which resulted from the exclusive emphasis on the germ component of the theory of contagious disease.

The fact that one can so readily conceive of many different logical unfoldings of Pasteur's work, all compatible with his endowments, his training, his imagination, and with the environment in which he lived, is perhaps the most eloquent and convincing index of the richness of his personality. True genius, according to Dr. Johnson, is a mind of large general powers, accidentally determined to some particular direction, ready for all things, but chosen by circumstances for one. Although the existence of these multiple potentialities, only a few of which find expression during the life of any individual, is most obvious and overpowering in the case of outstanding creators, it is not peculiar to genius. It is a characteristic of all men, indeed of all living things, and Pasteur had even encountered it in the microbial world. The Mucor that he had studied could exist as a long filamentous mold on the surface of culture media, but it could also grow yeastlike when submerged in a sugar solution. Pasteur's own student Duclaux was the first to show that the metabolic equipment of microorganisms, their enzymatic make-up, is dependent upon the composition of the medium in which they live; the specific adaptation of the cell to the chemical environment by the production of appropriate enzymes is now known to be a phenomenon of universal occurrence. Each cell, each living being, has a multipotential biochemical personality, but the physicochemical environment determines the one under which it manifests itself. In these terms, the dependence of the expression of individuality upon the environment appears as mere chemistry, of little relevance to human problems. And yet, just as the Mucor grows either as filamentous mold or as a yeast depending upon the impact of the environment upon it, similarly, there exists for each one of us the potentiality of revealing ourselves to the world as many different individuals, but circumstances allow us to live only one of the many lives that we could have lived.

It is often by a trivial, even an accidental decision that we direct our activities into a certain channel, and thus determine which one of the potential expressions of our individuality will become manifest. Usually, we know nothing of the ultimate orientation or of the outlet toward which we travel, and the stream sweeps us to a formula of life from which there is no returning. There is drama in the thought that every time we make a choice, turn right instead of left, pronounce one word instead of another, we favor one of our potential beings at the expense of all the rest of our personality – nay, we likely starve and smother to death something of us that could have continued to live and grow. Every decision is like a murder, and our march forward is over the stillborn bodies of all our possible selves that will never be. But such is the penalty of a productive life. For "Except a corn of wheat fall into the ground and die, it abideth alone: but if it die, it bringeth forth much fruit."

Very often, Pasteur must have looked over his past and wondered what his life would have yielded had he selected other hunting grounds in which to spend his energies and display his genius. He often returned in his conversations and writings to the problems of his youth, and lamented having abandoned them on the way – a symbol of the tragedy of choice. He had been, he thought, "enchained" to an inescapable logic. But in reality, it was not logic that had enchained him. It was the strange compulsion, the almost insane urge, which makes the born investigator become possessed and indeed hypnotized by the new problems arising out of his own observations. The logic that Pasteur followed was not inevitable, although he did not succeed in escaping it. He could have followed many other courses, as logical, as fruitful. Discoveries greater than pasteurization and vaccination could be attached to the memory of his adult years, had he elected to live some other of the many lives that were offered to him by the gods – the drama in the life of every aspiring man.

CHAPTER XIV

Beyond Experimental Science

> Faith is the substance of things hoped for, the evidence of things not seen.
>
> — EPISTLE TO THE HEBREWS

THERE was a time when meditation on the relation of man to nature, the expression of wonder or anguish at the splendor and mysteries of the universe, the discovery of objective facts concerning the physical world and the application of these facts to the welfare of man were all, equally, the privilege and duty of the inquiring soul. Until late in the eighteenth century, philosophy — the love of wisdom — embraced the whole field of knowledge and was concerned with all aspects of the physical and metaphysical world.

This belief in the essential unity of mental processes survives in the custom of the French Academy of Letters — l'Académie Française — of admitting within its membership men of many different callings — churchmen, statesmen, soldiers, engineers, scientists — who have contributed to the advancement of mankind or to the glory of France. For, said Ernest Renan, "in a well organized society, all those who devote themselves to beautiful and good causes are collaborators; everything becomes great literature when it is done with talent." Pasteur was elected to the Académie Française in 1882. His literary titles were few but he brought to the venerable institution his genius, "this common basis for the creation of beauty and truth, this divine flame that inspires science, literature and art." Many circumstances made his formal reception an event of peculiar glamour. He was at that time the most famous representative of French chemistry and biology, and the atmosphere of legend already surrounded his name. Pasteur, symbol of the power of exact sciences, was taking the seat of Ernest Littré, who had become the prophet of positivist philosophy and the advocate of the scientific approach to human affairs, particularly in sociology and history. Pasteur, the man of unbending convictions and fiery temperament, worshiper of the experimental method but believer in the teachings of the Roman Catholic Church, was to be welcomed into the Académie by the smiling and skeptical Ernest Renan – linguist, philosopher, historian – one who had written of Christ as a son of man, and of positive science as the religion of the future.

In addition to his supreme intelligence, circumstances had placed Renan in a favored situation to act as the spokesman of the men of letters to the world of science. He had first intended to become a priest but had lost the Catholic faith when, after a "conscientious search for the historical basis of Christianity," he had become convinced of its "scientific impossibilities." In 1845, at the age of twenty-two, he left the Great Seminary of Saint Sulpice to study letters and philosophy. It was at that time that he met Marcellin Berthelot, then twenty years old, who was soon to become one of the founders of organic chemistry, and a convinced apostle of the role of exact sciences in human affairs. Won over to the scientific faith by the unlimited confidence of the young chemist, Renan learned to admire science for revealing the beauty and interest of natural phenomena. In it he saw also the promise of intellectual freedom as well as of material wealth, and the hope that exact knowledge would "solve the great enigma . . . and reveal to man in a definite manner the significance of things." Not only could the universe be controlled and admired. The time had come at last when it could be understood.

Renan continued throughout his life a lively intellectual intercourse with Berthelot by means of an extensive correspondence. But the dogmatic attitude of the chemist was fundamentally incompatible with his own temperament, always inclined as he was to doubt that truth was on one side only, and always willing to listen to the views of his opponents. At the Collège de France, where he was professor of Sanskrit, he became closely associated with Claude Bernard, who held the chair of experimental medicine. From him he learned that the great experimental principle is Doubt, not a sterile skepticism but rather a philosophical doubt, which leaves freedom and initiative to the mind. "Whereas," said Bernard, "the scholastic . . . is proud, intolerant, and does not accept contradiction . . . , the experimenter, who is always in doubt and never believes that he has achieved absolute certainty, succeeds in becoming the master of phenomena, and in bringing nature under his power."

To Renan, the man with the skeptical smile and warm heart, who questioned the absolute validity of any system but was always preoccupied with the thought of improving the world, nothing could be more attractive than this experimental method, based on doubt at every step and yet deriving its power of action from this very doubt.

According to academic custom, Pasteur was expected to pronounce the eulogy of his predecessor Littré on taking possession of his seat at the Académie Française. But he did more. He seized the opportunity to take issue with positivist philosophy by affirming that the scientific method is applicable only where experimentation is possible, and that it cannot be of any use in the problems involving emotions and religious faith.

"Auguste Comte's [1] fundamental principle," said Pasteur, "is to eliminate all metaphysical questions concerning first and final causes, to attempt to account for all ideas and theories in terms of concrete facts, and to consider as valid and established only that which has been shown by experience. According to him . . . the conceptions of the human mind proceed through three stages: theological, metaphysical and scientific or positivist. . . .

"M. Littré was all enthusiasm for this doctrine and for its author. I confess that I have come to a very different conclusion.

[1] Auguste Comte was the founder of positivist philosophy.

The origin of this conflict in our views probably results from the very nature of the studies which have occupied his life and those which have monopolized mine.

"M. Littré's studies have dealt with history, philology, scientific and literary erudition. The subject matter of these studies is exclusively the past, to which nothing can be added and from which nothing can be removed. Its only tool is the method of observation which, in most cases, fails to give rigorous demonstrations. On the contrary, the very characteristic of the experimental method is to accept none but absolutely convincing demonstrations. . . .

"Both of them unfamiliar with the practice of experimentation, Comte and Littré . . . use the word 'experience,' with the meaning which it has in the conversations of society, a meaning very different from that of the word 'experiment' in scientific language. In the former case, experience is merely the simple observation of things with the induction which concludes, more or less legitimately, from what has been to what could be. In contrast, the true experimental method aims at reaching a level of proof immune to any objection.

"The conditions and the daily results of the scientist's work lead his mind to identify the idea of progress with that of invention. In order to evaluate positivist philosophy, therefore, my first thought was to search it for the evidence of invention, and I did not find it. One certainly cannot dignify as invention the so-called law of the three stages of the human mind, or the hierarchic classification of sciences, views which are at best crude approximations without much significance. Positivism, offering me no new idea, leaves me reserved and suspicious."

As director of the Académie for the occasion, Renan had had the privilege of reading Pasteur's speech in advance of the meeting and took a malicious pleasure in opposing his own broad and subtle philosophy to the convictions of his new colleague. He did, of course, praise the intellectual beauty and importance of Pasteur's achievements, and expressed much admiration for the strenuous and incessant labor required of those who attempt to decipher the secrets of nature. As he sat facing the sixty-year-old scientist, he could read on his worn and wrinkled face the efforts which each discovery had cost him, the struggles with facts, with men, with their convictions and conventions, and even more, perhaps, with his own intellectual and moral weaknesses. Science is not the product of lofty meditations and genteel behavior, it is fertilized by heartbreaking toil and long vigils – even if, only too often, those who harvest the fruit are but the laborers of the eleventh hour. "Nature is plebeian; she demands that one work; she prefers callused hands and will reveal herself only to those with careworn brows."

But at the same time, Renan could not refrain from being somewhat amused by Pasteur's assurance that he understood clearly the respective place of science and sentiment in the problems of human life. With ironical words, he suggested that, in philosophical matters, hesitation and doubt are more often successful than overconfidence in apprehending reality and truth.

"Truth, Sir, is a great coquette. She will not be won by too much passion. Indifference is often more successful with her. She escapes when apparently caught, but she yields readily if patiently waited for. She reveals herself when one is about to abandon the hope of possessing her; but she is inexorable when one affirms her, that is when one loves her with too much fervor."

While admitting the power of the exact sciences, Renan emphasized with consummate grace and persuasiveness that the experimental method does not constitute the only legitimate technique for the acquisition of knowledge. Sociologists, historians, and even philosophers also make use of scientific judgment in their studies, and human feelings and behavior as well as religious dogma are, like other areas of thought, amenable to scientific analysis. He pointed out that the scientific method in historical matters consists in the discovery, identification and evaluation of texts. Although less simple and less directly convincing than the experimental method, historical criticism is also a worthy instrument for the understanding and formulation of important truths.

"The human spirit would be far less developed without it; I

dare say, indeed, that your exact sciences . . . would not have come into being if there did not exist near them a vigilant guardian to keep the world from being devoured by superstition and delivered defenseless to all the assertions of credulity."

Renan might have also answered Pasteur's scornful remarks on the sciences of "observation" by noting that Galileo and Newton had founded much of modern physics while working on problems of astronomy — where direct experimentation with the subject material is not possible. He could have pointed to the growth of other sciences, such as epidemiology or psychology, which are largely based on observation and for which experimentation can at best only provide oversimplified models. It is surprising that Pasteur, who so often upheld the place of the historical method in the study of exact sciences, should have been unable to recognize in historical criticism a legitimate technique for the evaluation of data pertaining to human relationships. He should have been aware that social problems, which must be solved objectively by future generations, had first to be defined by what Renan called his "little conjectural sciences," sociology and history.

Like Pasteur, Berthelot doubted that sociologists would ever contribute anything of practical importance to human welfare, but he based his skepticism on very different reasons. He had such unlimited confidence in the power of physicochemical sciences that he saw in them the eventual solution of all human problems. For him, the problems of life were mere extrapolations of problems of matter, and demanded no special methods for their elucidation. Pasteur, on the contrary, saw a profound schism between the world of matter and those vast areas where emotions and feelings rule the affairs of mankind. "In each one of us there are two men: the scientist who . . . by observation, experimentation and reasoning, attempts at reaching a knowledge of nature; but also the sensitive man, the man of tradition, faith and doubt, the man of sentiment, the man who laments for his children who are no more; who cannot, alas! prove that he will see them again, but believes and hopes it, who does not want to die like a vibrio, who wants to be convinced that the strength which is in him will not be wasted and will find another life."

Speaking in 1874 at the graduation exercises of the College of Arbois, where he had been a student, he affirmed that religious convictions are founded on the impregnable rock of direct personal experience. "The man of faith . . . believes in a supernatural revelation. If you tell me that this is incompatible with human reason I shall agree with you, but it is still more impossible to believe that reason has the power to deal with the problems of origins and ends. Furthermore, reason is not all . . . ; the eternal strength of the man of faith lies in the fact that the teachings of his creed are in harmony with the callings of the heart. . . . Who, by the deathbed of a beloved one, does not hear an inner voice assuring him that the soul is immortal? To say with the materialist 'Death is the end of all' is to insult the human heart."

At the time of his reception in the Académie Française, he attempted to present a more intellectual justification of religious faith. According to him, positivism, while pretending to explain human behavior in scientific terms, fails to take into account the most important of all the positivist notions, that of infinity. Although an inescapable conclusion of human thinking, the notion of infinity is incomprehensible to human reason. Indeed, Pasteur felt, it is more incompatible with it than are all the miracles of religion.

"I see everywhere in the world the inevitable expression of the concept of infinity. It establishes in the depth of our hearts a belief in the supernatural. The idea of God is nothing more than one form of the idea of infinity. So long as the mystery of the infinite weighs on the human mind, so long will temples be raised to the cult of the infinite, whether God be called Brahma, Allah, Jehovah or Jesus."

Replying to Pasteur, Renan pointed out that such statements gave a certificate of credibility to many strange tales. He was ready to grant that, in the field of the ideal, where nothing can be proved, all forms of belief and faith are justified. But miracles are specific claims that certain events have occurred, at definite times and places. They are, therefore, subject to historical criticism; and Renan failed to find substantiating evidence for any of the particular facts of religious history that had been thoroughly investigated. It was not, however, on the miracles peculiar to each religious lore that Pasteur had based his criticism of positivist philosophy. He had merely claimed that the notion of infinity conditions human behavior, and has immense consequences in the life of societies. Auguste Comte and his followers had failed to recognize these deep and mysterious sources of inspiration through which the notion of infinity expresses itself in the hearts of men. Pasteur saw in them the spiritual link of humanity and the origin of man's nobility.

"The Greeks understood the mysterious power of the hidden side of things. They bequeathed to us one of the most beautiful words in our language – the word 'enthusiasm' – *En theos* – an Inner God.

"The grandeur of human actions is measured by the inspiration from which they spring. Happy is he who bears within himself a god, an ideal of beauty, and who obeys it; ideal of art, of science, of patriotism, of the virtues symbolized in the Gospel. These are the living sources of great thoughts and great acts. All are lighted by reflection from the infinite."

Pasteur's opponents have seen an evidence of the philosophical limitation of his mind in this unwillingness to accept the possibility that human emotions and religious faith could be amenable to scientific scrutiny. They have also regarded his attitude as an intellectual surrender due to his acceptance of the Catholic discipline. Before accepting this interpretation, it is well to remember that many of the greatest scientific minds of the nineteenth century – Davy, Faraday, Joule, Maxwell, Lord Kelvin, Helmholtz – to mention only a few of the masters of exact sciences in non-Catholic countries – have, like Pasteur, forcefully acknowledged their allegiance to the Christian faith and have dissociated their beliefs as men of sentiment from their behavior as experimental scientists. In the course of a lecture before the Royal Institution, Faraday once stated in words not very different from those often used by Pasteur that the concept of God came to his mind through channels as certain as those which led him to truths of physical order.

Many scientific workers possess the ability to follow two independent and apparently conflicting lines of thought: on the one hand, the acceptance of religious dogmas, on the other, an absolute confidence in the ability of experimental science to analyze and control the mechanisms of the physical world. This attitude is symbolized by Newton, who formulated the mechanical laws ruling over the operations of the universe while retaining faith in the existence and power of a Creator who first put those forces into motion. The divorce between Christian faith and positive science became most widespread among the French scientific philosophers of the eighteenth century, but it is certain that they arrived at intellectual agnosticism or atheism, not through the examination of scientific knowledge, but by the way of philosophy. As physicists, they accepted Newton's gravitational force; as philosophers, they saw no need of his hypothesis concerning the existence of God. They converted Newtonian science into a mechanical philosophy in which the past and the future were theoretically calculable and man was a mere machine. They did not heed Newton's caution that the cause or nature of the gravitational force was unknown, that his law described only the relations between bodies, not the ultimate nature of these bodies nor the forces that acted upon them.

In contrast to their French contemporaries many British philosophers and scientific workers arrived at the conviction that the ultimate truths of nature transcend human understanding and that the most man can hope is to recognize and describe the relationships between objects and events. While intellectually a defeatist philosophy, this point of view is a source of great effective power. It does not weaken man's conviction that he can learn through experience to control nature and use it for his own ends, and it directs his efforts towards objective and practical achievements instead of speculative problems. Thus most scientists came to dismiss from their conscious minds, if not from their subconscious, the pretense that they were about to come to grips with the ultimate nature of reality. But they retained the hope, or carefully nursed the illusion, that their efforts would help improve the lot of man on earth. Should even that satisfaction fail them, there remained, as alluring as ever, the enjoyment of search and the intoxication of discovery, and in this they found sufficient inducement to make of experimental science the dominating force of Western society.

The dissociation between religious faith and experimental science received philosophical sanction and dignity from Immanuel Kant. He assured the world that all matters concerning which no information can be obtained through the direct experience of the senses — the ultimate nature of the universe, the soul and God — fall outside the range of rational knowledge. We can maintain neither their existence nor their nonexistence, and we are therefore justified in continuing to believe without need of proof in the existence of God and in the immortality of the soul.

Kant's philosophy had a great appeal for many scientists. It absolved them from having to deal with the philosophical significance of reality, encouraged them to devote their efforts to tasks of practical importance and, by making all fundamental creeds immune from the attacks of science, it permitted them to retain their religious beliefs. The acceptance of Kant's philosophy explains how so many of the greatest experimentalists of the nineteenth century found it possible to follow positive science and religious faith simultaneously. They felt free to suspend judgment on the interrelationships between the two on the ground that there was not as yet, and perhaps never could be, any objective evidence to permit understanding of their deeper implications.

It is probable that the practice of this double standard by men such as Faraday, Maxwell, Helmholtz, Pasteur, appeared intellectually dishonest to those who had accepted what they considered the inevitable logical consequences of the scientific point of view, and who were willing to disavow any form of allegiance to Biblical teaching. In England, this attitude was proudly upheld by Tyndall, when he stated in his Belfast address: "We claim, and we shall wrest from theology, the entire domain of cosmological theory." In France, the extreme form of the materialist faith was expressed by Berthelot in the astonishing statement: "The world, today, has no longer any mystery for us."

Pasteur was far too conscientious and earnest to reject scientific materialism merely on the basis of emotions – on the longings of his heart or the appeal to him of the notion of infinity. He examined the question time and time again, attempting to restate, in terms of his scientific experience, those problems which have always compelled the thoughts of man and which so many philosophies and religions have sought to answer. It was particularly at the time of the controversy on spontaneous generation that he found it necessary to formulate, for himself, an opinion concerning the ability of experimental science to decipher the riddle of life. He reached the conclusion, as Claude Bernard had, that the mystery of life resides not in the manifestations of vital processes – all of which pertain to ordinary physicochemical reactions – but in the predetermined specific characters of the organisms which are transmitted through the ovum, through what he called the "germ."

"The mystery of life does not reside in its manifestations in adult beings, but rather and solely in the existence of the germ and of its becoming. . . .

"Life is the germ with its becoming, and the germ is life. . . .

"Once the germ exists, it needs only inanimate substances and proper conditions of temperature to obey the laws of its development . . . it will then grow and manifest all the phenomena that we call 'vital,' but these are only physical and chemical phenomena; it is only the law of their succession which constitutes the unknown of life. . . .

"This is why the problem of spontaneous generation is all-absorbing, and all-important. It is the very problem of life and of its origin. To bring about spontaneous generation would be to create a germ. It would be creating life; it would be to solve the problem of its origin. It would mean to go from matter to life through conditions of environment and of matter.

"God as author of life would then no longer be needed. Matter would replace Him. God would need be invoked only as author of the motions of the world in the Universe."

Like an obsession, there recurs time and time again through his writings, often in unpublished fragments, the statement that "Life is the germ and its becoming." The concept of "becoming" was obviously borrowed, perhaps unknown to Pasteur, from the *Werden* of the Hegelian doctrine. This taught a logical or dialectic development of things according to which the whole world – spiritual phenomena, man, together with all natural objects – was the unfolding of an act of thought on the part of a creative mind. There is some irony in seeing the great French patriot and champion of the power of the experimental method, struggling to express his philosophical view of life in the words of the German archenemy of the experimental scientists.

Pasteur was unquestionably sincere in affirming his willingness – nay, his eagerness – to believe in the spontaneous generation of life provided adequate proof was brought forward to demonstrate its occurrence. His religious faith was independent of scientific knowledge. Well aware of the limitations of the experimental method, he knew that his work had not proved that the generation of life *de novo* was impossible, and that he had done nothing more than show the fallacy of all known claims. "I do not pretend to establish that spontaneous generation does not occur. One cannot prove the negative." But by the same token he protested the assumption, for which no evidence is yet available, that spontaneous generation had been the origin of life in the universe.

"I have been looking for spontaneous generation during twenty years without discovering it. No, I do not judge it impossible. But what allows you to make it the origin of life? You place matter before life, and you decide that matter has existed for all eternity. How do you know that the incessant progress of science will not compel scientists . . . to consider that life has existed during eternity and not matter? You pass from matter to life because your intelligence of today . . . cannot conceive things otherwise. How do you know that in 10,000 years one will not consider it more likely that matter has emerged from life . . . ?"

Pasteur never published these remarks, written in 1878. He may have reserved them for some ulterior communication, since other posthumous fragments suggest that he had intended to return to the problem of spontaneous generation, a project he did not fulfill. Despite his conviction and self-assurance, he may have also feared the opposition that these unorthodox views would encounter in the scientific world. In reality, however, he was not alone in questioning the order of the relation of life to matter. At about the same time Fechner in Germany pointed out that everywhere the living generates not only the living, but also, and much more frequently, the inanimate, although we never see life develop *de novo* out of inorganic matter. Preyer also asked himself whether, instead of the living being evolved from dead matter, it is not the latter which is a product of the former. Contrary to common sense as these views appear, they foreshadowed some of the modern developments of the theory of knowledge. Common-sense realism gives us only a very limited view of the world. Our preceptions are no more than plane sections of the universe, from which we construct models of it to fit our practical needs and to help ourselves recognize some qualitative and quantitative relations between its component parts. But perceptions, and the models we derive from them, give us little if any understanding of the intrinsic nature of reality. The concepts of "life" and of "matter" probably correspond to two of these abstract models, and the human mind has not yet succeeded in tracing significant relationships between them.

If Pasteur saw no hope that the experimental method would ever reveal the origins and ends of the universe it was because he believed that "in good science, the word 'cause' should be reserved for the primary divine impulse which gave birth to the universe. We can observe nothing but correlations. It is only by stretching the true meaning of words unjustifiably that we speak of a cause and effect relationship when referring to one phenomenon which follows another in time and cannot occur without it." In much the same vein, Claude Bernard had also written: "The obscure concept of cause . . . has meaning only with reference to the origin of the universe . . . in science it must yield to the concept of relation or conditions. Determinism establishes the conditions of phenomena and permits us to predict their occurrence and even under certain conditions to provoke it. Determinism does not give us any account of nature, but renders us master of it. . . . Although we may think, or rather feel, that there is a truth which goes beyond our scientific caution, we are compelled to limit ourselves to determinism."

It is interesting to recognize that despite their differences in religious convictions, Bernard, who was probably an agnostic, and Pasteur, who was a practicing Roman Catholic, had arrived at essentially the same scientific philosophy. Both limited the role of experimental science in biology to the physicochemical determinism of living processes, but they accepted its power as supreme within this restricted field.

It is thus certain that many influences other than the Catholic dogma played a decisive part in shaping Pasteur's conviction that materialist doctrines are inadequate to account for the origin of life. Like most men, he believed that it was through spontaneous inner feelings and direct experience that he had arrived at the metaphysical and religious views which he expressed with such warmth and conviction. And indeed, sheltered as he was within the walls of his laboratory, coming into contact with the world almost exclusively through his dealings with objective scientific problems, he appeared protected from the whims and fluctuations of public thinking. But the currents of human thought are made of immensely diffusible stuff, and permeate the whole fabric of human societies; no walls are impermeable to them; they reach the peasant hearth as well as the inner rooms of scientific sanctuaries. Hegelian logic compounded with positive science, Catholic faith tempered by the intellectual philosophy of the eighteenth century, physicochemical interpretation of living processes colored with a touch of emergent evolution and *élan vital* — all these influences and probably many others had found their way into Pasteur's mind while he was working over his microscope and injecting his animals. It was through a mistaken illusion, such as he was so fond of detecting in others, that he came to regard his beliefs as spontaneous generations of his heart.

Pasteur did not intend to propound a philosophical doctrine when he emphasized with such intensity the limitations of experimental science. He meant only to state that Creation is more vast than what is revealed by our senses, even with the aid of scientific insight and instruments. The universe certainly transcends the concepts devised by the human mind to imagine that which cannot be seen, and it is because men perceive only a very small angle of reality that they often disagree so profoundly. In the search for truth, tolerance is no less essential than objectivity and sincerity. Pasteur will be remembered for having contributed his stone to the great edifice of human understanding, but it is for simpler reasons that he labored and that his name is now honored. If he devoted himself to science with so much passion, it was not only for the sake of interest in philosophical problems, but also because he found "enchantment" in the "serene peace of libraries and laboratories." It was there also that he satisfied the romantic urges of his enthusiasm, of that inner god which made him regard each experiment as a miracle, each conflict as a crusade.

With struggle, but with great success, he served through his eventful life many of the deities worshiped by thinking men. He used the experimental method to create for humankind the wealth, comfort and health which make our sojourn on earth more enjoyable. He tried to answer by the techniques of science some of the eternal questions which have been asked in so many different forms by all civilizations. He even dared to attempt to create life anew, or to modify it, by his own artifices. And yet, throughout all these bold ventures — where as much as any living man he manifested the glorious conceit of the human race — he retained, childlike, the creed and worshipful attitude of his ancestors. His life symbolizes the hope that a time will come when the infallibility of the experimental method can be reconciled with the changing but eternal dreams of the human heart.

Appendix: CHRONOLOGY OF EVENTS IN PASTEUR'S LIFE

REFERENCE NOTES TO THE INTRODUCTION

BIBLIOGRAPHY

INDEX

Appendix: Chronology of Events in Pasteur's Life

1822: Birth of Louis Pasteur at Dôle on December 27.

1827: Removal of his family to Arbois.

1838: Trip to Paris in October with plan to study at the Institution Barbet.
Return to Arbois with his father in November.

1839–1842: Secondary education at the Collège Royal de Besançon.

1842: First admission to the Ecole Normale Supérieure in Paris, and resignation in hope of achieving better rank.

1842–1843: Completion of secondary studies at the Lycée Saint-Louis, at the Sorbonne and at the Institution Barbet in Paris.

1843: Readmission to the Ecole Normale, fifth in rank.

1844–1846: Beginning of chemical and crystallographic studies, as a student at the Ecole Normale.
Discovery of molecular asymmetry.

1846: Appointment as assistant to Balard, at the Ecole Normale.

1847: Completion of requirements for doctorate *ès-sciences*.

1848: Appointment as professor of chemistry at the Université de Strasbourg.

1849: Marriage to Marie Laurent, daughter of the Rector of the University, on May 29.

1850: Birth of his daughter Jeanne.

1851: Birth of his son Jean-Baptiste.

1853: Birth of his daughter Cécile.
Prix de la Société de Pharmacie de Paris for the synthesis of racemic acid.
Award of Legion of Honor.

1854: Appointment as professor of chemistry and dean in the newly organized Faculté des Sciences at Lille.

1855: Beginning of studies on fermentation.

1857: Publication of the *Mémoire sur la fermentation appelée lactique*.
Rumford Medal from the Royal Society of London for his studies on crystallography.

Appointment as manager and director of scientific studies at the Ecole Normale Supérieure in Paris.

1858: Birth of his daughter Marie-Louise.

1859: Death of his daughter Jeanne in September at Arbois.
Beginning of studies on spontaneous generation.
Prix de Physiologie Expérimentale (Académie des Sciences).

1860: Publication of the *Mémoire sur la fermentation alcoolique.*
Two lectures before the Société Chimique de Paris on "*Recherches sur la dissymétrie moléculaire des produits organiques naturels.*"

1861: Discovery of anaerobic life.
Lecture before the Société de Chimie, "*Sur les corpuscules organisés qui existent dans l'atmosphère. Examen de la doctrine des générations spontanées.*"
Prix Jecker (Académie des Sciences) for studies on fermentations.

1862: Election in December as a member of the Paris Académie des Sciences in the section of mineralogy.
Studies on acetic acid fermentation.
Prix Alhumbert for his studies on spontaneous generation.

1863: Studies on wine.
Appointment as professor of geology, physics and chemistry at the Ecole des Beaux Arts.
Birth of his daughter Camille.

1864: Publication of the "*Mémoire sur la fermentation acétique.*"
Lecture at the Sorbonne on "*Des générations spontanées.*"
Controversy with Pouchet, Joly and Musset on spontaneous generation.
Establishment of a field laboratory for the study of wines in home of his school friend Jules Vercel in Arbois.

1865: Studies on pasteurization.
Beginning of studies on silkworm diseases, at Alais; continued until 1869.
Death of his father in June at Arbois.
Death of his youngest daughter Camille in September (burial at Arbois).

1866: Publication of the *Etudes sur le vin.*
Publication of an essay on the scientific achievements of Claude Bernard.
Death of his daughter Cécile (burial at Arbois).

1867: Lecture in Orléans on manufacture of vinegar.
Grand Prix of the Exposition Universelle of 1867 for method of preservation of wines by heating.
Appointment as professor of chemistry at the Sorbonne.
Resignation from his administrative duties at the Ecole Normale.

1868: Publication of *Etudes sur le vinaigre.*
Attack of paralysis (left hemiplegia) in October.
Enlargement of laboratory at the Ecole Normale.

1869: Resumption of studies on silkworm diseases, first at Alais, then on the estate of the Prince Impérial at Villa Vicentina (Austria).

1870: Publication of *Etudes sur la maladie des vers à soie.*
Return to Paris, then to Arbois. Franco–Prussian War.

1871: Trip from Arbois to Pontarlier in search of his son Jean-Baptiste of the French Army in retreat.
Sojourn for a few months at Clermont-Ferrand; beginning of the studies on beer in Duclaux's laboratory.

1871–1877: Studies on beer and on fermentation, in Paris at the Ecole Normale.

1873: Election as an associate member of the Académie de Médecine.

1874: Address on occasion of graduation exercises at the Collège d'Arbois, in August.

1875: Establishment of a field laboratory at Arbois (again in Jules Vercel's home) for studies on fermentation.

1876: Candidacy for election to the Senate. Defeat at the election.
Publication of *Etudes sur la bière.*

1877: Beginning of studies on anthrax.

1878: Controversies (especially with Colin) on etiology of anthrax.
Studies on gangrene, septicemia, childbirth fever.
Publication of the memoir *La théorie des germes et ses applications à la médecine et à la chirurgie.*
Discussion of a posthumous publication of Claude Bernard, and controversy with Berthelot on fermentation.

1879: Studies on chicken cholera. Discovery of immunization by means of attenuated cultures.
Marriage of his daughter Marie-Louise to René Vallery-Radot.
Marriage of his son Jean-Baptiste.

1880: Beginning of studies on rabies.
Publication of the memoir *Sur les maladies virulentes et en particulier sur la maladie appelée vulgairement choléra des poules.*

1881: Publication of studies on anthrax vaccination *De la possibilité de rendre les moutons réfractaires au charbon par la méthode des inoculations préventives.*
Field trial of anthrax vaccination at Pouilly le Fort.
Paper before the International Congress of Medicine in London on the studies on fowl cholera and anthrax—

"Vaccination in Relation to Chicken Cholera and Splenic Fever."

1882: Election to the Académie Française and reception by Ernest Renan.
Studies on cattle pleuropneumonia.
Paper before the Congress of Hygiene at Geneva on "*Atténuation des Virus.*"
Controversy with Koch on anthrax immunization.
Studies on rabies.

1883: Establishment of a laboratory at family home in Arbois.
In July, official celebration at his birthplace and Pasteur's address to the memory of his parents.
Vaccination against swine erysipelas.
Studies on cholera (Death of Thuillier in Egypt).
Lecture before the Société Chimique de Paris "*La dissymétrie moléculaire.*"

1884: Studies on vaccination against rabies.
Paper before the International Congress of Medicine in Copenhagen on "*Microbes pathogènes et vaccins.*"

1885: Treatment of Joseph Meister and Jean Baptiste Jupille against rabies.

1886: Establishment of kennels for the study of rabies in dogs at Garches (Villeneuve l'Etang).
International subscription for the foundation of an Institut Pasteur, devoted to the study and treatment of rabies and other microbiological problems (Pasteur Institute).
Controversies on rabies.
Convalescence for a few weeks at Villa Bischoffheim, Bordighera. Return to Paris to answer attacks against rabies treatment.

1887: Report of the English commission on rabies.

1888: Inauguration of the Pasteur Institute on November 14.

1892: Pasteur Jubilee at the Sorbonne on December 27.

1894: Last stay at Arbois (July to October).

1895: Death on September 28 at Villeneuve l'Etang.

Reference Notes to the Introduction

All quotations from Pasteur's writings in the Introduction are to be found in *Oeuvres de Pasteur, réunies par Pasteur Vallery-Radot* (Paris: Masson et Cie, 1933–1939), the volumes of which are listed by title in section II of the Bibliography. References to these volumes are cited here as *Oeuvres,* with the volume number and page number; in addition these notes include the publication data for the work in which the quotation originally appeared.

1. Pasteur Vallery-Radot, "Pasteur as I Remember Him," in *The Pasteur Fermentation Centennial 1857–1957: A Scientific Symposium* (New York: Chas. Pfizer & Co., Inc., 1958), p. 185.
2. *Oeuvres,* vol. IV, p. 244; *Etudes sur la maladie des vers à soie* (Paris: Gauthier-Villars, 1870).
3. *Oeuvres,* vol. V, p. 435; *Etudes sur la bière* (Paris: Gauthier-Villars, 1876).
4. *Oeuvres,* vol. IV, p. 245; *Etudes sur la maladie des vers à soie.*
5. *Oeuvres,* vol. VII, p. 54.
6. *Oeuvres,* vol. VI, p. 92; "Observations verbales à l'occasion de la communication de M. Alph. Guérin," *Comptes rendus de l'Académie des sciences* 78 (1874): 867–868.
7. *Oeuvres,* vol. VI, p. 337; "De l'atténuation des virus et de leur retour à la virulence," *Comptes rendus de l'Académie des sciences* 92 (1881): 429–435.
8. *Oeuvres,* vol. VI, p. 109; "Discussion: La théorie des germes et ses applications à la chirurgie," *Bulletin de l'Académie de médecine* 7 (1878): 139–167.
9. *Oeuvres,* vol. IV, p. 279; *Etudes sur la maladie des vers à soie.*
10. *Oeuvres,* vol. IV, pp. 240, 233; *Etudes sur la maladie des vers à soie.*
11. *Oeuvres,* vol. IV, p. 568; *Etudes sur la maladie des vers à soie.*
12. *Oeuvres,* vol. IV, p. 721; *Etudes sur la maladie des vers à soie.*
13. *Oeuvres,* vol. VI, p. 124; "La Théorie des germes et ses applications à la médecine et à la chirurgie," *Comptes rendus de l'Académie des sciences* 86 (1878): 1037–1043.
14. *Oeuvres,* vol. IV, p. 633; *Etudes sur la maladie des vers à soie.*
15. *Oeuvres,* vol. VI, p. 92; "Observations verbales à l'occasion du rapport de M. Gosselin," *Comptes rendus de l'Académie des sciences* 80 (1875): 87–95.

16. René Vallery-Radot, *The Life of Pasteur,* trans. Mrs. R. L. Devonshire (Garden City, N.Y.: Garden City Publishing Co., Inc., n.d.), p. 462.
17. L. Pasteur Vallery-Radot, *Pasteur Inconnu* (Paris: Flammarion, 1954), pp. 232–238. See also André George, *Pasteur* (Paris: Albin Michel, 1948).
18. *Oeuvres,* vol. VII, p. 338.

Bibliography

I

Detailed information concerning the events of Pasteur's life can be found in the following publications:

DUCLAUX, E.[1]: "*Le Laboratoire de M. Pasteur.*" In Centième Anniversaire de la Naissance de Pasteur. (Institut Pasteur, Paris: Hachette, 1922.)

DUCLAUX, MADAME E.: *La Vie de Emile Duclaux.* (Paris: L. Barneoud & Cie., 1906.)

LOIR, A.[2]: *A l'ombre de Pasteur.* In Mouvement Sanitaire. Vol. 14 (1937), pp. 43, 84, 135, 188, 269, 328, 387, 438, 487, 572, 619, 659; Vol. 15 (1938), pp. 179, 370, 503.

Pasteur, Correspondance réunie et annotée par Pasteur Vallery-Radot [3] (Paris: Bernard Grasset, 1940.)

Pasteur, Dessinateur et Pastelliste (1836–1842). Compiled by René Vallery-Radot. (Paris: Emile Paul, 1912.)

ROUX, E.[4]: "*L'Oeuvre Médicale de Pasteur.*" In Centième Anniversaire de la Naissance de Pasteur. (Institut Pasteur, Paris: Hachette, 1922.)

VALLERY-RADOT, R.[5]: *La Vie de Pasteur.* (Paris: Hachette, 1922.)

VALLERY-RADOT, R.: *Madame Pasteur.* (Paris: Flammarion, 1941.)

[1] Duclaux became Pasteur's assistant in 1862 and remained his close associate to the end. He was the first director of the Pasteur Institute after Pasteur's death.

[2] Loir was Pasteur's nephew and was his technical assistant between 1884 and 1888.

[3] Pasteur's grandson.

[4] Roux became Pasteur's assistant in 1876 and became director of the Pasteur Institute in 1904.

[5] Pasteur's son-in-law.

II

All of Pasteur's publications as well as a large number of unpublished manuscripts and notes have been collected in:

Oeuvres de Pasteur, réunies par Pasteur Vallery-Radot
(Paris: Masson et Cie., 1933–1939)

VOLUME
I. *Dissymétrie moléculaire*
II. *Fermentations et générations dites spontanées*
III. *Etudes sur le vinaigre et sur le vin*
IV. *Etudes sur la maladie des vers à soie*
V. *Etudes sur la bière*
VI. *Maladies virulentes, virus-vaccins et prophylaxie de la rage*
VII. *Mélanges scientifiques et littéraires*

The development of Pasteur's scientific work has been lucidly and critically analyzed in:

DUCLAUX, E.: *Pasteur. Histoire d'un esprit* (Sceaux: Charaire et Cie., 1896.)

III

The views expressed in the present volume concerning the influence of historical factors, and of the scientific and social environment, on the unfolding of Pasteur's career are based on information derived from the following sources:

ACKERKNECHT, E. H.: *Anticontagionism between 1821 and 1867.* (*Bull. of Hist. of Med.* 1948, *22*, 562–593.)

AYKROYD, W. R.: *Three Philosophers.* (London: William Heinemann, 1935.)

BALLANTINE, W. G.: *The Logic of Science.* (N. Y.: Thomas Y. Crowell Co., 1933.)

BÉDIER, J. et HAZARD, P.: *Histoire de la Littérature Française Illustrée.* (Paris: Larousse, 1924.)

BERNAL, J. D.: *The Social Function of Science.* (London: George Routledge & Sons, Ltd., 1939.)

BERNARD, C.: *An Introduction to the Study of Experimental Medicine.* (N. Y.: The Macmillan Co., 1927.)

BROWN, H.: *Scientific Organizations in Seventeenth-Century France.* (Baltimore: The Williams & Wilkins Co., 1934.)

BULLOCH, W.: *The History of Bacteriology.* (London: Oxford University Press, 1938.)

CASTIGLIONI, A.: *A History of Medicine.* (N. Y., A. A. Knopf, 1941.)

COHEN, I. B.: *Science, Servant of Man.* (Boston, Little, Brown and Co., 1948.)

CONANT, J. B.: *On Understanding Science: an Historical Approach.* (New Haven: Yale University Press, 1947.)

CROWTHER, J. G.: *British Scientists of the Nineteenth Century.* (London: George Routledge & Sons, 1935.)

———: *Famous American Men of Science.* (N. Y.: W. W. Norton & Co., 1937.)

CROWTHER, J. G.: *The Social Relations of Science.* (N. Y.: The Macmillan Co., 1941.)

DAMPIER, W. C. D.: *A History of Science and Its Relation with Philosophy and Religion.* (N. Y.: The Macmillan Co., 1930.)

DRACHMAN, J. M.: *Studies in the Literature of Natural Sciences.* N. Y.: The Macmillan Co., 1930.)

FRANKLAND, P. F. and FRANKLAND, MRS. P.: *Pasteur.* (N. Y. The Macmillan Co., 1898.)

GARRISON, F. H.: *An Introduction to the History of Medicine.* (Phila.: W. B. Saunders Co., 4th edition, 1929.)

GODLEE, R. J.: *Lord Lister.* (London, Macmillan & Co. Ltd., 1918.)

GOLDENWEISER, A.: *Robots or Gods.* (N. Y.: A. A. Knopf, 1931.)

GREENWOOD, M.: *Epidemics and Crowd-Diseases.* (London: Williams & Norgate Ltd., 1935.)

GREGORY, J. C.: *The Scientific Achievements of Sir Humphry Davy.* (London: Oxford University Press, 1930.)

HAGGARD, H. W.: *Devils, Drugs, and Doctors.* (N. Y.: Harper, 1929.)

HARDEN, A.: *Alcoholic Fermentation.* (London: Longmans, Green & Co., 3rd edition, 1923.)

HARDING, R. E. M.: *The Anatomy of Inspiration.* (Cambridge, England: W. Heffer & Sons, 2nd edition, 1942.)

HELMHOLTZ, H.: *Popular Lectures on Scientific Subjects.* Translated by E. Atkinson. First Series. (London: Longmans, Green & Co., 1898.)
Second Series. (London: Longmans, Green & Co., 1881.)

HOGBEN, L.: *Science for the Citizen.* (N. Y.: A. A. Knopf, 1938.)

HUME, E. E.: *Max von Pettenkofer.* (N. Y.: Paul B. Hoeber, Inc., 1927.)

HUXLEY, L.: *Life and Letters of Thomas Henry Huxley.* (N. Y.: D. Appleton Co., 2 vols., 1902.)

KRAMER, H. D.: *The Germ Theory and the Early Public Health Program in the United States.* (*Bull. Hist. Med.* 1948, 22, 233–247.)

LARGE, E. D.: *The Advance of the Fungi.* (London: Jonathan Cape, 1940.)

LAVOISIER, A. L.: *Oeuvres de Lavoisier.* (Paris: 4v. Imprimerie Impériale, 1862–1868.)

LEVY, H.: *Modern Science — A Study of Physical Science in the World Today.* (N. Y.: A. A. Knopf, 1939.)

MEAD, G. H.: *Movements of Thought in the Nineteenth Century,* edited by M. H. Moore (Chicago: University of Chicago Press, 1936.)

MEES, C. E. K.: *The Path of Science.* (N. Y.: John Wiley & Sons, Inc., 1946.)

MERZ, J. T.: *A History of European Thought in the Nineteenth Century.* (London: William Blackwood & Sons, ed. 4 vol. I, ed. 3 vol. II, 1923–1928.)

MURRAY, R. H.: *Science and Scientists in the Nineteenth Century.* (N. Y.: The Macmillan Co., 1925.)

NEWMAN, G.: *Interpreters of Nature.* (London: Faber & Gwyer, 1927.)

NEWSHOLME, A.: *Evolution of Preventive Medicine.* (Baltimore: Williams & Wilkins Co., 1927.)

NORDENSKIOLD, E.: *The History of Biology; a Survey.* (N. Y.: A. A. Knopf, 1932.)

OLMSTED, J. M. D.: *Claude Bernard, Physiologist.* (N. Y.: Harper, 1938.)

——: *François Magendie, Pioneer in Experimental Physiology and Scientific Medicine in XIX Century France.* (N. Y.: Henry Schuman, 1944.)

PACK, G. T. and GRANT, F. R.: *The Influence of Disease on History.* (*Bull. N. Y. Academy Medicine,* 1948, *24,* 523–540.)

PLEDGE, H. T.: *Science Since 1500.* (London: His Majesty's Stationery Office, 1939.)

PORTERFIELD, A. L.: *Creative Factors in Scientific Research.* (Durham: Duke University Press, 1941.)

ROBINSON, V.: *Pathfinders in Medicine.* (N. Y.: Medical Life Press, 2nd edition 1929.)

SARTON, G.: *The Life of Science.* (N. Y.: Henry Schuman, 1948.)

SHENSTONE, W. A.: *Justus von Liebig, His Life and Work.* (London: Cassell & Co., Ltd., 1901.)

SHRYOCK, R. H.: *The Development of Modern Medicine.* (N. Y.: A. A. Knopf, 1947.)

SINGER, C.: *A Short History of Medicine.* (N. Y.: Oxford University Press, 1928.)

STEARN, E. W. and STEARN, A. E.: *The Effect of Smallpox on the Destiny of the Amerindian.* (Boston: Bruce Humphries, Inc., 1945.)

THORNTON, J. E.: *Science and Social Change.* (Washington: Brookings Institution, 1939.)

TYNDALL, J.: *Fragments of Science.* (N. Y.: Appleton, 2v., 1896.)

TYNDALL, J.: *New Fragments.* (N. Y.: Appleton, 1896.)

WALLACE, A. R.: *The Wonderful Century. Its Successes and Its Failures.* (N. Y.: Dodd, Mead & Co., 1899.)

WEBSTER, L. T.: *Rabies.* (N. Y.: The Macmillan Co., 1942.)

WHITEHEAD, A. N.: *Science and the Modern World.* (N. Y.: The Macmillan Co., 1925.)

WIEGAND, W. B.: *Motivation in Research.* (*Chem. and Eng. News,* 1946, *24,* 2772–2773.)

WINSLOW, C. E.: *The Conquest of Epidemic Disease.* (Princeton: Princeton University Press, 1943.)

WOLF, A.: *A History of Science, Technology and Philosophy in the Eighteenth Century.* (London: Allen & Unwin Ltd., 1938.)

Index

ABOUT, EDMOND, 177.
Académie Française (Academy of Letters), 50, 333, 355, 377, 385, 387, 391.
Academy of Medicine, 48, 74–75, 185–186, 189, 248, 260–261, 263, 284–287, 301, 304, 324–325, 327, 335, 345–348.
Academy of Sciences, 6, 8–10, 14, 34, 37, 40, 42–43, 45, 73, 92, 108, 121, 165, 173–174, 176, 182, 198, 220, 357.
Acetic acid fermentation, 116, 137–141, 144, 161.
Adaptation, *see* Virulence, alterations of.
Adaptive enzymes, 383.
Aerobic life, 136. *See also* Respiration.
Alais, 30, 52, 215–232.
Alcoholic fermentation, 73, 115–126, 129, 148, 161, 182–186, 191–208, 372.
Aleppo boil, 300.
Alhumbert Prize, 166, 173.
Ampère, Jean-Jacques, 7, 12, 153.
Amyl alcohol, 115, 125, 126.
Anaerobic life, 45, 135, 136, 188–190, 257, 360, 372. *See also* Fermentation.
Analogy, use of, in scientific discovery, 138, 327, 373–374.
Anthrax, 48, 67, 68, 73, 81, 234, 240, 244, 249, 250–256, 258, 260–261, 266, 272–276, 284, 289, 309, 330–331, 337–343, 357, 380.
Antisepsis, 48, 128, 151, 245, 300–302.
Appert, François, 10.
Arago, Dominique-François, 16, 99.
Arbois, 21–22, 24, 28, 37, 45, 51, 68, 143–144, 146, 184, 332, 349, 391.
Aseptic techniques, 150, 301–302.
Asparagine, 104, 108.
Aspartic acid, 104, 108–109.
Assistants, *see* Pasteur's assistants.
Asymmetry, molecular, 40–41, 47, 88, 90–115, 373; relation to living processes of, 108, 378–379.
Attenuation of virulence, 281, 329–334, 355, 374.
Australia, 310, 312.
Autoclave, 179.
Auzias-Turenne, 324–326, 354.

BACILLUS BOMBYCIS, 225.
Bacon, Francis, 6, 10, 90, 188, 209, 313, 362, 366–367.
Bacteriological warfare, 312–313, 318–319.
Bacteriology, 49, 157, 189, 209; development of techniques, 186–187. *See also* Golden era of bacteriology.
Bacteriotherapy, 309–310, 380–381.
Bail, Karl, 191–192.
Balard, Antoine Jérôme, 31–32, 34–35, 42, 91, 97, 168–169.
Bassi, Agostino, 209, 240, 258.
Bastian, Henry Charlton, 72, 176, 178–179, 185.
Beer, studies on, 47, 68, 84, 147–150, 246, 361.
Berkeley, Miles Joseph, 259–260.
Bernard, Claude, 14, 16, 19, 30, 50, 61, 72, 181, 241, 269, 358, 387, 395, 398; posthumous publication, 182–186, 198–208.
Bert, Paul, 16, 182, 253, 256.
Berthelot, Marcellin, 72, 123, 182, 198–208, 386, 390, 395.
Berthollet, Claude Louis, 9.
Bertin-Mourot, Pierre Augustin, 28, 36, 60, 68, 148.

Berzelius, Jöns Jakob, 117, 119, 121, 123, 129, 137, 139, 155, 156, 198, 199, 203.
Bezançon, 17, 21, 36.
Billroth, Albert Christian T., 247.
Biochemical memory, 328–329.
Biochemical sciences, 9, 133, 187–208.
Biochemical unity of living processes, 48, 153, 188–208, 379.
Biological sciences and technology, 141, 144–145, 153, 155, 380.
Biological warfare, *see* Bacteriological warfare.
Biot, Jean Baptiste, 34–35, 37, 41, 43, 46, 90–92, 97, 102, 104, 108, 115, 166.
Bonn, diploma from University of, 51.
Botrytis bassiana, 209.
Boussingault, Jean Baptiste, 30.
Boyle, Robert, 237–238, 243.
Boylston, Zabdiel, 320.
Brauell, J. F., 251–252, 253.
Bretonneau, Pierre, 239.
Brücke, Ernst von, 243.
Büchner, Hans, 156, 201–202.
Budd, William, 305.
Butyric acid fermentation, 45, 116, 123, 134–136, 188–190, 195, 244, 247, 251, 257, 372.

Cagniard de la Tour, Charles, 120–121, 123, 203, 239.
California wine maker, 152.
Cancer, Pasteur's views on, 284.
Cantoni, Gaetano, 213, 216, 219.
Carlsberg laboratory, 148.
Carnot, Sadi, 7, 368.
Carriers of infection, 270, 278–279, 298.
Catalysis, 117, 119, 137.
Chadwick, Edwin, 295–297.
Chamberland, Charles Edouard, 61, 66, 69, 75, 178, 261, 338, 340–341.
Chance and scientific discovery, 19, 95, 100–101, 106, 201, 368. *See also* Pasteur's luck.
Chapin, Charles V., 307.
Chappuis, Charles, 18, 28–29, 33, 40, 43, 125.
Chartres, 276.
Chateaubriand, François René, 15.
Chemical processes of life, 9, 41, 108, 110–115, 123, 190–208; basis of disease, 289–291, 382; basis of immunity, 354–358.
Chemical vaccine, *see* Vaccinating substance.
Chemotherapy, 307–310. *See also* Bacteriotherapy.
Chevreul, Michel Eugène, 174.
Chicken cholera, 261, 272, 277–282, 289, 311, 327–330, 354–355, 371.
Childbirth fever, 261, 303.
Cholera, 241, 250, 268–272.
Clemenceau, Georges, 347–348.
Cohn, Ferdinand, 122, 179, 247, 250, 253, 280.
Colbert, Jean Baptiste, 6.
Coleridge, Samuel Taylor, 12.
Colin, G., 73, 74, 75, 284–287.
Collège de France, 35, 98, 387.
Columella, 236.
Commercial exploitation of discoveries, 69, 80–81, 341.
Common sense, use in science of, 248, 397.
Comte, Auguste, 387–388, 392.
Contagion, mechanisms of, 267–291.
Contagious diseases, 48, 69, 107, 209–316. *See also* Diseases.
Controversies, *see* Pasteur.
Cowpox, 300, 320–324, 327, 329.
Creighton, Charles, 323.
Crystal repair, analogy of, with wound healing, 125.
Crystallography, 33–34, 73, 90–115, 360, 370, 378.
Cuvier, Georges, 13–14, 16.
Cycle of matter in nature, 159, 360.

Daguerre, Louis, 8, 31.
Darwin, Charles, 14, 16–17, 76, 277, 292, 366–368.
Davaine, Casimir Joseph, 244, 251–254, 256, 274.
Davy, Humphry, 8, 12–13, 16, 83, 392.
Delafond, Onésime, 251.
Delafosse, Gabriel, 32, 91, 92.

Dessaignes, Victor, 108.
Deville, Sainte-Claire, *see* Sainte-Claire.
Dijon, 35.
Discovery, *see* Scientific discovery.
Disease, chemical basis of, 289–291; mechanism of, 267–291; study of, 46, 48, 361, 380.
Disease as physiological conflict, 189–191.
Diseases of fermentation, 144–154.
Diseases of filth, 295–297, 304–305.
Dôle, 3, 22, 24.
Domestication of microbial life, 116–158.
Du Bois-Reymond, Emil, 243.
Duclaux, Emile, 21, 59, 63, 70, 73, 81, 113–114, 148, 168, 193, 196, 208, 210, 213, 214, 217, 220, 223, 263, 382.
Dumas, Jean Baptiste, 16, 29, 30–31, 35, 41, 43, 46, 50, 58, 59, 81, 118, 166, 168, 203, 213.
Duruy, Victor, 15.
Dusch, Theodor von, 164.
Dust, as carrier of microorganisms, 179–180, 235–236, 245–246.

Earthworms, 14, 275–277, 371.
Ecole des Beaux Arts, *see* School of Fine Arts.
Ecole Normale Supérieure, 17, 26, 29, 31–32, 36, 42, 62, 69, 81, 90, 95, 216, 360.
Economy of matter, 159, 360.
Edelfeldt, 26, 334.
Edison, Thomas A., 19.
Egg-selection method, 68, 216–232.
Ehrenberg, Christian, 122.
Einstein, Albert, 364.
Encyclopedists, 6, 393.
Energy in biochemical processes, 194–197, 372.
English commission investigating rabies, 349–352.
Enthusiasm, 23, 392.
Environmental factors, influence on susceptibility to disease of, 314–316.
Enzymes, 199–208, 379.
Epidemic climate, 270.
Epidemics, 46, 209, 267–291, 294; empirical control of, 292–299.
Epidemiology, 49, 230–232, 234–235, 256, 278, 282.
Exhaustion theory of immunity, 353–354.
Experimental method, 56, 74, 131, 362, 376–377, 387–389, 396.

Faraday, Michael, 5, 7, 12–13, 16, 19, 50, 364–365, 367, 392–394.
Fechner, Gustav, 397.
Fermentation, 41, 48, 67, 73, 90, 106–107, 114, 116–117, 119, 160, 164, 187, 189–208, 247, 301, 360, 379; analogy of, with disease, 48, 214, 233, 237–239; and respiration, 194–208; diseases of, 144–154; mechanisms of, 118–159, 372; relation of, to muscle metabolism, 196. *See also* individual fermentations (acetic, alcoholic, butyric, lactic, tartaric).
Fermentation correlative with life, 43, 129–134, 189–208.
Fermentation in synthetic media, 130–133, 157, 205.
Fermentation is life without air, 188–208.
Ferments, 117, 127, 133, 190–208.
Filterable viruses, 225, 264–266.
Filth and disease, 295–297, 304–305.
Filtration and removal of germs, 164, 170, 178–179, 255, 290.
Flacherie, 212, 221–232, 247, 371.
Flaming, use of, for sterilization, 302.
Flourens, Marie Jean Pierre, 173.
Fortoul, Hippolyte, 18.
Fourcroy, Antoine François, 9.
Fracastoro, 237.
Franco-Prussian War, 47, 51, 77–78, 300.
Franklin, Benjamin, 32.

Galileo, 4, 41, 390.
Garches, 69.
Garrison, Fielding H., 295.
Gattine, 212, 221–232.
Gauss, Karl Friedrich, 364–365.
Gay-Lussac, Joseph Louis, 118, 120, 129, 171, 203.

Genius, 87–88, 382, 385.
Geoffroy Saint-Hilaire, Etienne, 5, 14, 16.
Germs in air, 170–187.
Germ theory, 41, 44, 72.
Germ theory of disease, 41, 48, 75, 107, 115, 161, 233–266; of fermentation, 41, 47, 72, 115–208, 161.
Gernez, Désiré Jean Baptiste, 53, 105, 217, 220.
Goethe, Johann Wolfgang von, 124, 198.
Golden era of bacteriology, 260–261, 314.
Grancher, Jacques Joseph, 62, 335, 345, 347.
Greenwood, Major, 322, 324.
Grosvenor, Robert, 323.
Grove, William Robert, 367.

Harvey, William, 5, 159.
Haüy, René Just, 91, 99.
Hay bacillus, 254.
Hay bacillus in anthrax vaccine, 341.
Heat, effects of, 120, 151, 164, 166, 168–169, 178–179.
Hegel, Georg, 396.
Helmholtz, Hermann von, 7, 13, 16, 122–123, 165, 168, 243, 248, 363, 368, 392, 394.
Hemihedral facets, 91.
Henle, Jacob, 240.
Henle-Koch postulates, 240, 266.
Henry, Joseph, 7, 10, 367.
Hereditary susceptibility to disease, 230–232, 314–316.
Herschel, John, 91, 92.
Hétérogénie, 166.
Hippocratic medicine, 236, 313–314.
Holmes, Oliver Wendell, 261, 303–304.
Host-parasite relationships, 273.
Humboldt, Friedrich von, 13.
Huxley, Thomas Henry, 14, 16, 76, 180, 366.
Hydrophobia, *see* Rabies.
Hygiene, 270–271, 292–316.

Immunity, 49, 261, 284, 299, 317–358, 361, 373; chemical basis of, 381; mechanisms of, 353–358.
Infection, carriers of, 180, 270.
Infectious diseases, *see* Contagious diseases.
Infinitely small, the, 17, 45, 162, 243.
Insect vectors, 298.
Interference phenomenon, 357.
Intuition in scientific discovery, 131, 133, 154, 219, 359–360, 362, 365–367, 371.
Isomerism, 98, 106.

Jacobsen, J. C., 148.
Jaillard, Pierre François, 252, 254, 256.
Jefferson, Thomas, 317, 323.
Jenner, Edward, 300, 317–324, 327–328, 331, 336, 373.
Johnson, Dr. Samuel, 382.
Joly, Nicolas, 173.
Joubert, Jules François, 60, 178, 233.
Joule, James Prescott, 7, 368, 392.
Jupille, Jean Baptiste, 49, 53, 69, 336, 345.
Juvenal, 233.

Kant, Immanuel, 394.
Kekule, Friedrich August, 363–364, 367.
Kelvin, Lord (William Thompson), 7, 19, 363, 367, 392.
Koch-Henle postulates, 240, 266.
Koch, Robert, 48, 73, 240, 249–250, 253–254, 257, 260–261, 269–271, 274, 280, 306, 324–325, 341–342, 380.
Kützing, Friederich Traugott, 120–121, 123, 139, 203, 239.

Lactic acid fermentation, 116, 123, 125–126, 133, 245.
Laënnec, René T. H., 15.
Laplace, Pierre Simon, 11.
Large, E. D., 260.
Laurent, Auguste, 91.
Laurent, Marie, *see* Pasteur, Mme. Louis.
Lavoisier, Antoine Laurent, 4, 8, 9, 118–119, 129–130, 160, 203.
Leblanc, Nicolas, 10.
Lechartier, Georges Vital, 195.
Leeuwenhoek, Anton van, 122, 238.

Leonardo da Vinci, 58.
Leplat, F., 252, 254, 256.
Leprosy, 293.
Liebig, Justus von, 9, 13, 30, 72, 81, 121–132, 137, 139, 140, 144, 154–157, 197–208, 243, 248, 287, 358.
Life, origin of, 40, 47, 111–115, 162, 166, 177, 181, 187, 395–396.
Lille, 18, 41–42, 112, 116, 124–125, 146.
Lindley, John, 259–260.
Lister, Joseph, 48, 244–247, 301.
Littré, Emile, 50, 386–388.
Living processes, chemical unity of, 48, 135, 188–208, 379.
Living processes and chemistry, 9, 30, 41, 108, 110–115, 119.
Logic and scientific discovery, 131, 138, 222, 359, 375.
Loir, Adrien, 60, 62, 65, 79, 81, 87, 302, 310–312, 326, 333–334, 375.
London, Great Stench of, 305–306; Pasteur's visit to brewery in, 148–149.
Ludwig, Karl, 243.

Macaulay, Thomas Babington, 317.
Magendie, François, 363.
Malaria, 294–295, 298.
Malic acid, 104–105, 109.
Marat, Jean Paul, 11.
Marggraf, Andreas Sigismund, 8.
Marx, Karl, 18.
Mather, Cotton, 320, 323.
Mathilde, Princess, 15.
Maxwell, Clerk, 15, 364–365, 392, 394.
Meister, Joseph, 49, 53, 69, 335–336, 344–345, 349.
Melun, 49, 67, 338.
Mendelejeff, Dmitrij, 368.
Mesotartaric acid, 104, 109.
Metchnikoff, Elie, 272.
Meyer, Lothar, 368.
Meze, 153.
Microbe, 188.
Microbiology, *see* Bacteriology.
Microorganisms, 41, 189; activities of, affected by environment, 145. *See also* Germ theory.
Mitscherlich, Eilhardt, 33, 92–97, 102.
Molecular configuration, 40, 90–115, 193, 360.
Monge, Gaspard, 10.
Montagu, Lady, 319.
Montaigne, Michel de, 317.
Morphology, Pasteur's interest in, 193–194, 263.
Morphology as guide to study of functions, 192–194.
Morse, Samuel F. B., 7.
Morts flats, 212, 221–232.
Motility of microorganisms, 134–135.
Mucor mucedo, 191–192.
Munich, 271.
Muscle metabolism, 196.
Musset, Charles, 173.
Mycoderma aceti, 138–141, 145–147, 157, 191; *vini,* 147.

Napoleon I, 10–11.
Napoleon III, 16–18.
National Academy of Sciences, 10.
Needham, John Turberville, 163.
Newcomen, Thomas, 7.
Newton, Isaac, 4, 40, 390, 393.
Nicolle, Charles, 339.
Niepce, Joseph Nicéphore, 8.
Nightingale, Florence, 248, 297.
Nocard, Edmond Isidore Etienne, 269.
Nosema bombyces, 222.
Nutritional concepts of parasitism, 283–284; requirements of microorganisms and higher organisms, 133, 145, 264–266, 353–354, 379; theories of disease and immunity, 325.

Obermeier, Otto Hugo Franz, 248.
Oersted, Jean Christian, 7, 18.
Office of Scientific Research and Development, 10.
Optical activity, crystal shape and physiological processes, 33–34, 39, 90–115, 125–126, 370.
Oriental sore, 299–300.
Origin of life, *see* Life.
Orléans, 68, 139.

Osimo, Marco, 213, 216.
Osteomyelitis, 263.

PAGET, STEPHEN, 24.
Parasitism, 239, 242, 267, 282–283, 288–291.
Paratartaric acid, 33–34, 40, 41, 93–113.
Pasteur Institute, 23, 53, 56, 70, 336, 344, 355.
Pasteur, Jean Joseph, 3, 24–25, 35, 51, 83–84, 125.
Pasteur, Madame Jean Joseph, 25.
Pasteur, Madame Louis, 24, 36–41, 46, 51–52, 66, 79, 88, 217, 227.
Pasteur, Louis, assistants of, 43, 59–60, 64–66, 79, 215; birth, 3; burial of, in chapel, 23, 57; character and temperament of, 13, 26–29, 37–38, 42, 44, 50–54, 63, 73, 75–76, 103, 154, 156, 171, 173, 183, 207, 214, 227–230, 337–338, 345, 369–370, 395–400; children, 51, 217; controversies, 44, 49, 67, 70–78, 153, 157, 173, 197–208, 228, 284–287, 341, 345–349; daily life, 66, 79; death, 3, 56; education, 25–26, 52; enthusiasm, 23, 53; experimenter, 44–45, 59–88, 101, 114–115, 166, 207, 219, 226, 342, 370–376; fame and honors, 16–17, 21–24, 40, 43, 51, 53–56, 83; family life, 25, 27, 38, 51, 63, 79; father, 3, 24–25, 35, 51, 83–84, 125; genius, 24, 87–88, 101, 266, 382; illnesses, 47, 53, 56, 64, 230, 347, 354, 357; interest in empirical practices, 143, 226, 231, 276; intuition, 359, 360, 361, 371; jubilee, 17, 54, 76, 80, 85; laboratories, 32, 40, 42, 43, 45, 47, 60–62, 64, 68–70, 81, 146, 215, 217, 227, 275–276, 370; legend, 21–57; letters, 27, 33–34, 37–38, 40, 43, 51, 67, 71, 73, 80, 103, 104, 125; logic, 360–362; luck, 100–101, 106, 219, 309, 327, 340, 342; mannerisms, 79–80; marriage, 36–38; mother, 25; motivation, 28–30, 80–89, 361; paintings, 21, 25–26, 370; patriotism and public spirit, 28, 50–51, 56, 77, 83–85; philosophy and religious faith, 50, 54, 85, 166, 376–377, 385–400; promoter of science, 45–47, 49–50, 59, 68–70, 140, 151–152, 227–230, 342–343; public debates and lectures, 15, 44–45, 50, 53–54, 58, 70–71, 73, 75, 78, 84, 87, 139, 177, 187, 377, 387, 391; sequence of discoveries, 190, 233, 359–384; teachers, 28–33; technique, 64–65, 101; travels, 40, 103, 148–149, 230; use of microscope by, 64, 65, 92, 135, 146, 149, 155, 170, 216–218, 222, 226, 263, 371; views on physicians as scientists, 61–62, 74–75, 249; views on theoretical and practical science, 18–19, 67, 88, 156; wife, 24, 51; will and testament, 55. *See also* Appendix outlining events of his life in chronological order, 403–406.
Pasteurization, 23, 47, 68, 80, 151–154, 361.
Pavlov, Ivan Petrovich, 328.
Pébrine, 212–232, 371.
Pelletier, Louise, 345.
Perkin, William Henry, 110.
Peter, Michel, 73, 246, 346, 348, 352.
Pettenkofer, Max Joseph von, 270–272, 297–298.
Philosophy, influence of, on biological theories, 172–173, 177, 347.
Phipps, James, 321.
Phylloxera, 310.
Physicians, Pasteur's views on, 61, 74–75, 249.
Physicochemical interpretation of living processes, 119, 122–124, 133–134, 139, 154, 197–208, 243, 358, 390, 395.
Physiological state and its effect on disease, 225, 230–232, 236, 272, 284–291, 302, 314–316, 382.
Plant diseases, 258–260.
Pneumococcus, 263, 375.
Poincaré, Henri, 364.
Polarization of light, 33–34, 90–115, 193, 360, 378.
Pont Gisquet, 217, 227.

Positivist philosophy, 50, 386–388, 391.
Potato blight, 259.
Potentialities of living things, multiple, 377–384.
Pouchet, Félix Archimède, 72, 165–177, 287.
Pouilly le Fort, 15, 49, 331, 333, 338–340, 342–343.
Practical *vs.* theoretical science, 4, 18, 19, 42, 47, 58, 67, 68, 86, 162, 190, 359.
Preconceived ideas and scientific discovery, 94, 96–97, 100, 109, 157, 362, 369, 376.
Preyer, W., 397.
Priority, concern with, 73, 82–83.
Prophylactic and therapeutic measures, difficulty of establishing evaluation of, 323–324, 344–353.
Provostaye, Frederic de la, 92–96.
Proust, Marcel, 328.
Public health, 49, 270–271, 292–316.
Puerperal fever, *see* Childbirth fever.
Putrefaction, 116–117, 122–136, 160, 164–165, 168, 180, 187–188, 301; analogy of, with fermentation and disease, 48, 237–239, 247.

QUARANTINE MEASURES, 298.

RABBITS, destruction of, by introduction of disease, 310–312.
Rabies, 49, 53, 62, 69–70, 73, 263–266, 298–299, 326, 329, 332–334, 344–358, 374.
Racemic acid, 93, 97, 106.
Raulin, Jules, 112, 133, 284.
Rayer, Roger, 244, 251.
Religion and science, 390–400.
Renan, Ernest, 333, 377, 385–391.
Reservoirs of infection, 270–278, 298.
Respiration and fermentation, 194–208.
Robespierre, Maximilien de, 11.
Robin, Charles, 240.
Rossignol, H., 337–338.
Roux, Emile, 39, 61, 63, 65, 75, 88, 201, 249, 261, 264–265, 269, 275, 334–335, 338–340, 346–347, 355.
Royal Institution, 6, 12–13, 392.
Royal Society, 6, 238, 320, 322.
Rubner, Max, 202.

SAINTE-CLAIRE DEVILLE, HENRI ETIENNE, 42, 241, 269.
Saint-Cloud, 22, 56, 69.
Salmon, David Elmer, 381.
Sand, George, 15.
Sanitarians, 295–297, 304–307.
Scarlet fever, 294.
School of Fine Arts, Pasteur teaches at, 50, 81, 270, 316.
Schröder, Heinrich, 164, 168, 248.
Schülze, Franz, 164.
Schwann, Theodor, 120–123, 164, 203, 239, 240.
Science, popularity of, 11.
Science, and country, 77, 84–85; and philosophy and religion, 85–86, 390–400; and society, 1–20, 77–78, 84–85; in war, 10, 77. *See also* Practical, Theoretical.
Scientific discovery, mechanisms of, 29, 87, 93, 159, 359–384.
Sédillot, Charles Emmanuel, 188.
Semmelweis, Ignaz Philipp, 261, 304.
Shelley, Percy Bysshe, 359
Silkworms, culture of, 210–212; studies on, 46, 53, 68, 209–232, 247, 308, 315, 361.
Smallpox, 249, 299, 317–324, 327, 329.
Smith, Stephen, 296.
Smith, Theobald, 381.
Snow, John, 241–242, 269, 322.
Social sciences, 389–390.
Sorbonne, 31, 34, 81; Pasteur's jubilee at, 17, 54, 76, 80, 85; Pasteur's lecture on spontaneous generation at, 15, 177, 187.
Spallanzani, Lazzaro, 163.
Specificity, 107; basis of germ theory, 128–129, 133–134, 145, 158, 166, 244, 264–266; of disease, 145, 239, 242.
Spectator, the, 23–24.
Spencer, Herbert, 14.

Spontaneous generation, 43–44, 72, 114, 159–187, 248, 347, 361, 395–397.
Spores, 179, 253–254, 256, 274, 277, 330.
Staphylococcus, 262–263.
Stearn, W. E., 318.
Stereochemistry, 90, 105.
Stereoisomerism, 378.
Strasbourg, 34, 36, 39, 103–104, 112.
Streptococcus and puerperal fever, 261, 303.
Streptococcus bombycis, 225.
Surgery, *see* Lister.
Swanneck flasks, 169, 175, 177.
Swift, Jonathan, 267.
Swine erysipelas, 68, 261, 272, 329, 331, 344.
Sydenham, Thomas, 239.
Symbiosis, 282–283.
Synthetic media, 130–133, 157, 205.
Syphilis, 237, 294, 324.

TARTARIC ACIDS, 33–34, 41, 73, 92–113, 116, 193, 266.
Thénard, Louis Jacques, 102, 118, 124, 129, 130, 132.
Theoretical *vs.* practical science, 4, 18, 19, 42, 58, 67, 86, 162, 190, 359.
Thermodynamics and living processes, 195.
Thompson, William, *see* Kelvin.
Thomson, Arthur, 86.
Thoreau, Henry D., 21.
Thucydides, 326.
Thuillier, Louis Ferdinand, 269, 331, 340.
Toxins, 273, 289–290, 355, 381.
Transformations of microorganisms, 191–193, 280, 328–333; of organic matter, 41, 43; of virulence, *see* Attenuation.
Traube, Moritz, 205.
Tuberculosis, 13, 250, 258, 293, 315.
Turpin, Pierre Jean François, 121, 123, 203.
Tyndall, John, 13, 16, 165, 176, 178–179, 181, 186, 354, 362, 394.
Tyndallization, 180.
Typhoid fever, 271, 279, 306.
Typhus, 295, 298.

VACCINATING SUBSTANCE, 355–358.
Vaccination, 67–68, 273, 317–358, 361.
Vallery-Radot, Pasteur, 21.
Vallery-Radot, René, 21, 75.
Variolation, 320–321.
Vibrion septique, 252, 256–257, 266, 287–288, 303, 341.
Villemin, Jean Antoine, 347.
Villeneuve l'Etang, 3, 56.
Vinegar studies, 45, 68, 136–141, 145–146, 246, 361.
Virchow, Rudolph, 243.
Virulence, alterations of, 280–282.
Vitalistic theories, 121–122, 139, 154–156, 189–208, 243, 358.
Voltaire, 159, 163.
Vulpian, Edme Félix Alfred, 335, 346.

WALLACE, ALFRED RUSSEL, 17, 20, 328, 368.
Waterhouse, Benjamin, 323.
Watt, James, 7.
Wheatstone, Charles, 7.
Wilberforce, Bishop, 14.
Wine studies, 45, 68, 141–153, 183, 246–247, 361.
Wöhler, Friedrich, 8, 121, 123, 139.
Wonderful Century, The, 17, 20.
Working hypotheses, 62, 93, 101, 107, 109, 128, 361–367.
Wound healing, analogy of, with crystal repair, 125.
Wound infections, 300.
Würtz, 60.

YEAST, 48, 117, 120–121, 146, 149, 159–187, 189–208.
Yellow fever, 298.

ZYMASE, 201–202.

About the Author

René Dubos, professor emeritus at The Rockefeller University in New York City, is a microbiologist and experimental pathologist who pioneered the discovery of germ-fighting drugs from microbes. For his scientific contributions, Dr. Dubos has received many awards; most recently he was the recipient of the first Institut de la Vie prize for his work devoted to environmental problems. He is also well known as an author and a lecturer; his books include *So Human an Animal* (1969 Pulitzer Prize winner); *A God Within; Reason Awake! Science for Man; Only One Earth* (with Barbara Ward); *Beast or Angel? Choices That Make Us Human; The Torch of Life; The Unseen World; The Dreams of Reason; The White Plague* (with Jean Dubos); *The Mirage of Health;* and *Man Adapting.*